Foundations of Bilevel Programming

Nonconvex Optimization and Its Applications

Volume 61

The titles published in this series are listed at the end of this volume.

Foundations of
Bilevel Programming

1 921 449

90-02
90C

by

Stephan Dempe

Freiberg University of Mining and Technology,
Freiberg, Germany

KLUWER ACADEMIC PUBLISHERS

DORDRECHT / BOSTON / LONDON

A C.I.P. Catalogue record for this book is available from the Library of Congress.

ISBN 1-4020-0631-4

Published by Kluwer Academic Publishers,
P.O. Box 17, 3300 AA Dordrecht, The Netherlands.

Sold and distributed in North, Central and South America
by Kluwer Academic Publishers,
101 Philip Drive, Norwell, MA 02061, U.S.A.

In all other countries, sold and distributed
by Kluwer Academic Publishers,
P.O. Box 322, 3300 AH Dordrecht, The Netherlands.

Printed on acid-free paper

Printed in the Netherlands.

Contents

Preface

The central topic of this monograph are bilevel programming problems – problems whose first formulation dates back to 1934 when they have been formulated by H.v. Stackelberg in a monograph on market economy. One special type of bilevel programming problems, called Stackelberg games, has been considered over the years within economic game theory.

Bilevel programming problems have been introduced to the optimization community in the seventies of the 20th century. After that moment a rapid development and intensive investigation of these problems begun both in theoretical and in applications oriented directions. Contributions to its investigation have been delivered by mathematicians, economists and engineers and the number of papers within this field is ever growing rapidly.

Bilevel programming problems are hierarchical ones – optimization problems having a second (parametric) optimization problem as part of their constraints. In economics, this reflects the right of subaltern parts of large economic units to find and select their own, best decisions depending on the "environmental data" posed by the management. In engineering and in natural sciences this "inner" or "lower level" problem can be used to find a correct model for nature. In all situations, the "outer" or "upper level" problem is used to reflect our aim for reaching a certain goal. And the sense of bilevel programming is that we cannot realize this without considering the reaction of the subaltern part or nature on our decisions.

From the mathematical point of view, bilevel programming problems are complicated problems: they are NP-hard; their formulation has inherent difficulties even with respect to the notion of a solution; for many of its reformulations as one-level optimization problems regularity conditions cannot be satisfied at any feasible point.

It is the aim of this monograph to present the theoretical foundations of bilevel programming. Focus is on its optimistic and pessimistic (or weak and strong) solution concepts. Different reformulations as one level optimization problems will be introduced and the relations between the original and the reformulated problems are highlighted. A larger part of this monograph is dedicated to optimality conditions, a smaller part to solution algorithms. In the last chapter the discrete bilevel programming problem is considered and I aim to give some useful directions for investigating it.

I have included a thorough overview of used results in parametric nonlinear optimization.

The book addresses several types of readers. The primal audience are applied mathematicians working in mathematical optimization, operations research or economic modelling. It is intended that graduated and advanced undergraduate student should use it to get a deeper insight into a rapidly developing topic of mathematical optimization. I have separated the proofs of the theorems from the theorems itself and hope that this is helpful for the readers to get an overview over the theory more quickly. A larger number of applications should show the wide fields of possible applications. With both these I hope to initiate or intensify the reader's interest in own contributions to the theory and applications of bilevel programming.

During my work in bilevel programming I had fruitful and helpful discussions on the topic with many colleagues. It is impossible to mention all of them but I can only single some of them out. The first is Klaus Beer who introduced this topic to me. I want to thank Jonathan Bard, Evgenij Gol'stein, Diethard Pallaschke, and Danny Ralph for joint work on related topics. My gratitude is to Diethard Klatte, Jiri Outrata, Stefan Scholtes, and Heiner Schreier. And I want also to thank my Ph.D. students Diana Fanghänel, Heidrun Schmidt, Steffen Vogel for their contributions to the topic and for their helpful comments on parts of the manuscript.

Stephan Dempe

Chapter 1

INTRODUCTION

Bilevel programming problems are mathematical optimization problems where the set of all variables is partitioned between two vectors x and y and x is to be chosen as an optimal solution of a second mathematical programming problem parameterized in y. Thus, the bilevel programming problem is hierarchical in the sense that its constraints are defined in part by a second optimization problem. Let this second problem be introduced first as follows:

$$\min_{x}\{f(x,y) : g(x,y) \leq 0,\ h(x,y) = 0\}, \qquad (1.1)$$

where $f : \mathbb{R}^n \times \mathbb{R}^m \to \mathbb{R}, g : \mathbb{R}^n \times \mathbb{R}^m \to \mathbb{R}^p, h : \mathbb{R}^n \times \mathbb{R}^m \to \mathbb{R}^q, g(x,y) = (g_1(x,y), \ldots, g_p(x,y))^\top, h(x,y) = (h_1(x,y), \ldots, h_q(x,y))^\top$. This problem will also be referred to as the *lower level* or the *follower's problem*. Let $\Psi(y)$ denote the solution set of problem (1.1) for fixed $y \in \mathbb{R}^m$. Then Ψ is a so-called *point-to-set mapping* from $\mathbb{R}^m$ into the power set of $\mathbb{R}^n$, denoted by $\Psi : \mathbb{R}^m \to 2^{\mathbb{R}^n}$.

Denote some element of $\Psi(y)$ by $x(y)$ and assume for the moment that this choice is unique for all possible y. Then, the aim of the bilevel programming problem is to select that parameter vector y describing the "environmental data" for the lower level problem which is the optimal one in a certain sense. To be more precise, this selection of y is conducted so that certain (nonlinear) equality and/or inequality constraints

$$G(x(y), y) \leq 0,\ H(x(y), y) = 0 \qquad (1.2)$$

are satisfied and an objective function $F(x(y), y)$ is minimized, where $F : \mathbb{R}^n \times \mathbb{R}^m \to \mathbb{R}, G : \mathbb{R}^n \times \mathbb{R}^m \to \mathbb{R}^k, H : \mathbb{R}^n \times \mathbb{R}^m \to \mathbb{R}^l$. Throughout the book we will assume that all functions F, G, H, f, g, h are sufficiently

1

smooth, i.e. that all the gradients and Hessian matrices of these functions
exist and are smooth. Clearly this assumptions can be weakened at
many places but it is not our intention to present the results using the
weakest differentiability assumptions. In most cases, the generalization
of the presented material to the case of, say, locally Lipschitz functions
is straightforward or can be found in the cited literature.

The problem of determining a best solution y^* can thus be described as
that of finding a vector y^* of parameters for the parametric optimization
problem (1.1) which together with the response $x(y) \in \Psi(y)$ proves to
satisfy the constraints (1.2) and to give the best possible function value
for $F(x(y), y)$. That is

$$\text{``}\min_{y}\text{''}\{F(x(y), y) : G(x(y), y) \le 0, \ H(x(y), y) = 0, \ x(y) \in \Psi(y)\}.$$

$$(1.3)$$

This problem is the *bilevel programming problem* or the *leader's problem*.
The function F is called the *upper level objective* and the functions G and
H are called the *upper level constraint functions*. Strongly speaking, this
definition of the bilevel programming problem is valid only in the case
when the lower level solution is uniquely determined for each possible y.
The quotation marks have been used to express this uncertainity in the
definition of the bilevel programming problem in case of non-uniquely
determined lower level optimal solutions. If the lower level problem has
at most one (global) optimal solution for all values of the parameter,
the quotation marks can be dropped and the familiar notation of an
optimization problem arises.

The bilevel programming problem (1.3) is a generalization of several well-known optimization problems: It is a realization of a *decomposition approach* to optimization problems [88] if $F(x, y) = f(x, y)$ on
$\mathbb{R}^n \times \mathbb{R}^m$. *Minimax problems* of mathematical programming [65] arise if
$F(x, y) = -f(x, y)$ for each x and y. In both instances, $F(x(y), y)$ is independent of the particular choice of $x(y) \in \Psi(y)$. In such circumstances
"min" can be replaced by ordinary minimization in the formulation of
(1.3). If the dependency of both problems in y is dropped, problem
(1.3) can be interpreted as one of finding the best point in the set of optimal solutions of problem (1.1) with respect to the upper level objective
function. In this case we obtain one approach for attacking *bicriteria
optimization* problems [130]. What distinguishes bilevel optimization
from bicriteria optimization is that in the latter both objective functions F and f are considered jointly. In such cases the aim is to find
a best compromise between the objectives. Such a compromise yields
a solution that, in general, is not a feasible solution to the bilevel programming problem: Considered as a bilevel problem, the lower level

problem has to be solved, i.e., f is minimized over the feasible set, and F is then minimized over the resulting set of optimal solutions. An example illustrating this difference is given in Chapter 3. If the function $f(x, y) = -F(x, y) = \alpha(x) + y^\top \beta(x)$ is equal to the Lagrangean of a mathematical optimization problem $\min\{\alpha(x) : \beta(x) = 0\}$, then problem (1.3) is equivalent to its *dual*. Moreover, one way of treating so-called *improper optimization problems* [90] uses bilevel programming.

Bilevel programming problems can be converted into ordinary (one-level) optimization problems by at least three different approaches:

- Implicit function theorems can be applied to derive a local description of the function $x(y) : \mathbb{R}^m \to \mathbb{R}^n$. If this description is inserted into problem (1.3) an optimization problem with implicitly defined objective and/or constraint functions arises [67, 226]. Based on the stability results in Chapter 4 we will discuss these ideas in Chapters 5 and 6.

- The lower level problem can be replaced by its Karush-Kuhn-Tucker conditions, by a variational inequality or by a semi-infinite system of inequalities. This results in a typical one-level mathematical program [20, 24, 85] which is often called *Mathematical Program with Equilibrium Constraints* or MPEC [23, 188, 224]. Recently, the resulting problem aroused a lot of interest. But note that, even under certain regularity assumptions, the resulting problem is in general not equivalent to the bilevel problem. It is also only possible to apply this approach to the so-called optimistic position of bilevel programming and there seems to be no efficient way to use it in the pessimistic one. A more detailled investigation of this approach will be given in Chapters 5 and 6.

- The lower-level objective in problem (1.1) can be replaced by an additional non-differentiable equation $f(x, y) = \varphi(y)$, where

$$\varphi(y) = \min_x \{f(x, y) : g(x, y) \leq 0, \ h(x, y) = 0\}.$$

The function $\varphi : \mathbb{R}^m \to \mathbb{R}$ is locally Lipschitz continuous under assumptions which are not too restrictive [153]. This approach has been used for deriving optimality conditions for bilevel programming [297] and designing algorithms for finding global optimal solutions [262]. We discuss these ideas in Chapter 5.

The bilevel programming problem demonstrates that applications in economics, in engineering, medicine, ecology etc. have often inspired mathematicians to develop new theories and to investigate new mathematical models. The bilevel programming problem in its original formulation goes back to H.v. Stackelberg [267] who, in 1934, introduced a special case of such problems when he investigated real market situations. This particular formulation is called a Stackelberg game which we discuss in Chapter 2. Even though bilevel programming problems were first introduced to the optimization community by J. Bracken and J. McGill [47, 48] in 1973/74, it was not until about five years later that intensive investigation of bilevel programming problems began (cf. e.g. [2, 19, 27, 41, 54, 99, 263, 285]). Since that time there has been a rapid development and broad interests both from the practical and the theoretical points of view (cf. e.g. the bibliography [278], the three monographs on mathematical programs with equilibrium constraints [188, 224] and bilevel programming [26] and the two edited volumes on bilevel programming [6, 205]).

Some words seem to be in order to distinguish the present monograph from previous work. The edited volumes [6, 205] collect many interesting papers reflecting selected applications of bilevel programming, solution algorithms, and theoretical properties. Both monographs [188, 224] are devoted to MPECs. On the one hand MPECs are slightly more general than bilevel programming problems in the cases when the lower level optimal solution is uniquely determined for all parameter values or when the optimistic position can be used. On the other hand, a reduction of the bilevel problem to an MPEC is not possible in each situation. This is as well reflected in neglecting the pessimistic position as in the recognition that not every feasible solution to an MPEC is also feasible to the corresponding bilevel programming problem. The recent monograph [26] gives a comprehensive overview over all the "classical" results in bilevel programming especially with respect to algorithms for the search for a global optimal solution. In almost all topics in [26] a unique lower level optimal solution for all parameter values is assumed. All three monographs are very interesting sources of results also for the bilevel programming problem showing the rich theory developed by their authors.

In distinction to the previous work it is our aim to reflect the theoretical foundation of bilevel programming as well as the new developments using nondifferentiable approaches for bilevel programming problems. Special attention will be given to the difficulties arising from non-unique lower level solutions.

The outline of the book is as follows. After a Chapter on applications of bilevel programming we will use the linear case for introducing the bilevel problem and illustrating its geometric nature in Chapter 3. Relations to standard optimization problems are discussed and equivalence to the Boolean linear optimization problem is shown. Many authors have verified the $\mathcal{NP}$-hardness of bilevel programming. We include one such result showing strong $\mathcal{NP}$-hardness. Optimality conditions and some algorithms for computing optimal solutions of the bilevel problem conclude Chapter 3.

Chapter 4 will collect some theoretical results from parametric optimization which will be helpful in the sequel. The formulation of existence results of (global) optimal solutions relies on continuity results for the point-to-set mapping $\Psi(\cdot)$. Optimality conditions as well as solution algorithms can rely on directional derivatives resp. generalized Jacobians of the function $x(\cdot)$ describing (local) optimal solutions of the lower level problem.

Chapter 5 is devoted to necessary as well as sufficient optimality conditions. Here we will discuss different approaches for the formulation of such conditions: approaches based on directional derivatives of lower level solutions, ideas using Clarke's generalized Jacobian of the function $x(\cdot)$ and results derived using equivalent formulations of the bilevel programming problem. One of the main topics in this Chapter is concerned with the different notions of optimality in case of non-uniquely determined lower level optimal solutions. Then, at least two different approaches to attack the problem are discussed in the literature: the optimistic and the pessimistic approaches. In the optimistic approach the leader supposes that he is able to influence the follower such that the latter selects that solution in $\Psi(y)$ which is best suited for him. Using the pessimistic point of view the leader tries to bound the damage resulting from the worst possible selection of the follower with respect to the leader's objective function. Results about the relations of bilevel programing problems to its various reformulations as well as the question of satisfiability of constraint qualifications for these reformulations can be found here.

Many attempts for constructing algorithms solving bilevel programming problems have been made in the last few years. Some of these will be described in Chapter 6. However it is not our intention to provide extensive numerical comparison between the different algorithms since such a comparison would be formidable at best and most likely inconclusive. The recent books [26, 224] as well as the papers [100, 131, 135, 225, 269] can be used to get a feeling about the numerical behavior of the various algorithms.

Chapter 7 is devoted to bilevel programming problems with non-unique lower level optimal solutions. We will discuss the stability of bilevel programming problems with respect to perturbations as well as different approaches related to the computation of optimal solutions. One of the mail topics here is also to give some material concerning the possibilities to compute optimal solutions for the bilevel programming problem not using the reformulation of the lower level problem via the Karush-Kuhn-Tucker conditions or variational inequalities. Most results which can be found so far in the literature are devoted to the optimistic position. Some ideas in Chapter 7 can also be used to attack the pessimistic bilevel problem. The latter one is a very complicated problem, but in our opinion, it deserves more attacks in research.

The last Chapter 8 is used to introduce the discrete bilevel programming problem which can only seldom be found in the literature. But saying that this results from low practical relevance is absolutely wrong. Bilevel programming problems are difficult to attack and this is even more true for discrete ones. Here we give attempts into three directions and hope that our ideas can be used as starting points for future investigations: We present a cutting plane algorithm applicable if the constraints of the lower level problem are parameter-independent, some ideas for the computation of bounding functions for branch-and-bound algorithms as well as an idea for the application of parametric discrete optimization to bilevel programming.

It is our intention to write a book which can be used in different directions:

- If all proofs, which are concentrated at the ends of the respective chapters, are omitted, Chapters 2, 3, 5, 6, 7, and 8 give an overview about recent results in bilevel programming. If the reader is interested in a quick and comprehensive introduction into bilevel programming we refer to Chapters 2 and 3. The material in Chapters 5, 6, 7, 8 will give a broad and deep insight into the foundations and into the different attempts to attack bilevel programming problems.

- The Chapter 4 can be used as complementary source in parametric optimization. Here we have also included a number of proofs for basic results which can be found in different original sources. One of the main topics is an introduction to piecewise continuously differentiable functions and its application to the solution function of smooth parametric optimization problems.

- The proofs of the results in all chapters are concentrated in the respective sections at the end of the chapters. This is done to make the reading of the monography more fluently and to obtain a quicker and easier grip to the main results.

Chapter 2

APPLICATIONS

The investigation of bilevel programming problems is strongly motivated by (real world) applications. Many interesting examples can be found in the monographs [26, 188, 224] and also in the edited volumes [6, 205]. We will add some of them for the sake of motivation.

2.1 STACKELBERG GAMES

2.1.1 MARKET ECONOMY

In his monograph about market economy [267], H.v. Stackelberg used by the first time an hierarchical model to describe real market situations. This model especially reflects the case that different decision makers try to realize best decisions on the market with respect to their own, generally different objectives and that they are often not able to realize their decisions independently but are forced to act according to a certain hierarchy. We will consider the simplest case of such a situation where there are only two acting decision makers. Then, this hierarchy divides the two decision makers in one which can handle independently on the market (the so-called leader) and in the other who has to act in a dependent manner (the follower). A leader is able to dictate the selling prices or to overstock the market with his products but in choosing his selections he has to anticipate the possible reactions of the follower since his profit strongly depends not only on his own decision but also on the response of the follower. On the other hand, the choice of the leader influences the set of possible decisions as well as the objectives of the follower who thus has to react on the selection of the leader.

It seems to be obvious that, if one decision maker is able to take on an independent position (and thus to observe and utilize the reactions of the dependent decision maker on his decisions) then he will try to make good use of this advantage (in the sense of making higher profit). The problem he has to solve is the so-called *Stackelberg game*, which can be formulated as follows: Let X and Y denote the set of admissible strategies x and y of the follower and of the leader, respectively. Assume that the values of the choices are measured by means of the functions $f_L(x,y)$ and $f_F(x,y)$, denoting the utility functions of the leader resp. the follower. Then, knowing the selection y of the leader the follower has to select his best strategy $x(y)$ such that his utility function is maximized on X:

$$x(y) \in \Psi(y) := \operatorname*{Argmax}_{x} \{f_F(x,y) : x \in X\}.$$

Being aware of this selection, the leader solves the Stackelberg game for computing his best selection:

$$\text{``}\operatorname*{max}_{y}\text{''}\{f_L(x,y) : y \in Y,\ x \in \Psi(y)\}.$$

If there are more than one person on one or both levels of the hierarchy, then these are assumed to search for an equilibrium (as e.g. a Nash or again a Stackelberg equilibria) between them [110, 260, 263].

Bilevel programming problems are more general than Stackelberg games in the sense that both admissible sets can also depend on the choices of the other decision maker.

2.1.2   COURNOT-NASH EQUILIBRIA

Consider the example where n decision makers (firms) produce one homogeneous product in the quantities $x_i, i = 1, \ldots, n$ [224]. Assume that they all have differentiable, convex, non-negative cost functions $f_i(x_i)$ and get a revenue of $x_i p \left(\sum_{j=1}^{n} x_j \right)$ when selling their products on the common market. Here $p : \mathbb{R} \to \mathbb{R}$ is a so-called inverse demand function describing the dependency of the market price on the offered quantity of that product. Assume that the function p is continuously differentiable, strictly convex and decreasing on the set of positive numbers $\mathbb{R}_{++} = \{z : z > 0\}$, but the function $g(x) = xp(x)$ is concave there. Let $Y_i = [a_i, b_i] \subset \mathbb{R}_{++}$ be a bounded interval where firm i believes to have a profitable production. Then, for computing an optimal quantity yielding a maximal profit the firm i has to solve the problem

$$\max_{x_i} \left\{ x_i p \left(\sum_{j=1}^{n} x_j \right) - f_i(x_i) : x_i \in Y_i \right\} \qquad (2.1)$$

which has an optimal solution $x_i(x_{-i})$, where the abbreviation x_{-i} denotes $x_{-i} = (x_1, \ldots, x_{i-1}, x_{i+1}, \ldots, x_n)^\top$. Now, consider the situation where one of the firms (say firm 1) is able to advantage over the others in the sense that it can fix its produced quantity x_1 first and that all the other firms will react on this quantity. Then, firms $2, \ldots, n$ compute a Nash equilibrium between them by solving the $n - 1$ problems (2.1) for $i = 2, \ldots, n$ simultaneously. Let

$$p\left(\sum_{i=2}^{n} b_i\right) - f_1(0) > 0.$$

This implies that the firm 1 will produce a positive quantity [224]. Suppose that the Nash equilibrium $(x_2(x_1), \ldots, x_n(x_1))^\top$ between the firms $2, \ldots, n$ is uniquely determined for each fixed x_1. Then, firm 1 has to solve the following problem in order to realize its maximal profit:

$$\max_{x_1} \left\{ x_1 p\left(\sum_{j=1}^{n} x_j\right) - f_1(x_1) : x_1 \in Y_1, x_i = x_i(x_1), i = 1, \ldots, n \right\},$$

which can also be posed in the following form:

$$x_1 p\left(\sum_{j=1}^{n} x_j\right) - f_1(x_1) \to \max_{x_1}$$

where $x_1 \in Y_1$

and $x_i \in \operatorname*{Argmax}_{x_i} \left\{ x_i p\left(\sum_{j=1}^{n} x_j\right) - f_i(x_i) : x_i \in Y_i \right\}, i = 2, \ldots, n.$

This is a simple example of a Stackelberg game.

2.1.3 PRINCIPAL-AGENCY PROBLEMS

In modern economics a generalization of this model is often treated within *Principal-Agency Theory*. In the following we will describe one mathematical problem discussed within this theory taken from [231]. Possible generalizations of this model to more than one follower and/or more than one decision variable in both of the upper and lower level problems are obvious and left to the reader.

In the problem discussed in [231] one decision maker, the so-called *principal* has engaged the other, the *agent* to act for, on behalf of, or as representative for him. Both decision makers have made a contract where it is fixed that the principal delegates (some part of) jurisdiction to the agent thus giving him the freedom to select his actions (more or less) according to his own aims only. Hence, having only an expectation about the results of his actions and using an utility function $G : \mathbb{R} \times A \to \mathbb{R}$ for measuring the value of the reward $s(x)$ from the principal against the effort for his action a, the agent tries to maximize the expected utility of his action $a \in A$:

$$\int_X G(s(x), a)g(x|a)dx \to \max_{a \in A}, \qquad (2.2)$$

where X is the set of possible results of the actions $a \in A$ of the agent. The density function $g(x|a)$ is used to describe the probabilities of realizing the result $x \in X$ if the agent uses action $a \in A$. The reward $s(x)$ is paid by the principal to the agent if result x is achieved. The function $s : X \to \mathbb{R}$ is also part of the contract made by both parties. From the view of the principal, the function s describes a system of incentives which is used to motivate the agent to act according to the aims of the principal. Thus, the principal has to select this function such that he achieves his goals as best as possible. Assuming that the principal uses the utility function $H : \mathbb{R} \to \mathbb{R}$ to measure his yield $x - s(x)$ resulting from the activities of the agent and that he uses the same density function g to evaluate the probabilities for realizing the result x, he will maximize the function

$$\int_X H(x - s(x))g(x|a')dx \to \max_{s \in S}, \qquad (2.3)$$

where S is a set of possible systems of incentives and a' solves (2.2) for a fixed function $s(\cdot)$. The model (2.2), (2.3) for describing the principal-agency relationship is not complete without the condition

$$\int_X G(s(x), a')g(x|a')dx \geq c, \qquad (2.4)$$

where a' maximizes (2.2). If this inequality is not satisfied, the agent will not be willing to sign the contract with the principal. Inequality (2.4) is a constraint of the type (1.2) which is used to decide about feasibility of the principal's selection after the follower's reply.

Summing up, the principal's problem is to select $s \in S$ maximizing the function (2.3) subject to the condition that $a' \in A$ solves (2.2) for fixed function $s \in S$ and (2.4) is satisfied. This problem is an example for bilevel programming problems.

In modern economics, this model and its generalizations will be used to describe a large variety of real-life situations, cf. e.g. [92, 93, 112, 145, 166, 167, 231, 237, 238].

2.2   OPTIMAL CHEMICAL EQUILIBRIA

In producing substances by chemical reactions we have often to answer the question of how to compose a mixture of chemical substances such that

- the substance we like to produce really arises as a result of the chemical reactions in the reactor and

- the amount of this substance should clearly be as large as possible or some other (poisonous or etching) substance is desired to be vacuous or at least of a small amount.

It is possible to model this problem as a bilevel optimization problem where the first aim describes the lower level problem and the second one is used to motivate the upper level objective function.

Let us start with the lower level problem. Although the chemists are technically not able to observe in situ the single chemical reactions at higher temperatures, they described the final point of the system by a convex programming problem. In this problem, the entropy functional $f(x, p, T)$ is minimized subject to the conditions that the mass conservation principle is satisfied and masses are not negative. Thus, the obtained equilibrium state depends on the pressure p and the temperature T in the reactor as well as on the masses y of the substances which have been put into the reactor:

$$\sum_{i=1}^{N} c_i(p, T) x_i + \sum_{i=1}^{G} x_i \ln \frac{x_i}{z} \to \min_x$$

$$z = \sum_{j=1}^{G} x_j, \quad Ax = \overline{A}y, \quad x \geq 0,$$

where $G \leq N$ denotes the number of gaseous and N the total number of reacting substances. Each row of the matrix A corresponds to a chemical element, each column to a substance. Hence, a column gives the amount of the different elements in the substances; x is the vector of the masses of the substances in the resulting chemical equilibrium whereas y denotes the initial masses of substances put into the reactor; $\overline{A}$ is a submatrix of A consisting of the columns corresponding to the initial substances. The value of $c_i(p, T)$ gives the chemical potential of a substance which depends on the pressure p and the temperature T [265]. Let $x(p, T, y)$ denote the unique optimal solution of this problem. The variables p, T, y can thus be considered as parameters for the chemical reaction. The problem is now that there exists some desire about the result of the chemical reactions which should be reached as best as possible, as e.g. the goal that the mass of one substance should be as large or as small as possible in the resulting equilibrium. To reach this goal the parameters p, T, y are to be selected such that the resulting chemical equilibrium satisfies the overall goal as best as possible [221]:

$$\langle c, x \rangle \;\to\; \min_{p,T,y}$$
$$(p, T, y) \in Y, \quad x = x(p, T, y).$$

2.3   ENVIRONMENTAL ECONOMICS
2.3.1   WASTE MINIMIZATION

The conditions for the production of some decision maker are often influenced by the results of the economic activities of other manufacturers even if they are not competitors on the market. This will become more clear in an environmental setting when one producer pollutes the environment with by-products of his production and the other manufacturer needs a clean environment as a basis for his own activities. A very simple example for such a situation is a paper producing plant situated at the upper course of some river influencing the natural resource water of a fishery at the lower course of the same river. A higher productivity of the paper plant implies a lower water quality in the river since then more waste is led into the river by the plant. This implies a decreasing stock of fish in the river and hence a decreasing income for the fishery. If the paper plant does not recover for the pollution of the environment, the fishermen only have to pay for the resulting damage, i.e. for the negative external effect caused by the waste production of the plant. Let $g_P(x_1)$, $g_F(x_1, x_2)$ denote the profit of the paper plant and of the fishery, resp., depending on their respective economical efforts x_1 and x_2. Here, the common profit obtained by the fishery depends on the effort of the paper plant and will be decreasing with increasing x_1. Without loss of generality, the functions $g_P(\cdot)$ and $g_F(x_1, \cdot)$ are assumed to be concave, while $g_F(\cdot, x_2)$ is decreasing on the space of non-negative arguments. If both parties try to maximize their respective profits, the market will fail since the fishery will be destroyed by the waste produced by the plant.

Now assume that the government is interested in saving the fishery. Then it has to correct the failure of the market e.g. by a tax which has to be paid by the producer of the external effect [232]. In this case, the external effect will be internalized.

Let for simplicity this tax depend linearly on x_1. Then, the profit function of the paper plant changes to $g_P(x_1) - rx_1$, where r is the tax rate determined by the government and the paper plant will now maximize its profit depending on r. Denote the optimal effort by $x_1(r)$. The larger r the smaller $x_1(r)$ of the paper plant will be. Hence, the damage for the fishery will be decreased by increasing r. This can indeed save the fishery. Consequently, the optimal effort of the fishery $x_2(x_1(r))$ also depends on the tax rate r.

Now, it is the government's task to determine the tax rate. For doing so, e.g. a welfare function can be used measuring the social welfare in dependence on the paper and the fish produced by both industries and on the value of the tax paid by the paper plant. Let, for simplicity, this function be given by $W(x_1, x_2, r)$. Then, the government can decide to determine r such that $W(x_1(r), x_2(x_1(r)), r)$ is maximized where $x_1(r)$ and $x_2(x_1(r))$ are the optimal decisions of the paper plant and the fishery induced by a fixed value of r. In a very natural way, this leads to a formulation as a bilevel programming problem (cf. e.g. [248]):

$$W(x_1, x_2, r) \to \max_{r \geq 0}$$

subject to the conditions that x_1 solves the problem

$$g_P(x_1) - rx_1 \to \max_{x_1 \geq 0},$$

x_2 solves

$$g_F(x_1, x_2) \to \max_{x_2 \geq 0}$$

given r, x_1, and the upper level constraint

$$g_F(x_1, x_2) \geq 0$$

is also satisfied. Clearly, as formulated here, this is rather a problem with three levels than a bilevel one. For multilevel problems the reader is referred e.g. to [4, 22, 41, 56, 287]

For a more detailed discussion of this approach to environmental economy, the interested reader is referred to [283].

In [72, 239], a similar problem has been considered where some authority asks for an optimal waste disposal price and two followers have to solve a two stage economic order quantity model describing the manufacturing of new and the repair of used products in a first shop and the employment of the products in a second one. Here, the waste disposal rate (and hence the repair rate) clearly depends on the waste disposal price.

Some arguments for the necessity of ecological investigations together with some implications of product recovery management for the production can also be found in [273]. These can be used in a similar way to formulate bilevel problems which are helpful to find reasonable decisions.

2.3.2   BIOFUEL PRODUCTION

The following problem has been described in [29]. Led by the high subsidies for the agricultural sector and the need to reduce the environmental pollution associated with automobile emissions, the French government decided to explore the possibilities to encourage the petrochemical industry to use farm crops for producing biofuel. Different nonfood crops as wheat, corn, rapseet, and sunflower can be used for this purpose. Unfortunately, industry's cost for producing biofuel from nonfood crops is significantly higher than it is when hydrocarbon-based raw materials are used to produce fuel. To encourage the industry to use farm output, government incentives in form of tax credits are necessary.

For developing one possible model, the industry is assumed to be neutral and to produce any profitable biofuel. The agricultural sector is represented by a subset of the farms in some region of France. We will describe the model for only one farm, it can easily be enlarged by adding more farms. The farm can either let some part of its land unused or use it for nonfood crops. In each case it will have some revenue either in form of set-aside payments from the government for leaving part of the land fallow or in form of subsidies from the European Union plus income for selling the nonfood crops to the industry. By maximizing the total profit, the farm decides by itself how much of the land will be used to produce nonfood crops and how much will be left fallow. Let $x_f \in \mathbb{R}, x_n \in \mathbb{R}^p, x_c \in \mathbb{R}^q$ denote the amount of land left fallow, used for the different nonfood crops, and used for various kinds of food crops by some farmer. Let e^p denote the summation vector $e^p = (1, \ldots, 1)^\top \in \mathbb{R}^p$. Than, the farmer's problem consists of maximizing the income of the farm

$$\langle p_c, x_c \rangle + \langle p_n + s - c_n, x_n \rangle + \gamma x_f \to \max_{x_f, x_c, x_n} \tag{2.5}$$

$$\langle e^p, x_n \rangle + \langle e^q, x_c \rangle + x_f \leq t \tag{2.6}$$

$$\langle e^p, x_n \rangle + x_f = \sigma_1 t \tag{2.7}$$

$$\langle e^p, x_n \rangle + \langle e^q, x_c \rangle \leq \sigma_2 t \tag{2.8}$$

$$x_c \leq t', x_n \geq 0, x_c \geq 0, x_f \geq 0 \tag{2.9}$$

subject to constraints describing the total amount of arable land of the farm (2.6), restrictions posed by the European Union (2.7) on the percentage of the land either left fallow or used for non-food crops, and such reflecting agronomic considerations (2.8). The first inequality in (2.9) is a special bound for sugar beet production. p_c, p_n denote the income and the price vector for food resp. nonfood crops, s is the vector of subsidies for nonfood crops paid by the European Union, c_n are the farmer's costs for producing nonfood crops and γ is the unit set-aside payment for unused land.

The government plays the role of the leader in this problem. Its aim is to minimize the total value of tax-credits given to the petro-chemical industry minus the savings from the set-aside payments for the land left fallow. Let k_r denote the vector of the unit amounts of biofuel $r, r = 1, \ldots, \omega$, produced from one unit of the different nonfood crops and τ_r be the variable government tax credit given to industry for biofuel r. Then, the government has to solve the problem

$$\sum_{r=1}^{\omega} \tau_r \langle k_r, x_n \rangle - \gamma \langle e^p, x_n \rangle \to \text{``} \min_{\tau} \text{''} \tag{2.10}$$

$$\langle e^p, x_n \rangle \geq t'' \tag{2.11}$$

$$\sum_{r=1}^{\omega} \langle k_r, x_n \rangle \leq H \tag{2.12}$$

$$p_n = p_n(\tau) \geq 0, \tau \geq 0 \tag{2.13}$$

and x_n, x_c, x_f solve (2.5)–(2.9) $\tag{2.14}$

subject to constraints on the available land (2.11), the amount of biofuel produced (2.12) and the price paid by industry for nonfood crops (2.13). In the paper [29], the industry is modeled as a neutral element which means that the prices paid by industry are fixed functions $p_n = p_n(\tau)$ of the tax credits which must not be negative and guarantee some profit for the industry. The model can be enlarged by a third level to include the industry's problem.

2.4 DISCRIMINATION BETWEEN SETS
2.4.1 THE MAIN IDEA

In many situations as e.g. in robot control, character and speech recognition, in certain finance problems as bank failure prediction and credit evaluation, in oil drilling, in medical problems as for instance breast cancer diagnosis, methods for discriminating between different sets are used for being able to find the correct decisions implied by samples having certain characteristics [82, 118, 193, 194, 258, 264]. In doing so, a mapping $\mathcal{T}_0$ is used representing these samples according to their characteristics as points in the input space (usually the n-dimensional Euclidean space) [193]. Assume that this leads to a finite number of different points. Now, these points are classified according to the correct decisions implied by their originals. This classification can be considered as a second mapping $\mathcal{T}_1$ from the input space into the output space given by the set of all possible decisions. This second mapping introduces a partition of the

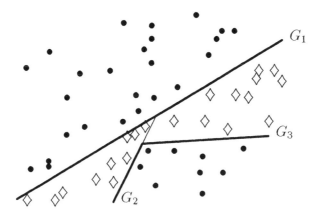

Figure 2.1. Splitting of R^2 into three subsets each containing points of one of the sets $\mathcal{A}$ respectively $\mathcal{B}$ only

input space into a certain number of disjoint subsets such that all points in one and the same subset are mapped to the same decision (via its inverse mapping). For being able to determine the correct decision implied by a new sample we have to find that partition of the input space without knowing the mapping $\mathcal{T}_1$.

Consider the typical case of discriminating between two disjoint subsets $\mathcal{A}$ and $\mathcal{B}$ of the input space $\mathbb{R}^n$ [193]. Then, for approximating this partition, piecewise affine surfaces can be determined separating the sets $\mathcal{A}$ and $\mathcal{B}$ (cf. Fig. 2.1 where the piecewise affine surfaces are given by the bold lines). For the computation of these surfaces an algorithm is given in [193] which starts with the computation of one hyperplane (say G_1) separating the sets $\mathcal{A}$ and $\mathcal{B}$ as best as possible. Clearly, if both sets are separable, then a separating hyperplane is constructed. In the other case, there are some misclassified points. Now, discarding all subsets containing only points from one of the sets, the remaining subsets are partitioned in the same way again, and so on. In Fig. 2.1 this means that after constructing the hyperplane G_1 the upper-left half-space is discarded and the lower-right half-space is partitioned again (say by G_2). At last, the lower-right corner is subdivided by G_3.

2.4.2   MISCLASSIFICATION MINIMIZATION

This algorithm reduces this problem of discriminating between two sets to that of finding a hyperplane separating two finite sets $\mathcal{A}$ and $\mathcal{B}$

of points as best as possible. In [194] an optimization problem has been derived which selects the desired hyperplane such that the number of misclassified points is minimized. For describing that problem, let A and B be two matrices the rows of which are given by the coordinates of the s and t points in the sets $\mathcal{A}$ and $\mathcal{B}$, respectively. Then, a separating hyperplane is determined by an n-dimensional vector w and a scalar γ as $H = \{x \in \mathbb{R}^n : \langle w, x \rangle = \gamma\}$ with the property that

$$Aw > \gamma e^s, \; Bw < \gamma e^t$$

provided that the convex hulls of the points in the sets $\mathcal{A}$ and $\mathcal{B}$ are disjoint. Up to normalization, the above system is equivalent to

$$Aw - \gamma e^s - e^s \geq 0, \; -Bw + \gamma e^t - e^t \geq 0. \tag{2.15}$$

Then, a point in $\mathcal{A}$ belongs to the correct half-space if and only if the given inequality in the corresponding line of the last system is satisfied. Hence, using the step function a_* and the plus function a_+ which are component-wise given as

$$(a_*)_i = \begin{cases} 1 & \text{if } a_i > 0 \\ 0 & \text{if } a_i \leq 0 \end{cases}, \; (a_+)_i = \begin{cases} a_i & \text{if } a_i > 0 \\ 0 & \text{if } a_i \leq 0 \end{cases}$$

we obtain that the system (2.15) is equivalent to the equation

$$e^{s\mathsf{T}}(-Aw + \gamma e^s + e^s)_* + e^{t\mathsf{T}}(Bw - \gamma e^t + e^t)_* = 0. \tag{2.16}$$

It is easy to see that the number of misclassified points is counted by the left-hand side of (2.16). In [194], for $a, c, d, r, u \in \mathbb{R}^l$, the step function is characterized as follows:

$$r = a_*, u = a_+ \iff \begin{cases} \begin{pmatrix} r \\ u \end{pmatrix} = \begin{pmatrix} r - u + a \\ r + u - e^l \end{pmatrix}_+ \\ \text{and } r \text{ is minimal in case of uncertainty} \end{cases}$$

and we have

$$c = d_+ \iff c - d \geq 0, \; c \geq 0, \; c(c - d) = 0.$$

Using both relations, we can transform the problem of minimizing the number of misclassified points or, equivalently, the minimization of the left-hand side function in (2.16) into the following optimization problem

[194]

$$e^{s\top}r + e^{t\top}s \;\to\; \min_{w,\gamma,r,u,p,v}$$

$$
\begin{aligned}
u + Aw - \gamma e^s - e^s &\geq 0 & v - Bw + \gamma e^t - e^t &\geq 0 \\
r &\geq 0 & p &\geq 0 \\
r^\top(u + Aw - \gamma e^s - e^s) &= 0 & p^\top(v - Bw + \gamma e^t - e^t) &\geq 0 \\
-r + e^s &\geq 0 & -p + e^t &\geq 0 \\
u &\geq 0 & v &\geq 0 \\
u^\top(-r + e^s) &= 0 & v^\top(-p + e^t) &= 0
\end{aligned}
$$

This problem is an optimization problem with linear complementarity constraints, a generalized bilevel programming problem. In [193] it is shown that the task of training neural networks can be modeled by a similar problem.

2.5   FURTHER APPLICATIONS

The list of reported applications of bilevel programming is very long and quickly increasing. In the following some more applications will be shortly touched. Of course, this list is far from being complete and is only included to give an impression of the many different fields where bilevel programming applications can be found.

- An hierarchical optimization problem motivated by a variety of defense problems is formulated in [48]. Especially they investigated problems such as strategic offensive and defensive force structure design, strategic bomber force structure and basing as well as allocation of tactical aircraft to missions.

- The Bracken–McGill problem [49] for computing optimal production and marketing decisions subject to the constraint that the firm's minimum share function for each product is not less than some given constant is an example for a principal-agency relationship where the agent's variables do not appear in the upper level problem.

- If a firm is organized in a hierarchical manner with one superior unit and several subordinate units where each subordinate unit is assumed to control a unique set of variables and tries to maximize its own objective function over jointly dependent strategy sets, an example of bilevel programming arises if the superior unit for example wants to optimally allocate the resources between the subordinate units [21, 91, 149, 247, 281].

- An equilibrium facility location problem attempts to find the best location for a new facility of a firm, to compute the production level of this new facility and to plan the shipping patterns such that the locating firm's profit is maximized. The main feature of the equilibrium facility location problem is it that this model accounts for the changes in the market prices and production levels at each of the competing firms resulting from the increase of the overall supply of the products resulting from the production of the new facility [207, 208].

- Different traffic planning or transportation problems with congestion can be found in [35, 198, 200, 204]. An example is the following: A superior unit wants to optimally balancing the transportation, investment and maintenance costs of a traffic network where the users behave according to Wardrop's first principle of traffic equilibrium [198]. A related urban transport problem is considered in [95] where road taxing and transit ticket prices are to be determined.

- In [5] bilevel programming is applied to conflict resolution in international river management. India and Bangladesh share water from the Ganges river and obtain hydroelectric power, irrigation, and flood protection by use of a series of dams in both countries. Both countries make investments in reservoirs and decide about the size of dams and groundwater storage, levels of water use for irrigation and hydroelectric power. Situations where either India or Bangladesh is the leader as well as an arbitrator (the UN) is involved are investigated. The related problem of the Indus basin model is considered in [44].

- Optimum operating configuration of an aluminum plant is investigated in [219]. Here, in the lower level costs are minimized resulting from activities and raw material consumption in the rodding and anode areas of the aluminium smelter. The objective in the upper level maximizes the output of aluminium.

- Several optimum shape design problems can be found e.g. in [224]. One example of such problems is the following: Consider a rigid obstacle and an elastic membrane above it. A rigid obstacle does not give in as a result of the contact pressure of the membrane on it. We are interested in the shape of the membrane above the obstacle. Assume that the obstacle can be described by a function $\chi : \text{cl } \widehat{\Omega} \to \mathbb{R}$, where $\widehat{\Omega} = (0, c_2) \times (0, 1)$ for some positive constant $0 < c_1 \leq c_2$. The function $\chi \in H^2(\text{cl } \widehat{\Omega})$ is assumed to belong to the Sobolev space $H^2(\text{cl } \widehat{\Omega})$ defined on $\text{cl } \widehat{\Omega}$ with

$$\chi \leq 0 \text{ on } \partial\widehat{\Omega} \cup ((c_1, c_2) \times (0, 1)),$$

where $\partial\Omega(\alpha)$ denotes the boundary of $\Omega(\alpha)$. Then, the membrane is given as

$$\Omega(\alpha) = \{(\xi_1, \xi_2) : 0 < \xi_1 < \alpha(\xi_2) \ \forall \ \xi_2 \in (0, 1)\}$$

with $\alpha \in U_{ad} \subseteq C^{0,1}([0, 1])$ belonging to a subset of the space of Lipschitz continuous functions over the interval $[0, 1]$,

$$U_{ad} = \{\alpha \in C^{0,1} : 0 < c_1 \leq \alpha(\xi_2) \leq c_2, \left|\frac{d}{d\xi_2}\alpha(\xi_2)\right| < c_3 \text{ a.e. in } (0, 1)\}$$

with a constant $c_3 > 0$. The boundary-value problem with a rigid obstacle consists in computing a function

$$u \in H^1(\Omega(\alpha)) : -\Delta u \geq f, \ u \geq \chi \text{ in } \Omega(\alpha)$$
$$(\Delta u + f)(u - \chi) = 0 \text{ in } \Omega(\alpha) \qquad (2.17)$$
$$u = 0 \text{ on } \partial\Omega(\alpha)$$

where $f \in L_2(\Omega(\alpha))$ is a force that is perpendicularly applied to the membrane $\Omega(\alpha)$ and Δu abbreviates the Laplace operator. Note that this problem can equivalently be written as a variational inequality in an infinite dimensional space.

The packaging problem with rigid obstacle aims to minimize the area of the domain $\Omega(\alpha)$ under the condition that a part of the membrane has contact with the obstacle throughout a given set Ω_0. This problem (and its generalizations) has various applications in the formulation of other problems as e.g. filtration of liquids in porous media, the lubrication problem, elastic-plastic torsion problems etc.

- In [119] different bilevel models are presented which can be used to minimize greenhouse-gas emissions subject to optimally behaving energy consumers. In one model the tax for greenhouse-gas producing technologies is significantly increased. In a second model, this tax is complemented by subsidies for the development and introduction of new ecological technologies. Comparing numerical calculations show that both approaches can be used to reduce the greenhouse-gas emissions significantly but that the second approach imposes much less additional costs on the consumers.

The linear bilevel programming problem is the one which has mostly been considered in application oriented papers. Among others, such applications are the network design problem [34, 37, 38, 197, 198, 200, 271] and the resource allocation problem [56, 91, 216, 247, 286]. Nonlinear resource allocation problems have been investigated in [145, 281].

Chapter 3

LINEAR BILEVEL PROBLEMS

With this chapter the investigation of bilevel programming problems (1.1), (1.3) starts. We begin with the simplest case, the linear bilevel programming problem. This problem will be illustrated by Figure 3.1 in the one-dimensional case. After that we will investigate the relations between bilevel and other mathematical programming problems, describe properties of and possible solution algorithms for linear bilevel programming problems. The chapter also gives different optimality conditions for linear bilevel problems.

3.1 THE MODEL AND ONE EXAMPLE

Let us start our considerations with the linear bilevel programming problem which is obtained from (1.1), (1.3) if all functions defining this problem are restricted to be affine. At almost all places throughout the book we will consider only problems where the upper level constraints (1.2) depend only on y.

Let for the moment the lower level problem be given as

$$\min_{x}\{\langle c, x\rangle : A^1 x \leq a - A^2 y, x \geq 0\}, \tag{3.1}$$

where $x \in \mathbb{R}^n$, $y \in \mathbb{R}^m$, $a \in \mathbb{R}^p$ and the matrices A^1, A^2 as well as the vector c are of appropriate dimensions. Note that the description of the parametric linear optimization problem in this form is not an essential restriction of the general case provided that we have linear right-hand side perturbations only. The case when the objective function of the lower level problem depends on the parameter as well as the case when both the right-hand side and the objective function are linearly perturbed can be treated in a similar way. If we have nonlinear perturbations or if the

21

matrix A^1 depends on y, then the material in this chapter cannot be applied directly and we refer to the consideration of nonlinear bilevel programming problems. Let for the moment

$$\Psi_L(y) = \underset{x}{\text{Argmin}} \ \{\langle c, x \rangle : A^1 x \le a - A^2 y, \ x \ge 0\}$$

denote the set of optimal solutions of problem (3.1). Then, the bilevel programming problem can be stated as

$$\text{``}\min_{y}\text{''}\{\langle d^1, x \rangle + \langle d^2, y \rangle : A^3 y = b, y \ge 0, x \in \Psi_L(y)\}, \qquad (3.2)$$

where $b \in \mathbb{R}^l$ and all other dimensions are determined such that they match with the above ones. Note again that the formulation with the quotation marks is used to express the uncertainity in the definition of the bilevel problem in case of non-uniquely determined lower level optimal solutions.

Example: Consider the problem of minimizing the upper level objective function

$$3x + y \to \min$$

subject to $1 \le y \le 6$ and x solving the lower level problem

$$\min_{x}\{-x : x + y \le 8, \ 4x + y \ge 8, \ 2x + y \le 13\}.$$

The set of all pairs (x, y) satisfying the constraints of both the lower and the upper levels, denoted by M, as well as the minimization directions of the objective functions of both the lower and the upper level problems, marked by the arrows, are depicted in Figure 3.1.

The feasible set of the lower level problem for a fixed value of y is just the intersection of the set M with the set of all points above the point $(0, y)$ on the y-axis. Now, if the function $f(x, y) = -x$ is minimized on this set, we arrive at a point on the thick line which thus is the optimal solution of the lower level problem:

$$x(y) = \begin{cases} 6.5 - 0.5y & \text{for } 1 \le y \le 3, \\ 8 - y & \text{for } 3 \le y \le 6, \end{cases}$$

In doing this for all values of y between 1 and 6, all points on the thick lines are obtained. Hence, the thick lines give the set of feasible solutions to the upper level problem. On this set, the upper level objective function $F(x, y)$ is to be minimized:

$$F(x(y), y) = \begin{cases} 19.5 - 0.5y & \text{for } 1 \le y \le 3, \\ 24 - 2y & \text{for } 3 \le y \le 6. \end{cases}$$

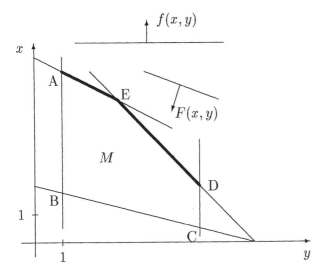

Figure 3.1. The linear bilevel programming problem

The global optimal solution is found at the point $D = (2; 6)^\top$ with an optimal function value of 12. □

From Figure 3.1 we see that, even in its simplest case of linear functions, the bilevel programming problem is a nonconvex and nondifferentiable optimization problem. Hence, the occurrence of local optimal and/or stationary solutions is possible.

It should be mentioned that the bilevel programming problem can be considered as a special case of multilevel programming problems (cf. e.g. [5, 22, 39]). These problems can be modelled as follows:

$$f_1(x_1, \ldots, x_p) \to \min_{x_1}$$
$$g_1(x_1, \ldots, x_p) \leq 0$$
where $x_2, \ldots, x_p$ solves

$$\begin{cases} f_2(x_1, \ldots, x_p) \to \min_{x_2} \\ g_2(x_1, \ldots, x_p) \leq 0 \\ \vdots \\ \text{where } x_p \text{ solves} \\ \begin{cases} f_p(x_1, \ldots, x_p) \to \min_{x_p} \\ g_p(x_1, \ldots, x_p) \leq 0 \end{cases} \end{cases}$$

Here we have a hierarchy of not only two but of p levels and the constraint set of the l-th level problem is given by explicit inequality and equality constraints as well as the solution set mapping of the $l+1$-st level problem. Here, the $l+1$-st level problem has the variables of the first l levels as parameters and is used to choose an optimal decision for the variables of the $l+1$-st level. Hence, these problems are a sequence of (parametric) optimization problems each but one of them having the next problem as part of its constraints. Multilevel linear optimization problems have been investigated e.g. in the papers [22, 39, 41, 214, 284, 287]. We will not consider multilevel programming problems in what follows but restrict ourselves to bilevel problems.

Problems with multiple leaders [259] and multiple followers [21, 145, 294] have been considered in the literature. In most of the cited papers different decision makers on the same level of hierarchy are assumed to behave according to a Nash equilibrium in that level (cf. also [110]).

3.2   THE GEOMETRIC NATURE OF LINEAR BILEVEL PROGRAMMING

Figure 3.1 suggests that the feasible set of the linear bilevel programming problem could be composed by the union of faces of the set M. We will show that this is a general property for the linear bilevel problem.

DEFINITION 3.1 *A point-to-set mapping* $\Gamma : \mathbb{R}^p \to 2^{\mathbb{R}^q}$ *is called polyhedral if its graph*

$$\text{grph } \Gamma := \{(x, y) \in \mathbb{R}^q \times \mathbb{R}^p : x \in \Gamma(y)\} \qquad (3.3)$$

is equal to the union of a finite number of convex polyhedral sets.

Here, a *convex polyhedral set* is the intersection of a finite number of halfspaces [241].

THEOREM 3.1 *The point-to-set mapping* $\Psi_L(\cdot)$ *is polyhedral.*

The proofs of this and other theorems are given in Section 3.7 at the end of this Chapter.

Using Theorem 3.1 the linear bilevel programming can be solved by minimizing the objective function on each of the components of grph $\Psi_L(\cdot)$ subject to the upper level constraints of problem (3.2). Each of these subproblems is a linear optimization problem. Hence, as a corollary of Theorem 3.1 we obtain

COROLLARY 3.1 *If the optimal solution of the lower level problem (3.1) is uniquely determined for each value of the parameter y, then there is*

an optimal solution of the problem (3.2) which is a vertex of the set
$\{(x, y) : A^1 x + A^2 y \leq a, A^3 y = b, x \geq 0, y \geq 0\}$.

This result has been formulated also in [23, 43, 55].

REMARK 3.1 *A generalization to problems with lower level problems having linear constraints and a quasiconcave resp. fractional objective function can be found in [52, 53].*

Using the proof of Theorem 3.1 it is also easy to see that the graph of $\Psi_L(\cdot)$ is connected. This implies that the feasible set of the linear bilevel programming problem (3.2) is also connected. The following example yet shows that this is not longer valid if constraints are added to the upper level problem which depend on the lower level optimal solution.

Example: Consider the problem of minimizing the upper level objective function

$$3x + y \rightarrow \min$$

subject to $0 \leq y \leq 8$, $x \leq 5$ and x solving the lower level problem

$$\min_x \{-x : x + y \leq 8, \; 4x + y \geq 8, \; 2x + y \leq 13, \; 2x - 7y \leq 0\}.$$

Figure 3.2 illustrates the feasible set of this example. The optimal

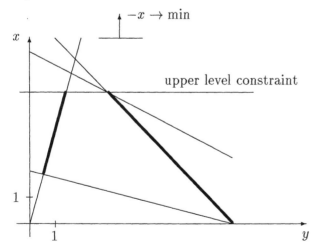

Figure 3.2. The linear bilevel programming problem with disconnected feasible set

solution of the lower level problem is equal to

$$
x(y) = \begin{cases}
3.5y & \text{if } \frac{8}{15} \leq y \leq \frac{13}{8} \\[2mm]
6.5 - 0.5y & \text{if } \frac{13}{8} \leq y \leq 3 \\[2mm]
8 - y & \text{if } 3 \leq y \leq 8
\end{cases}
$$

but only for $y \in \left[\frac{8}{15}, \frac{10}{7}\right] \cup [3, 8]$ the inequality $x \leq 5$ holds.   □

It should be noted that the feasible set of the problem in the last example extremely changes if the upper level constraint $x \leq 5$ is moved into the lower level problem. Then, the feasible set of the upper level problem is again connected and equal to the set of all points $(x(y), y)$ with

$$
x(y) = \begin{cases}
3.5y & \text{if } \frac{8}{15} \leq y \leq \frac{10}{7} \\[2mm]
5 & \text{if } \frac{10}{7} \leq y \leq 3 \\[2mm]
8 - y & \text{if } 3 \leq y \leq 8
\end{cases}
$$

Also note that the position of constraints is not arbitrary from the practical point of view. A constraint placed in the lower level problem restricts the feasible decisions of the follower. If the same constraint is placed in the upper level problem it restricts the decisions of the leader in the sense that the feasibility of his selections is investigated after the followers choice, i.e. it is an implicit constraint to the leaders task. This problem is even more difficult if the follower's selection is not uniquely determined for some values of the parameter since in this case some of the follower's selections $x \in \Psi(\tilde{y})$ can imply that the leader's selection $\tilde{y}$ is feasible but others can reject the same value $\tilde{y}$. It has been shown in [250] that an upper level constraint involving the follower's reply can be moved into the lower level problem provided that there is at least one optimal solution of the lower level problem which is not affected by this move for each value of the parameter.

In general, the optimal solution of the bilevel programming problem is neither optimal for the leader if the follower's objective function is dropped [55] nor is it the best choice for both the leader and the follower if the order of play is deleted. This can be seen in the Example on page 31 and was also the initial point for the investigations in [288]. But, if the game is an hierarchical one then it is not allowed to assume that both decision makers can act simultaneously. This means that the leader has to anticipate that the follower has the right and the possibility to choose an optimal solution under the conditions posed by the leader's selection. Pareto optimal solutions are in general not optimal for minimizing either of the objective functions. In this connection it is necessary to point out that the solution strongly depends on the order of play [27]. This is illustrated in Figure 3.3 where (x^j, y^j) is the respective optimal solution of the problem if the j-th player has the first choice.

— —feasible set 2nd player moves first

– – –feasible set 1st player moves first

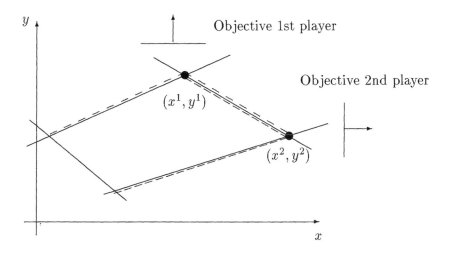

Figure 3.3.   Importance of the order of the play

3.3   EXISTENCE OF OPTIMAL SOLUTIONS

Consider now the linear bilevel programming in a slightly more general setting where the lower level programming problem is replaced by

$$\Psi_L(y) = \underset{x}{\text{Argmin}} \ \{\langle y^1, x \rangle : A^1 x + A^2 y^2 \leq a, x \geq 0\} \qquad (3.4)$$

for $y = (y^1, y^2)^\top \in \mathbb{R}^{n+m}$. Then again, the point-to-set mapping $\Psi_L(\cdot)$ is polyhedral. If the lower level optimal solution is uniquely determined for all parameter values then it is possible to use Weierstraß' Theorem to prove the existence of optimal solutions for the bilevel programming problem. By use of parametric linear programming [220] it is easy to see that if problem (3.4) has a unique optimal solution for all parameter values then these solutions define a continuous function $x : \mathbb{R}^{n+m} \to \mathbb{R}^n$ with $\{x(y)\} = \Psi_L(y)$ for all y for which problem (3.4) has an optimal solution. Then, if this solution is inserted into the bilevel programming problem (3.2), we get the problem

$$\min_y \{\langle d^1, x(y) \rangle + \langle d^2, y \rangle : A^3 y = b, y \geq 0\} \qquad (3.5)$$

with well-defined minimization with respect to y. This is a continuous nondifferentiable optimization problem. Using Weierstraß' Theorem we get

THEOREM 3.2 *If the set $\overline{M} := \{y \geq 0 : A^3 y = b\}$ is not empty and compact, the lower level problem (3.4) has at most one optimal solution for all $y \in \overline{M}$ and the feasible set of problem (3.2) is not empty, then this problem has at least one optimal solution.*

Such (global) optimal solution and also all local optimal solutions can be found at vertices of some polyhedral sets. These sets are the projection of the solution set of one of the following systems of linear (in)equalities onto $\mathbb{R}^n \times \mathbb{R}^m$. Each if these systems corresponds to two index sets $I \subseteq \{1, \ldots, n\}$, $J \subseteq \{1, \ldots, p\}$ and is determined as

$$
\begin{aligned}
(A^1 x + A^2 y^2 - a)_i &= 0, \quad \lambda_i \geq 0, \quad \text{for } i \in J \\
(A^1 x + A^2 y^2 - a)_i &\leq 0, \quad \lambda_i = 0, \quad \text{for } i \notin J \\
(A^{1\mathsf{T}} \lambda + y^1)_j &= 0, \quad x_j \geq 0 \quad \text{for } j \in I \\
(A^{1\mathsf{T}} \lambda + y^1)_j &\geq 0, \quad x_j = 0 \quad \text{for } j \notin I \\
A^3 y &= b, \quad y \geq 0.
\end{aligned}
\qquad (3.6)
$$

If the objective function of the lower level problem does not depend on the choice of the leader, then this implies that optimal solutions of problem (3.1), (3.2) can be found at vertices of the set $\{(x, y) \geq 0 : A^1 x + A^2 y \leq a, A^3 y = b\}$ which is similar to the result of Corollary 3.1 (see also [42] where the result first appears).

The uniqueness assumption of Theorem 3.2 cannot be satisfied at least in the case of linear lower level optimization problems having a parameter in the objective function but no parametrization of the feasible set unless the optimal solution is constant over the set $\overline{M}$. Then, since the leader (or the upper level decision maker) has no control about the real choice of the follower, he will be hard pressed to evaluate his objective function value before he is aware of the follower's real choice. In the literature several ways out of this situation can be found, each requiring some assumptions about the level of cooperation between the players. We will discuss only two of them.

In the first one the leader assumes that he is able to influence the follower to select in each case that solution out of $\Psi_L(y)$ which is the best one for the leader. This results in the so-called *optimistic* or *weak bilevel programming problem*:

$$\min_{x,y}\{\langle d^1, x\rangle + \langle d^2, y\rangle : A^3y = b, y \geq 0, x \in \Psi_L(y)\}, \qquad (3.7)$$

in which the objective function is minimized with respect to both the upper and the lower level variables. An optimal solution of problem (3.7) is called *optimistic optimal solution* of the bilevel programming problem (3.2), (3.4). It has been shown in [43] that this approach is possible e.g. in the case if the follower can participate in the profits realized by the leader. Note that, if this problem has an optimal solution, it can equivalently be posed as

$$\min_{y}\{\varphi_o(y) + \langle d^2, y\rangle : A^3y = b, y \geq 0\}, \qquad (3.8)$$

where

$$\varphi_o(y) = \min_{x}\{\langle d^1, x\rangle : x \in \Psi_L(y)\}$$

[187]. If this way out is not possible the leader is forced to choose an approach bounding the damage resulting from an unfavourable selection of the follower. This is reflected in the so-called *pessimistic* or *strong bilevel problem* :

$$\min_{y}\{\varphi_p(y) + \langle d^2, y\rangle : A^3y = b, y \geq 0\}, \qquad (3.9)$$

where

$$\varphi_p(y) = \max_{x}\{\langle d^1, x\rangle : x \in \Psi_L(y)\}.$$

A *pessimistic optimal solution* of the bilevel programming problem (3.2), (3.4) is defined to be an optimal solution of the problem (3.9).

For the optimistic bilevel problem we can use the polyhedrality of the point-to-set mapping $\Psi_L(\cdot)$ to investigate the existence of optimal solutions. Recall that this implies that the optimistic bilevel programming can be decomposed into a finite number of linear optimization problems the best optimal solution of which solves the original problem. This implies

THEOREM 3.3 *Consider the optimistic bilevel programming problem (3.7) and let the set $M := \{(x,y) \geq 0 : A^1x + A^2y^2 \leq a, A^3y = b\}$ be nonempty and bounded. Then, if the feasible set of the problem (3.7) is not empty, problem (3.7) has at least one optimal solution.*

Here again we can restrict the search for solutions to the vertices of convex polyhedral sets (3.6). If the pessimistic bilevel problem is used then the existence of optimal solutions is not guaranteed in general as can be seen using the following

Example: Consider the problem

$$-x_2 - 10y_1 - 10y_2 \to \underset{y}{\text{``min''}}$$

subject to $0 \le y_i \le 1$, $i = 1, 2$ and

$$x \in \Psi_L(y) := \underset{x}{\text{Argmin}} \ \{-\langle y, x \rangle : \ x_1 + x_2 \le 3, \ -x_1 + x_2 \le 1,$$
$$x_1 - x_2 \le 1, \ x \ge 0\}.$$

The feasible set and the minimization direction for the objective function of this problem are shown in Figure 3.3. Then,

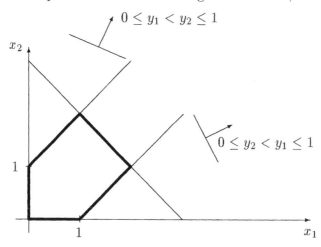

Figure 3.4.   Lower level feasible set and minimization directions

$$\Psi_L(y) = \begin{cases} \left\{ \begin{pmatrix} 2 \\ 1 \end{pmatrix} \right\} & \text{if } 0 \le y_2 < y_1 \le 1 \\[2ex] \left\{ \begin{pmatrix} 1 \\ 2 \end{pmatrix} \right\} & \text{if } 0 \le y_1 < y_2 \le 1 \\[2ex] \text{conv} \left\{ \left\{ \begin{pmatrix} 2 \\ 1 \end{pmatrix} \right\}, \left\{ \begin{pmatrix} 1 \\ 2 \end{pmatrix} \right\} \right\} & \text{if } 0 < y_1 = y_2 \le 1 \\[2ex] M(0,0) & \text{if } 0 = y_1 = y_2, \end{cases}$$

where $M(0,0)$ denotes the feasible set of the follower's problem. Hence, for $y_2^k > y_1^k$, $\lim\limits_{k \to \infty} y_1^k = \lim\limits_{k \to \infty} y_2^k = 1$ we have

$$\lim_{k \to \infty} \varphi_p(y_1^k, y_2^k) = \lim_{k \to \infty} (-10y_1^k - 10y_2^k - x_2(y_1^k, y_2^k)) = -22.$$

Since $x_2 \leq 2$, the optimal objective function value of the bilevel problem cannot be below -22, but for $y_1 = y_2 = 1$ we have $\varphi_p(1,1) = -1$ and the upper level objective function value is equal to -21. Hence, the infimal value of $\varphi_p(\cdot)$ is equal to -22 and it is not attained. The pessimistic bilevel problem has no optimal solution.   □

Here, the infimal value of the objective function in (3.2) again corresponds to a vertex $(\tilde{x}, \tilde{y})$ of some convex polyhedral set. This vertex is optimal if it is feasible for the problem (3.9) which is the case at least if $\Psi_L(\tilde{y})$ contains a unique point.

3.4   RELATIONS TO OTHER MP PROBLEMS

As already shown in the introduction, bilevel programming is closely related to other problems in mathematical programming. Here we will give two more connections which are not so obvious.

3.4.1   MULTICRITERIA OPTIMIZATION

The relations between bilevel programming and bicriteria optimization can easily be illustrated considering Figure 3.1. Here, the line between the points A and B constitutes the set of all Pareto optimal (cf. e.g. [130]) points of the problem

$$\text{``}\min_{x,y}\text{''} \left\{ \begin{pmatrix} c & 0 \\ d^1 & d^2 \end{pmatrix} \begin{pmatrix} x \\ y \end{pmatrix} : A^1 x + A^2 y \leq a,\ A^3 y = b,\ x, y \geq 0 \right\},$$

where the same notations have been used as in formulae (3.1), (3.2). Note, that the only efficient point which is feasible for the bilevel programming problem is point A which has the worst possible function value for the upper level objective function. Hence, in general it is not possible to use methods of multi-criteria optimization for solving bilevel programming problems directly (cf. also [66, 199, 288]). It is easy to see that an (optimistic or pessimistic) optimal solution of the bilevel programming problem is Pareto optimal for the corresponding bicriteria optimization problem if $c = \alpha d^1$ for some $\alpha \geq 0$. But this is in general not true if c is not parallel to d^1 [201]. A related result is also given in [204]. This is also illustrated by the following example:

Example: Consider the linear bilevel programming problem:

$$2x_2 - y \to \min_y$$
$$0 \le y \le 3$$
$$x \in \Psi(y),$$

where

$$\Psi(y) = \operatorname*{Argmin}_x \{-x_1 : 100x_1 - x_2 \le 1, \ x_2 \le y, \ x \ge 0\}.$$

Then, minimization of $-x_1$ implies

$$100x_1 = 1 + x_2 = 1 + y$$

or

$$\Psi(y) = \left\{ \begin{pmatrix} (1+y)/100 \\ y \end{pmatrix} \right\} \text{ for } 0 \le y \le 3.$$

Inserting this into the upper level objective function leads to

$$2x_2 - y = 2y - y = y \to \min : 0 \le y \le 3.$$

Thus,

$$y^* = 0, x_1^* = 1/100, x_2^* = 0$$

is the unique (global) optimal solution of the linear bilevel programming problem.

On the other hand, this solution is not efficient (Pareto optimal) and also not weakly efficient for the bicriteria optimization problem

$$\begin{pmatrix} 2x_2 - y \\ -x_1 \end{pmatrix} \to \text{ `` } \min_{y,x_1,x_2} \text{ ''}$$
$$0 \le y \le 3, \ 100x_1 - x_2 \le 1$$
$$x_2 \le y, \ x \ge 0$$

since it is dominated by

$$\hat{y} = 3, \hat{x}_1 = 1/100, \hat{x}_2 = 0$$

and also by each point

$$\tilde{y} = 3 - \varepsilon, \tilde{x}_1 = (1+\varepsilon)/100, \tilde{x}_2 = \varepsilon$$

for each sufficiently small $\varepsilon > 0$:

The vector-valued objective function values for the bicriteria optimization problem for the three points $(x^*, y^*), (\hat{x}, \hat{y}), (\tilde{x}, \tilde{y})$ are

$$\begin{pmatrix} 2x_2^* - y^* \\ -x_1^* \end{pmatrix} = \begin{pmatrix} 0 \\ -1/100 \end{pmatrix} \gneqq \begin{pmatrix} 2\hat{x}_2 - \hat{y} \\ \hat{x}_1 \end{pmatrix} = \begin{pmatrix} -3 \\ -1/100 \end{pmatrix}$$

resp.

$$\begin{pmatrix} 2x_2^* - y^* \\ -x_1^* \end{pmatrix} = \begin{pmatrix} 0 \\ -1/100 \end{pmatrix} > \begin{pmatrix} 2\tilde{x}_2 - \tilde{y} \\ \tilde{x}_1 \end{pmatrix} = \begin{pmatrix} -3 + 3\varepsilon \\ -(1 + \varepsilon)/100 \end{pmatrix}$$

for sufficiently small $\varepsilon > 0$. This shows that the optimal solution of a linear bilevel programming problem need not to be efficient for the corresponding vectorial optimization problem. $\qquad\square$

In [102] it has been shown that the relations between linear bilevel and linear multicriteria programming problems are closer than the above observation suggests. Namely, for each linear bilevel programming problem, there is some linear multicriteria problem such that the global optimal solution of the first problem and an optimal solution for minimizing the upper level objective function on the set of Pareto optimal points of the second problem coincide. And, vice versa, the problem of minimizing a linear function on the set of Pareto optimal points of some linear multicriteria optimization problem can be transformed into a bilevel programming problem.

Let us explain the ideas in [102] showing these relations.

First, consider a linear bilevel programming problem (3.1), (3.2) and construct a multicriteria problem such that the bilevel problem is the same as finding a best efficient solution of this problem. Let A be a (quadratic) submatrix of full row rank

$$\operatorname{rg} A = \operatorname{rg} \left(A^2, A^3, -E \right)^{\mathsf{T}}$$

composed of the rows of the matrix $\left(A^2, A^3, -E \right)^{\mathsf{T}}$. Construct the multicriteria optimization problem

$$\begin{pmatrix} 0 & A \\ 0 & -e^{m\mathsf{T}}A \\ c^{\mathsf{T}} & 0 \end{pmatrix} \begin{pmatrix} x \\ y \end{pmatrix} \to \text{``}\min_{x,y}\text{''} \tag{3.10}$$
$$A^1 x + A^2 y \leq a, \ A^3 y = b, \quad x, y \geq 0.$$

THEOREM 3.4 ([102]) *A point $(\overline{x}, \overline{y})$ is feasible to (3.2) with the lower level problem (3.1) if and only if it is a Pareto optimal solution of the problem (3.10).*

COROLLARY 3.2 ([102]) *The bilevel programming problem (3.1), (3.2) is equivalent to the problem of minimizing the upper level objective function $\langle d^1, x \rangle + \langle d^2, y \rangle$ over the set of Pareto optimal solutions for the multicriteria problem (3.10).*

Now, we show how to come back from the problem of finding a best Pareto optimal solution in some multicriteria optimization problem to a bilevel one. Let

$$\text{"min"}\{Cx : Ax \leq a\} \tag{3.11}$$

be a multicriteria optimization problem having k objective functions. Then, for any point y with $Ay \leq a$ every optimal solution of the following problem is a Pareto optimal solution for (3.11):

$$\min_x \{\langle e^k, Cx \rangle : C(y - x) \geq 0, \ Ax \leq a\}. \tag{3.12}$$

Moreover, by variation of the point y over the feasible set of problem (3.11) we can compute the whole set of Pareto optimal solutions for (3.11).

Let there be a measure $\langle d, x \rangle$ for the quality of some Pareto optimal solution of (3.12) which is to be minimized. Then, the question of finding the best Pareto optimal solution of (3.12) can be restated as

$$\min_{x,y}\{\langle d, x \rangle : Ay \leq a, \ x \text{ solves problem (3.12) for fixed } y\}. \tag{3.13}$$

THEOREM 3.5 ([102]) *For $\overline{x} \in \mathbb{R}^n$ the following statements are equivalent:*

- *$\overline{x}$ is a Pareto optimal solution of the problem (3.11),*

- *$\overline{x}$ is an optimal solution of the problem (3.12) for $y = \overline{x}$,*

- *there exists a point $\overline{y}$ such that $(\overline{x}, \overline{y})$ is a feasible solution to (3.13).*

In [291] the relations between the set of Pareto optimal points (for the problem where the objective functions of all decision makers are handled equally) and the multilevel programming problem is also investigated. It is shown that in the bilevel case there is always a point which is at the same time Pareto optimal for the bicriteria problem and feasible for the bilevel one. Such a point need not to exist in the multilevel case. In [288] the question is posed if the solution of a linear bilevel problem is more likely to be accepted by the practitioners than an Pareto optimal solution which gives both the follower and the leader a higher profit. Then, various procedures are proposed which can be used to "improve" the bilevel optimal solution.

3.4.2   LINEAR 0-1 PROGRAMMING

The linear bilevel programming problem (3.4), (3.7) in the optimistic approach can be equivalently transformed into a one-level problem with complementarity constraints

$$\langle d^1, x \rangle + \langle d^2, y \rangle \to \min_{x,y}$$
$$A^3 y = b, \; y \geq 0,$$
$$A^1 x + A^2 y^2 - a \leq 0, \; x \geq 0, \; \lambda \geq 0, \; A^{1\top}\lambda + y^1 \geq 0,$$
$$\langle A^1 x + A^2 y^2 - a, \lambda \rangle = 0, \; \langle x, A^{1\top}\lambda + y^1 \rangle = 0.$$

There are several ways to treat the complementarity conditions. One is the use of zero-one variables to transform them into linear conditions [99] which will be outlined in the following. Let $M > 0$ be a sufficiently large constant. Then, the conditions

$$A^1 x + A^2 y^2 - a \leq 0, \; \lambda \geq 0, \; \langle A^1 x + A^2 y^2 - a, \lambda \rangle = 0$$

are equivalent to

$$-Mz \leq A^1 x + A^2 y^2 - a \leq 0, \; M(e^p - z) \geq \lambda \geq 0, \; z_i \in \{0, 1\}, \; i = 1, \ldots, p$$

[158]. If problem (3.4), (3.7) has a bounded optimal solution, this can be used to compose the equivalent linear mixed-integer programming problem:

$$\langle d^1, x \rangle + \langle d^2, y \rangle \to \min_{x,y}$$
$$A^3 y = b, \; y \geq 0,$$
$$-Mz^1 \leq A^1 x + A^2 y^2 - a \leq 0, \; M(e^p - z^1) \geq \lambda \geq 0,$$
$$Mz^2 \geq A^{1\top}\lambda + y^1 \geq 0, \; M(e^n - z^2) \geq x \geq 0,$$
$$z_i^1 \in \{0, 1\}, \; i = 1, \ldots, p, \; z_j^2 \in \{0, 1\}, \; j = 1, \ldots, n.$$

There are different attempts to go the opposite way [11, 279]. These ideas are not only interesting from its own but also for clarifying the complexity of the algorithmic solution of the linear bilevel programming problem [25, 36, 116, 133]. The approaches in [11, 279] use upper level linear constraints depending on the follower's variables, resulting in a problem which is a little bit more general than problem (3.4), (3.7). Let

$$\langle c, x \rangle + \langle d, u \rangle \to \min_{x,u}$$
$$Ax + Bu \leq b, \; x \geq 0, \; u \in \{0, 1\}^p \tag{3.14}$$

be a mixed 0-1 optimization problem. The restriction to 0-1 variables u_i is of no lost of generality. The transformation is based on the knowledge that the optimal value of the simple program

$$\min_v \{-v : v \leq u, \ v \leq (1 - u)\} = -\min\{u, 1 - u\}$$

which results in $u \in \{0, 1\}$ if the optimal objective function value is restricted (in the upper level) to zero. This leads to the following problem equivalent to (3.14) [11]:

$$\begin{aligned}
&\langle c, x \rangle + \langle d, u \rangle \to \min_{x,u} \\
&Ax + Bu \leq b, \ x \geq 0, \ 0 \leq u \leq 1, \ v = 0 \\
&\text{where } v \in \operatorname*{Argmin}_v \ \{-\langle e^p, v \rangle : v \leq u, \ v \leq e^p - u\}.
\end{aligned} \qquad (3.15)$$

It is an important implication of this result that not only the problems (3.14) and (3.15) are equivalent but that also certain algorithms solving the problems compute the same sequence of subproblems.

3.5   OPTIMALITY CONDITIONS

In the following we will give three kinds of optimality conditions for linear bilevel programming problems. They all use equivalent optimization problems to (3.4), (3.7) as starting point. The first one is a nondifferentiable equivalent to the problem having the Karush-Kuhn-Tucker conditions in place of the lower level problem. Applying the optimality conditions for locally Lipschitz continuous problems to the letter one, we get F. John respectively Karush-Kuhn-Tucker type necessary optimality conditions for the bilevel programming problem. Note that this approach is valid only in the optimistic case. The second approach applies both to the optimistic and the pessimistic formulations. It uses the notion of a region of stability for linear parametric optimization where the optimal basis of the problem remains optimal. Here, local optimality for the bilevel problem means that the objective function value cannot decrease on each of the regions of stability related to optimal basic solutions of the lower level problem for the parameter under consideration. The last approach is a condition for a global optimal solution of the bilevel programming problem and uses linear programming duality to formulate a parametric nonconvex quadratic optimization problem. It is shown that the optimal function value of this problem is equal to zero if and only if an optimal solution corresponds to a feasible solution for the bilevel problem. Thus, global optimality for the bilevel problem is related to the lower bound for the parameter values leading to zero optimal function values of the quadratic problem.

3.5.1 KKT CONDITIONS

The necessary and sufficient optimality conditions for the linear lower level problem (3.4)

$$\Psi_L(y) = \operatorname*{Argmin}_x \; \{\langle y^1, x \rangle : A^1 x + A^2 y^2 \leq a, x \geq 0\}$$

are

$$A^1 x + A^2 y^2 - a \leq 0, \; x \geq 0, \; \lambda \geq 0, \; A^{1\top}\lambda + y^1 \geq 0,$$
$$\langle A^1 x + A^2 y^2 - a, \lambda \rangle = 0, \; \langle x, A^{1\top}\lambda + y^1 \rangle = 0$$

which can equivalently be written as

$$\min\{a - A^1 x - A^2 y^2, \lambda\} = 0, \; \min\{x, A^{1\top}\lambda + y^1\} = 0,$$

where the 'min' operator is understood component-wise. Denote

$$F(x, y, \lambda) = \begin{pmatrix} a - A^1 x - A^2 y^2 \\ A^{1\top}\lambda + y^1 \end{pmatrix}; \; G(x, y, \lambda) = \begin{pmatrix} \lambda \\ x \end{pmatrix}$$

This implies that the linear bilevel programming problem in its optimistic formulation (3.7) can be equivalently posed as

$$\begin{aligned} \langle d^1, x \rangle + \langle d^2, y \rangle &\to \min_{x,y} \\ A^3 y = b, y &\geq 0, \\ \min\{F(x, y, \lambda), G(x, y, \lambda)\} &= 0 \end{aligned} \tag{3.16}$$

Problem (3.16) is a nondifferentiable optimization problem with Lipschitz continuous constraints. Necessary optimality conditions can be obtained by use of Clarke's generalized derivative [61]. Note that the generalized derivative for a pointwise minimum $\min\{u(t), v(t)\}$ of two differentiable functions $u, v : \mathbb{R} \to \mathbb{R}$ is included in the convex hull of the derivatives of the functions if both are active [61]:

$$\partial \min\{u(t), v(t)\} \begin{cases} = \{u'(t)\} & \text{if } u(t) < v(t) \\ \subseteq \operatorname{conv} \{u'(t), v'(t)\} & \text{if } u(t) = v(t) \\ = \{v'(t)\} & \text{if } u(t) > v(t) \end{cases}$$

Thus, the following Lagrangian seems to be appropriate for describing the optimality conditions for the problem (3.16) [249]:

$$L(x, y, \lambda, \kappa_0, \kappa, \eta, \gamma, \xi) = \kappa_0(\langle d^1, x \rangle + \langle d^2, y \rangle)$$
$$+ \; \langle A^3 y - b, \kappa \rangle + \langle F(x, y, \lambda), \eta \rangle + \langle G(x, y, \lambda), \gamma \rangle - \langle y, \xi \rangle.$$

The following theorem gives F. John type conditions. The theorem together with its proof originates from an application of a result in [249] to the linear bilevel programming problem. Besides the derivatives of the Lagrangian with respect to the variables x, y, λ, complementarity conditions will appear in the necessary optimality conditions with respect to all constraints. In short, the theorem shows the existence of a vector $(\kappa_0, \kappa, \eta, \gamma, \xi)$ satisfying the following system:

$$\nabla_x L(x^0, y^0, \lambda^0, \kappa_0, \kappa, \eta, \gamma, \xi) = 0$$
$$\nabla_y L(x^0, y^0, \lambda^0, \kappa_0, \kappa, \eta, \gamma, \xi) = 0$$
$$\nabla_\lambda L(x^0, y^0, \lambda^0, \kappa_0, \kappa, \eta, \gamma, \xi) = 0$$
$$F_i(x^0, y^0, \lambda^0)\eta_i = G_i(x^0, y^0, \lambda^0)\gamma_i = 0, \quad i = 1, \ldots, n+p$$
$$\langle y^0, \xi \rangle = 0, \quad \kappa_0 \geq 0, \quad \xi \geq 0$$
$$\eta_i \gamma_i \geq 0, \quad i = 1, \ldots, n+p.$$

Here and in what follows, the symbol $\nabla z(x)$ denotes the gradient of the function $z : \mathbb{R}^p \to \mathbb{R}$ at the point x. The gradient is a row vector. We will use $\nabla_x z(x, y)$ to indicate that the gradient is taken only with respect to x while keeping y fixed.

Surprisingly, in these equations we do not have non-negativity constraints for the multipliers η and γ but only, that they both have the same sign in their components. This comes from the nondifferentiability of the last constraint in (3.16).

THEOREM 3.6   *If* (x^0, y^0) *is a local minimizer of problem (3.7) then there exist* $\lambda^0 \in \mathbb{R}^p$ *and a nonvanishing vector* $(\kappa_0, \kappa, \eta^1, \eta^2, \gamma^1, \gamma^2, \xi) \in \mathbb{R} \times \mathbb{R}^l \times \mathbb{R}^p \times \mathbb{R}^n \times \mathbb{R}^p \times \mathbb{R}^n \times \mathbb{R}^m$ *satisfying*

$$\kappa_0 d^1 - A^{1\top}\eta^1 + \gamma^2 = 0$$
$$\kappa_0 d^2 + A^{3\top}\kappa - \xi - \begin{pmatrix} -A^{2\top}\eta^1 \\ \eta^2 \end{pmatrix} = 0$$
$$A^1 \eta^2 + \gamma^1 = 0$$
$$(a - A^1 x^0 - A^2 y^{20})_i \eta_i^1 = 0, \quad \lambda_i^0 \gamma_i^1 = 0, \quad \forall i = 1, \ldots, p,$$
$$(A^{1\top}\lambda^0 + y^{10})_i \eta_i^2 = 0, \quad x_i^0 \gamma_i^2 = 0, \quad \forall i = 1, \ldots, n$$
$$\eta_i^1 \gamma_i^1 \geq 0, \quad \forall i : (a - A^1 x^0 - A^2 y^{20})_i = \lambda_i^0 = 0$$
$$\eta_i^2 \gamma_i^2 \geq 0, \quad \forall i : (A^{1\top}\lambda^0 + y^{10})_i = x_i^0 = 0$$
$$\xi \geq 0, \quad y_i^0 \xi_i = 0 \; \forall i = 1, \ldots, m, \quad \lambda \geq 0, \quad \kappa_0 \geq 0.$$

Note that for a feasible point (x^0, y^0) the relation $F_i(x^0, y^0, \lambda) = 0 < G_i(x^0, y^0, \lambda)$ together with $G_i(x^0, y^0, \lambda)\gamma_i = 0$ implies that $\eta_i \gamma_i = 0$. Hence, the inequality $\eta_i \gamma_i \geq 0$ is indeed satisfied for all $i = 1, \ldots, n+p$ in Theorem 3.6.

Theorem 3.6 is a Fritz John type necessary optimality condition. To obtain a Karush-Kuhn-Tucker type condition, a regularity assumption is necessary which is given below. Let (x^0, y^0, λ^0) be feasible for (3.16). The following relaxation is connected with problem (3.16):

$$\langle d^1, x \rangle + \langle d^2, y \rangle \;\to\; \min_{x,y}$$

$$
\begin{aligned}
&A^3 y = b, y \geq 0, \\
&F_i(x, y, \lambda) = 0, &&\text{if } F_i(x^0, y^0, \lambda^0) = 0, \\
&F_i(x, y, \lambda) \geq 0, &&\text{if } F_i(x^0, y^0, \lambda^0) > 0, \\
&G_i(x, y, \lambda) = 0, &&\text{if } G_i(x^0, y^0, \lambda^0) = 0, \\
&G_i(x, y, \lambda) \geq 0, &&\text{if } G_i(x^0, y^0, \lambda^0) > 0.
\end{aligned}
\tag{3.17}
$$

We will need the following generalized Slater's condition for this problem:

DEFINITION 3.2 *The generalized Slater's condition is satisfied for problem (3.17) if the gradients of all the equality constraints (with respect to (x, y, λ)) are linearly independent and there exists a feasible point $(\tilde{x}, \tilde{y}, \tilde{\lambda})$ satisfying all the inequality constraints as proper inequalities.*

Any feasible solution of problem (3.17) is also feasible for problem (3.16). Hence, a local optimal solution of problem (3.16) is also a local optimal solution of problem (3.17). The opposite statements are in general not true. The generalized Slater's condition can be assumed to be satisfied for problem (3.17) but not for problem (3.16) as long as the strict complementarity slackness condition does not hold (see Theorem 5.11). The F. John necessary optimality conditions for problem (3.17) are weaker than the conditions of Theorem 3.6 since they do not contain the conditions $\eta_i \gamma_i \geq 0$ for i with $F(x^0, y^0, \lambda^0) = G(x^0, y^0, \lambda^0) = 0$.

The original version of the following result for optimization problems with variational inequality constraints can be found in [249].

THEOREM 3.7 *If (x^0, y^0) is a local minimizer of problem (3.7) and the generalized Slater's condition is satisfied for problem (3.17) then there exist $\lambda^0 \in \mathbb{R}^p$ and a vector $(\kappa, \eta^1, \eta^2, \gamma^1, \gamma^2, \xi) \in \mathbb{R} \times \mathbb{R}^l \times \mathbb{R}^p \times \mathbb{R}^n \times \mathbb{R}^p \times \mathbb{R}^n \times \mathbb{R}^m$ satisfying*

$$d^1 - A^{1\top} \eta^1 + \gamma^2 = 0$$

$$d^2 + A^{3\top} \kappa - \xi - \begin{pmatrix} -A^{2\top} \eta^1 \\ \eta^2 \end{pmatrix} = 0$$

$$A^1 \eta^2 + \gamma^1 = 0$$

$$(a - A^1 x^0 - A^2 y^{20})_i \eta_i^1 = 0, \quad \lambda_i^0 \gamma_i^1 = 0, \quad \forall i = 1, \ldots, p,$$

$$(A^{1\top} \lambda^0 + y^{10})_i \eta_i^2 = 0, \quad x_i^0 \gamma_i^2 = 0, \quad \forall i = 1, \ldots, n$$

$$\eta_i^1 \gamma_i^1 \geq 0, \quad \forall i : (a - A^1 x^0 - A^2 y^{20})_i = \lambda_i^0 = 0$$

$$\eta_i^2 \gamma_i^2 \geq 0, \quad \forall i : (A^{1\top} \lambda^0 + y^{10})_i = x_i^0 = 0$$

$$\xi \geq 0, \quad y_i^0 \xi_i = 0 \;\; \forall i = 1, \ldots, m, \quad \lambda \geq 0.$$

3.5.2   NONEXISTENCE OF DESCENT DIRECTIONS

3.5.2.1   THE OPTIMISTIC CASE

Starting here we will change the notion of an optimistic optimal solution of the bilevel programming problem a little bit to one which is more oriented at our initial point that the leader has control over y only. If problem (3.7) has an optimal solution, it is equivalent to problem (3.8). But with respect to (3.8) we can define the optimality only with respect to the variable y and can add some corresponding $x \in \Psi(y)$ with $\langle d^1, x \rangle = \varphi_o(y)$ as some corresponding realization of the follower. In a way similar to the pessimistic optimal solution, from now on a point $(\overline{x}, \overline{y})$ is called a *(local) optimistic optimal solution* of the bilevel problem if $\overline{y}$ is a (local) optimal solution of the problem (3.8) and $\overline{x} \in \Psi(\overline{y})$ is such that $\langle d^1, \overline{x} \rangle = \varphi_o(\overline{y})$.

Consider an optimal vertex $\overline{x}$ of the lower level problem

$$\Psi_L(y) = \operatorname*{Argmin}_{x} \ \{ \langle y^1, x \rangle : A^1 x + A^2 y^2 = a, x \geq 0 \} \qquad (3.18)$$

for $y = \overline{y}$. Note that the concentration to equality constraints is of no loss of generality, but the ideas below are easier to describe for this model. Let $\mathbf{B}$ be a corresponding basic matrix, i.e. a quadratic submatrix of A^1 with rg $\mathbf{B} = $ rg A^1 and let the corresponding basic solution be given as $\overline{x} = x^{\mathbf{B}}(\overline{y}) = (x_B^{\mathbf{B}}(\overline{y}), x_N^{\mathbf{B}}(\overline{y}))^\top$ with the formulae

$$x_B^{\mathbf{B}} = \mathbf{B}^{-1}(a - A^2 \overline{y}^2) \geq 0, \ \ x_N^{\mathbf{B}} = 0.$$

Also, for simplicity we assume that rg $A^1 = p$. If this is not the case, all investigations are done in some affine submanyfold of $\mathbb{R}^m$ determined by solvability of the linear system of equations $A^1 x + A^2 y^2 = a$. This manyfold gives additional upper level constraints on y. Then, optimality of the basic matrix $\mathbf{B}$ is guaranteed if

$$\overline{y}_B^{1\top} \mathbf{B}^{-1} A^1 - \overline{y}^{1\top} \leq 0,$$

where $\overline{y}_B^1$ is the subvector of $\overline{y}^1$ corresponding to basic variables. This basic matrix need not be uniquely determined.

DEFINITION 3.3 *The region of stability* $\mathcal{R}(\mathbf{B})$ *is the set of all parameters y such that $\mathbf{B}$ remains an optimal basic matrix:*

$$\mathcal{R}(\mathbf{B}) = \{ y : \mathbf{B}^{-1}(a - A^2 y^2) \geq 0, \ y_B^{1\top} \mathbf{B}^{-1} A^1 - y^{1\top} \leq 0 \}. \qquad (3.19)$$

The number of different regions of stability is finite, each of them is a convex polyhedral set. For each sufficiently small perturbation $\tilde{y}$ of $\overline{y}$ such that problem (3.18) has a solution there exists at least one basic matrix $\mathbf{B}$ of (3.18) at $(\overline{x}, \overline{y})$ such that both $\overline{y}, \tilde{y} \in \mathcal{R}(\mathbf{B})$ (cf. Figure 3.5). Due to the vertex property of optimal solutions in Corollary 3.1 this implies that, in order to verify (local) optimality of a feasible point $(\overline{x}, \overline{y})$ for problem (3.7) we have to check if $(x^{\mathbf{B}}(\overline{y}), \overline{y})$ is an optimal solution of the problem (3.7) restricted to each of the regions of stability. Let

$$\varphi_o(\overline{y}) = \min_x \{\langle d^1, x \rangle : x \in \Psi_L(\overline{y})\}$$

denote the upper level objective function value of an optimistic optimal solution of the lower level problem and consider the set $\mathcal{M}(\overline{y})$ of all basic matrices corresponding to the basic solutions $(\tilde{x}, \overline{y})$ of this problem:

$$\mathcal{M}(\overline{y}) = \{\mathbf{B} : \overline{y} \in \mathcal{R}(\mathbf{B}), \ \langle d_B^1, \mathbf{B}^{-1}(a - A^2 \overline{y}^2) \rangle = \varphi_o(\overline{y})\},$$

where d_B^1 denotes the basic components of d^1. The set $\mathcal{M}(\overline{y})$ can contain more than one matrix and the matrices in $\mathcal{M}(\overline{y})$ can correspond to more than one optimal solution of the lower level problem. In this case, however, not only the follower but also the leader is indifferent between all these different solutions.

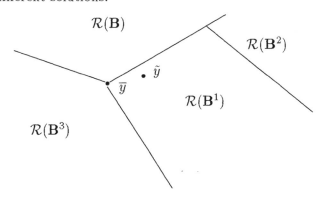

Figure 3.5.   The regions of stability

Let $\overline{y}, \tilde{y} \in \mathcal{R}(\mathbf{B})$ for some basic matrix $\mathbf{B}$. Then, for the corresponding optimal solutions $x^{\mathbf{B}}(\overline{y}), x^{\mathbf{B}}(\tilde{y})$, the relations

$$x_B^{\mathbf{B}}(\overline{y}) - x_B^{\mathbf{B}}(\tilde{y}) = \mathbf{B}^{-1} A^2 (\overline{y}^2 - \tilde{y}^2), \ x_N^{\mathbf{B}}(\overline{y}) - x_N^{\mathbf{B}}(\tilde{y}) = 0$$

hold. Thus, for the difference of the upper level objective function values the relation

$$\varphi_o(\overline{y}) + \langle d^2, \overline{y} \rangle - \varphi_o(\tilde{y}) - \langle d^2, \tilde{y} \rangle = \langle d_B^1, \mathbf{B}^{-1} A^2 (\overline{y}^2 - \tilde{y}^2) \rangle + \langle d^2, \overline{y} - \tilde{y} \rangle$$

is true. This leads to the following optimality condition:

THEOREM 3.8 *Let $(\overline{x}, \overline{y})$ be feasible for the optimistic linear bilevel programming problem (3.7) with $\Psi_L(y)$ given by (3.18), this means, let $\overline{x} \in \Psi_L(\overline{y})$, $A^3\overline{y} = b$, $\overline{y} \geq 0$. $(\overline{x}, \overline{y})$ is not a local optimistic optimal solution of the bilevel problem (3.2), (3.18) if and only if one of the following two conditions is satisfied:*

- *$\langle d^1, \overline{x} \rangle > \varphi_o(\overline{y})$, or*

- *there exist $\tilde{y}$ and $\mathbf{B} \in \mathcal{M}(\overline{y})$ such that $A^3\tilde{y} = b$, $\tilde{y} \geq 0$, $\overline{y}, \tilde{y} \in \mathcal{R}(\mathbf{B})$ and*
$$\langle d^1_B, \mathbf{B}^{-1}A^2(\tilde{y}^2 - \overline{y}^2) \rangle < \langle d^2, \overline{y} - \tilde{y} \rangle.$$

This Theorem is a consequence of the results in [77] (cf. also Chapter 5). Clearly, the first case corresponds to optimal solutions of the lower level problem which are not the best ones with respect to the upper level objective function, hence they are not feasible from the optimistic point of view.

It should be noted that all the conditions in Theorem 3.8 can be checked by solving linear optimization problems. For computing the set $\mathcal{M}(\overline{y})$ the family of all vertices of a convex compact polyhedron is to be computed which can be done e.g. with an algorithm in [83].

3.5.2.2   THE PESSIMISTIC CASE

Consider now the pessimistic linear bilevel programming problem (3.9). This consists of minimizing the generally not lower semicontinuous, piecewise affine-linear function $\varphi_p(y) + \langle d^2, y \rangle$ (cf. Figure 3.6). Each of the affine-linear pieces corresponds to one basic matrix and at a kink of the function $\varphi_p(\cdot)$ more than one basic matrix is optimal for the lower level problem (3.18). Let $\overline{y}$ be a (local) pessimistic optimal solution of the bilevel programming problem (3.2), (3.18) or equivalently a local optimal solution of the minimization problem (3.9)

$$\varphi_p(y) + \langle d^2, y \rangle \to \min_y$$
$$A^3 y = b, \ y \geq 0$$

with $\Psi_L(y)$ determined by (3.18).

$\overline{y}$ can be a point where the objective function has a kink or a jump. We remark that an optimal solution of this problem need not to exist, i.e. the infimal value of the objective function can correspond to a jump where the function value is not situated at the smallest function value.

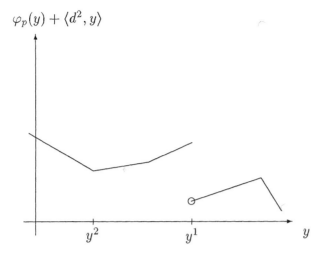

Figure 3.6.   The objective function in (3.9)

Nevertheless, since the function $\langle d^1, x \rangle$ is maximized over a convex polyhedral set for computing the value of

$$\varphi_p(y) = \max_x \{ \langle d^1, x \rangle : x \in \Psi_L(y) \}$$

each objective function value in (3.9) corresponds to a basic solution in the lower level problem. Hence, we can follow the lines in the previous subsection for investigating (local) optimality for (3.9) of some candidate $\bar{y}$.

Points $\bar{y}$ where $\varphi_p(\cdot)$ has a jump are characterized by $\Psi_L(\cdot)$ being not lower semicontinuous. Since upper semicontinuity of $\Psi_L(\cdot)$ implies that $\limsup_{y \to \bar{y}} \varphi_p(y) \leq \varphi_p(\bar{y})$, at a jump of $\varphi_p(\cdot)$ this inequality is satisfied as a strong inequality:

$$\lim_{k \to \infty} \varphi_p(y^k) < \varphi_p(\bar{y})$$

for at least one sequence $\{y^k\}_{k=1}^\infty$ converging to $\bar{y}$. This implies that we have to do two things:

- Given a candidate $(\bar{x}, \bar{y})$ with $\langle d^1, \bar{x} \rangle = \varphi_p(\bar{y})$ and $\bar{x} \in \Psi_L(\bar{y})$ we have to check if the function $\varphi_p(y) + \langle d^2, y \rangle$ has a jump at the point $\bar{y}$ or not. If it has no jump there (cf. the point y^2 in Figure 3.6) then we can use the ideas of the previous subsection to test if we have found a (local) pessimistic optimal solution.

- If the function has a jump at $\bar{y}$ (as at the point y^1 in Figure 3.6), then $\bar{y}$ cannot be a (local) pessimistic optimal solution.

Let again $\mathcal{M}(\overline{y})$ denote the set of all basic matrices corresponding to basic optimal solutions in $\Psi_L(\overline{y})$ and assume that the linear lower level problem (3.18) is a generic one, i.e. that, for each $\mathbf{B} \in \mathcal{M}(\overline{y})$, there exists a parameter value $\tilde{y} = \tilde{y}^{\mathbf{B}}$ such that the optimal solution and the optimal basis of problem (3.18) are uniquely determined for that $\tilde{y}^{\mathbf{B}}$. Denote the set of all points $\tilde{y}$ with this property by $\mathcal{R}^0(\mathbf{B})$:

$$\mathcal{R}^0(\mathbf{B}) = \{\tilde{y} \in \mathcal{R}(\mathbf{B}) : |\Psi_L(\tilde{y})| = 1, \mathcal{M}(\tilde{y}) = \{\mathbf{B}\}\}.$$

Then, by convexity of $\mathcal{R}(\mathbf{B})$, cl $\mathcal{R}^0(\mathbf{B}) = \mathcal{R}(\mathbf{B})$.

THEOREM 3.9 *Let* $(\overline{x}, \overline{y})$ *be feasible for the pessimistic linear bilevel problem (3.9) with* $\Psi_L(y)$ *defined by (3.18), that is, let* $A^3y = \overline{b}$, $\overline{y} \geq 0$, $\overline{x} \in \Psi_L(\overline{y})$, $\langle d^1, \overline{x} \rangle = \varphi_p(\overline{y})$. *Let the above genericity assumption be satisfied. Then,* $\overline{y}$ *is not a pessimistic local optimum of the bilevel programming problem (3.2), (3.18) in any of the following two cases:*

- *There exist a basis* $\mathbf{B} \in \mathcal{M}(\overline{y})$ *and a point* $\hat{y}$ *with* $\hat{y} \in R^0(\mathbf{B})$, $A^3\hat{y} = b$, $\hat{y} \geq 0$ *and* $\langle d^1, x^{\mathbf{B}}(\overline{y}) \rangle < \varphi_p(\overline{y})$.

- $\varphi_p(\overline{y}) = \varphi_o(\overline{y})$ *and there exist a basis* $\mathbf{B} \in \mathcal{M}(\overline{y})$ *and a point* $\hat{y}$ *with* $\overline{y} \in \mathcal{R}(\mathbf{B})$, $\hat{y} \in \mathcal{R}^0(\mathbf{B})$, $A^3\hat{y} = b$, $\hat{y} \geq 0$, *and*

$$\langle d^1_B, \mathbf{B}^{-1}A^2(\hat{y}^2 - \overline{y}^2) \rangle < \langle d^2, \overline{y} - \hat{y} \rangle.$$

Under the first condition of the theorem, the (by $\hat{y} \in R^0(\mathbf{B})$) feasible points $(x^{\mathbf{B}}(t\hat{y} + (1-t)\overline{y}), t\hat{y} + (1-t)\overline{y})$ have objective function values sufficiently close to $\langle d^1, x^{\mathbf{B}}(\overline{y}) \rangle + \langle d^2, \overline{y} \rangle < \varphi_p(\overline{y}) + \langle d^2, \overline{y} \rangle$ but the point $(x^{\mathbf{B}}(\overline{y}), \overline{y})$ is infeasible. If the second condition is satisfied then the feasible points $(x^{\mathbf{B}}(t\hat{y} + (1-t)\overline{y}), t\hat{y} + (1-t)\overline{y})$ again have better function values than $(x^{\mathbf{B}}(\overline{y}), \overline{y})$.

Note that the first case in Theorem 3.9 implies that $\varphi_o(\overline{y}) < \varphi_p(\overline{y})$ but that the satisfaction of this strong inequality does in general not imply that the function $\varphi_p(\cdot)$ has a jump at $y = \overline{y}$.

The following example shows that the restriction to points $\hat{y} \in R^0(\mathbf{B})$ is necessary in Theorem 3.9.

Example: Consider the small problem

$$-x_1 + y_2^2 \to \text{``} \min_y \text{''}$$

$$0 \leq y_1 \leq 1, y_2 \geq 0$$

$$x \in \text{Argmin}_x \{y_2 x_1 : x_1 + x_2 = y_1, -x_1 + x_3 = y_1, x_2 \geq 0, x_3 \geq 0\}$$

at the point $\overline{x} = (0,0,0)^\top$, $\overline{y} = (0,0)^\top$. Then, for $y \geq 0$

$$\Psi_L(y) = \begin{cases} \{x : x_1 \in [-y_1, y_1], x_2 = y_1 - x_1, x_3 = x_1 + y_1\}, & \text{if } y_2 = 0 \\ \{(-y_1, 2y_1, 0)^\top\}, & \text{if } y_2 > 0. \end{cases}$$

Thus, $\varphi_p(y) = y_1 + y_2^2$ and we have the optimal solution $(\overline{x}, \overline{y})$. This corresponds to the basic matrix

$$\mathbf{B}^0 = \begin{pmatrix} 1 & 1 \\ -1 & 0 \end{pmatrix}$$

with $\mathcal{R}(\mathbf{B}^0) = \{y : y_1 \geq 0, y_2 \geq 0\}$. But, if for $y_2 = 0$ the other basic matrix

$$\mathbf{B}^1 = \begin{pmatrix} 1 & 0 \\ -1 & 1 \end{pmatrix}$$

is taken, then we get $\mathcal{R}(\mathbf{B}^1) = \{y : y_1 \geq 0, y_2 \leq 0\}$, $x_1^{\mathbf{B}^1}(y) = y_1$ and there exists no point $y \in \mathcal{R}^0(\mathbf{B}^1)$, $y \geq 0$. For this basic matrix we get a direction of descent $\hat{y} - \overline{y} = (1,0)^\top$. But note, that in this example the point $\hat{x}$ with $\hat{x}_1 = y_1$ is the pessimistic solution in $\Psi_L(y)$ for no y with $y_1 \neq 0$. Hence, this basic matrix $\mathbf{B}^1$ cannot be used to verify optimality. $\square$

The following theorem gives a sufficient optimality condition:

THEOREM 3.10 *Let the assumptions of Theorem 3.9 be satisfied. If both conditions*

- *for all $x \in \Psi_L(\overline{y})$ the equation $\langle d^1, x \rangle = \varphi_p(\overline{y})$ hold and*

- *for all $\mathbf{B} \in \mathcal{M}(\overline{y})$ and all $\hat{y} \in R^0(\mathbf{B})$ with $A^3\hat{y} = b$, $\hat{y} \geq 0$ the inequality*

$$\langle d_B^1, \mathbf{B}^{-1}A^2(\hat{y}^2 - \overline{y}^2) \rangle \geq \langle d^2, \overline{y} - \hat{y} \rangle$$

  is satisfied,

then $(\overline{x}, \overline{y})$ is a local optimal solution of (3.9), (3.18) (or a local pessimistic optimal solution of the bilevel problem (3.2), (3.18)).

The proof of this theorem follows from a more general result in Theorem 5.25. Note, than in distinction to that result, we do not need a strong inequality in the second condition by piecewise linearity of $\varphi_p(\cdot)$. Note that this theorem is more restrictive than Theorem 3.9 since $\varphi_o(\overline{y}) = \varphi_p(\overline{y})$ is assumed here. In Theorem 3.9 this condition is weakened to

$$\forall \mathbf{B} \in \mathcal{M}(\overline{y}) \text{ either } \langle d^1, x^{\mathbf{B}}(\overline{y}) \rangle = \varphi_p(\overline{y}) \text{ or } R^0(\mathbf{B}) = \emptyset.$$

3.5.3   TWO AUXILIARY PROBLEMS

To investigate it, the linear bilevel programming problem is replaced by one of several problems. Most widely used is the problem arising if the lower level problem is replaced by its Karush-Kuhn-Tucker conditions. Let $\overline{y}$ be a fixed parameter vector. Then, $\overline{x}$ is an optimal solution of the problem (3.4), $\overline{x} \in \Psi_L(\overline{y})$, if and only if there exists $\overline{\lambda} \in \mathbb{R}^p$ such that $(x, y, \lambda) = (\overline{x}, \overline{y}, \overline{\lambda})$ satisfies

$$
\begin{aligned}
&A^1 x + A^2 y^2 \leq a, \ x \geq 0, \\
&A^{1\top} \lambda + y^1 \geq 0, \ \lambda \geq 0, \\
&\langle A^1 x + A^2 y^2 - a, \lambda \rangle = 0, \ \langle A^{1\top} \lambda + y^1, x \rangle = 0.
\end{aligned}
\tag{3.20}
$$

This leads to a first problem replacing the optimistic formulation (3.7) of the bilevel programming problem:

$$
\begin{aligned}
&\langle d^1, x \rangle + \langle d^2, y \rangle \to \min_{x,y} \\
&A^3 y = b, \ y \geq 0, \\
&A^1 x + A^2 y^2 \leq a, \ x \geq 0, \\
&A^{1\top} \lambda + y^1 \geq 0, \ \lambda \geq 0, \\
&\langle A^1 x + A^2 y^2 - a, \lambda \rangle = 0, \ \langle A^{1\top} \lambda + y^1, x \rangle = 0.
\end{aligned}
\tag{3.21}
$$

Both problems are equivalent if (local) optimistic optimality of the bilevel programming problem is investigated via problem (3.7). But this is no longer true if the approach in problem (3.8) is taken. We will come back to this phenomenon in Section 5.5.

In [67, 188], this problem has been used to construct optimality conditions by use of the tangent cone to the Karush-Kuhn-Tucker set mapping of the lower level problem. This tangent cone is in general not convex but the union of finitely many convex polyhedral cones. The resulting optimality conditions have a combinatorial nature: To check optimality the minimal optimal value of a finite family of linear optimization problems is to be compared with zero. We will come back to this in Chapter 5.

An alternative approach has been used in [134]. By linear programming duality, the point $\overline{x}$ is optimal for (3.4) at $y = \overline{y}$ if and only if there exists $\overline{\lambda} \in \mathbb{R}^p$ such that $(x, y, \lambda) = (\overline{x}, \overline{y}, \overline{\lambda})$ satisfies

$$
\begin{aligned}
&A^1 x + A^2 y^2 \leq a, \ x \geq 0, \\
&A^{1\top} \lambda + y^1 \geq 0, \ \lambda \geq 0, \\
&\langle y^1, x \rangle = \langle a - A^2 y^2, \lambda \rangle.
\end{aligned}
\tag{3.22}
$$

Moreover, for arbitrary feasible points $\tilde{x}, \tilde{\lambda}$ of the primal and the dual problems we have $\langle y^1, \tilde{x} \rangle \geq \langle a - A^2 y^2, \tilde{\lambda} \rangle$ by weak duality. Then, the following problem is posed:

$$\langle y^1, x \rangle - \langle a - A^2 y^2, \lambda \rangle \to \min_{x,y,\lambda}$$
$$A^1 x + A^2 y^2 \le a, \; x \ge 0,$$
$$A^{1\top} \lambda + y^1 \ge 0, \; \lambda \ge 0, \tag{3.23}$$
$$A^3 y = b, \; y \ge 0,$$
$$\langle d^1, x \rangle + \langle d^2, y \rangle \le h.$$

Let $\varphi(h)$ denote the optimal value of problem (3.23). A feasible solution of the problem (3.23) with zero objective function value is a feasible solution for problem (3.2), (3.4) with an upper level objective function value not larger than h. Hence, searching for the lowest possible value of h such that the global optimal function value of problem (3.23) is equal to zero is equivalent to computing an optimistic global optimal solution of the linear bilevel programming problem (3.2), (3.4).

THEOREM 3.11 ([134]) *A point $(\overline{x}, \overline{y})$ is a (global) optimal solution of (3.7) if and only if for any $h < \overline{h} = \langle d^1, \overline{x} \rangle + \langle d^2, \overline{y} \rangle$ the relation $\varphi(h) > 0$ holds.*

Problem (3.23) is a nonconvex quadratic optimization problem. Since the objective function is bounded from below by linear programming duality, it has a (global) optimal solution provided the feasible set is not empty [164]. Moreover, by the special perturbation, the optimal objective function value of the problem is not decreasing for decreasing h.

3.6 SOLUTION ALGORITHMS

In this section different algorithms for solving linear bilevel programming problems will be outlined. We start with some results about the complexity of the (linear) bilevel programming problem which gives a first imagination about the difficulties connected with the solution of bilevel programming problems on the computer.

Bilevel problems are nonconvex and nondifferentiable optimization problems (see Fig. 3.1). This poses the question of searching for a global optimum which is discussed in Subsection 3.6.5. Global optimization is often associated with enumeration algorithms which are outlined in Subsection 3.6.2. Other classes of algorithms do not intend to compute global but rather local optima as e.g. descent algorithms (see Subsection 3.6.3) or penalty methods (cf. Subsection 3.6.4). We don't want to go into the details of the algorithms but rather outline the basic ideas. A comparison of the numerical behavior of codes realizing them can be found in [26]. Also, proofs for theorems on correctness of algorithms are omitted since the development of solution algorithms is not a central topic of this book.

3.6.1   NUMERICAL COMPLEXITY

Complexity theory [104] tries to use relations between different problems for a classification of the problems into the "easy" and the "difficult" ones. For the easy problems, solution algorithms are known which solve arbitrary instances of the problem (obtained by fixing the data) with a number of elementary operations (multiplication, addition, division, subtraction, comparison) which can be bounded by a polynomial in the number of digits necessary to store the problem data in the computer memory (e.g. by use of binary encoding schemes). Such problems are the linear programming problem, the minimum spanning tree problem etc. The family of these problems is the class $\mathcal{P}$ of the polynomially solvable problems. A nondeterministic algorithm decomposes into two parts: a first one in which a guess is made for a solution of the problem and a second one in which this point is tested if it really solves the problem. In a polynomial nondeterministic algorithm both the storage of the guessed point as well as the test in the second part can be done with a numerical effort which can be bounded by a polynomial in the length of the input data provided that the guessed point really solves the problem. There is no bound on the time needed for finding a solution in practice on the computer. Clearly, the class $\mathcal{P}$ is a subclass of the family $\mathcal{NP}$ of problems solvable by polynomial nondeterministic algorithms. Optimization problems do often not belong to the class $\mathcal{NP}$ due to nonexistence of (polynomially testable) optimality criteria. For a minimization problem the corresponding decision problem asks for the existence of a feasible solution with an objective function value not larger than a given constant. For the traveling salesman problem e.g., the decision problem asks for the existence of an Hamiltonian circuit of length bounded by a given constant and not for the Hamiltonian circuit itself. Decision problems related to optimization problems often belong to the class $\mathcal{NP}$. The most difficult problems in the class $\mathcal{NP}$ form the subclass $\mathcal{NPC}$ of $\mathcal{NP}$–complete problems. The construction of the class $\mathcal{NPC}$ is done in such a way that if one problem in the class $\mathcal{NPC}$ proves to be polynomially solvable than both classes $\mathcal{P}$ and $\mathcal{NP}$ coincide. It is the most important question in this circumstance if the classes $\mathcal{NP}$ and $\mathcal{P}$ coincide or not with the negative answer assumed to be true by almost all scientists. The optimization problems corresponding to the decision problems in the class $\mathcal{NPC}$ are members of the class $\mathcal{NPH}$ of $\mathcal{NP}$–hard problems. The 0/1 knapsack and the traveling salesman problems are $\mathcal{NP}$–hard, their decision variants are $\mathcal{NP}$–complete. A large number of

$\mathcal{NP}$–complete and $\mathcal{NP}$–hard problems can be found in the monograph [104].

For showing that some problem Π is $\mathcal{NP}$–complete another problem Ω has to be found which is known to be $\mathcal{NP}$–complete and to prove that this problem is a "subproblem" of Π. This means that we have to describe a polynomial algorithm transforming an arbitrary instance B of Ω into some instance A of Π such that the answer to A is yes if and only the answer to B is yes.

THEOREM 3.12 ([81]) *For any $\varepsilon > 1$ it is $\mathcal{NP}$–hard to find a solution to the linear bilevel programming problem (3.1), (3.2) with not more than ε times the optimal value of the bilevel problem.*

It can be seen easily in the proof of this theorem, that the lower level problem has a unique optimal solution for all values of the parameters. Thus, a distinction between optimistic vs. pessimistic optimal solutions is not necessary here. The Theorem 3.12 implies that it is not likely to find a polynomial algorithm for globally solving the linear bilevel programming problem.

The proof of this Theorem (given in the Section 3.7) uses a transformation of the problem 3SAT to the bilevel programming problem. Since 3SAT is no number problem, Theorem 3.12 even shows that the linear bilevel programming problem is $\mathcal{NP}$–hard in the strong sense [103]. A fully polynomial approximation scheme is a solution algorithm parameterized in the accuracy of the computed solution which, for each given accuracy ε provides an ε–optimal solution in time polynomial in the length of the problem and in $1/\varepsilon$ [103]. The proof of Theorem 3.12 shows that the subproblems equivalent to 3SAT have nonnegative integer optimal function values zero or one for all instances. This implies that there cannot be a fully polynomial approximation scheme for the bilevel programming problem (unless $\mathcal{P} = \mathcal{NP}$) [103].

Other proofs for $\mathcal{NP}$–hardness of the linear bilevel programming problem can be found in [25, 36, 116, 133]. The result in [133] is the strongest one since it shows an even higher complexity of the multi-level linear programming problem.

3.6.2   ENUMERATIVE ALGORITHMS

3.6.2.1   SEARCH WITHIN THE VERTICES OF THE FEASIBLE SET

We consider the optimistic formulation of the linear bilevel programming problem (3.4), (3.7). As a result of Theorem 3.1 we can restrict the search for a (global) optimal solution to the vertices of the convex polyhedral components of the graph of $\Psi_L(\cdot)$. As given in Section 3.3

each such vertex is determined by two sets I, J of active constraints:

$$(A^1x + A^2y^2 - a)_i = 0, \ \lambda_i \geq 0, \ \text{ for } i \in J$$
$$(A^1x + A^2y^2 - a)_i \leq 0, \ \lambda_i = 0, \ \text{ for } i \notin J$$
$$(A^{1\top}\lambda + y^1)_j = 0, \ x_j \geq 0 \ \text{ for } j \in I$$
$$(A^{1\top}\lambda + y^1)_j \geq 0, \ x_j = 0 \ \text{ for } j \notin I$$
$$A^3y = b, \ y \geq 0.$$

Hence, we can reformulate the linear bilevel programming problem as the problem of finding such index sets I, J for which the following linear optimization problem has the smallest optimal function value:

$$\langle d^1, x \rangle + \langle d^2, y \rangle \to \min_{x,y}$$
$$(A^1x + A^2y^2 - a)_i = 0, \ \lambda_i \geq 0, \ \text{ for } i \in J$$
$$(A^1x + A^2y^2 - a)_i \leq 0, \ \lambda_i = 0, \ \text{ for } i \notin J$$
$$(A^{1\top}\lambda + y^1)_j = 0, \ x_j \geq 0 \ \text{ for } j \in I \qquad (3.24)$$
$$(A^{1\top}\lambda + y^1)_j \geq 0, \ x_j = 0 \ \text{ for } j \notin I$$
$$A^3y = b, \ y \geq 0.$$

This search can be realized by an enumeration over all the different possible sets I, J which of course results in a large and fortunately unnecessary amount of computations. Different ideas for minimizing the computational effort by use of intelligent search rules for new sets I, J which with high probability lead to improved solutions have been proposed [27, 55]. In [177] it is shown that this approach can lead to a polynomial algorithm for the linear bilevel programming problem if the follower controls a fixed number of variables only. The reason is that then only a polynomial number of different basic matrices for the lower level problem exist and that the resulting linear optimization problems (3.24) can be solved in polynomial time [148].

Solving the problem (3.1), (3.7) with the K–th best algorithm of [42], the enumeration of vertices is realized in the following simple way: Starting with the optimal solution of

$$\min_{x,y}\{\langle d^1, x \rangle + \langle d^2, y \rangle : A^1x + A^2y \leq a, A^3y = b, x, y \geq 0\}$$

a set M is constructed containing all the neighboring vertices of the set

$$\{(x, y) : A^1x + A^2y \leq a, A^3y = b, x, y \geq 0\}$$

to the already investigated vertices. In each step the best point with respect to the upper level objective function in the set M is checked for feasibility for the bilevel programming problem. If it is feasible then it is a global optimal solution of (3.1), (3.7). Otherwise it is excluded from M and all of its neighboring vertices are included in M.

REMARK 3.2 *An algorithm for the three-level linear programming problem based on the "k"-th best algorithm of [42] can be found in [287].*

3.6.2.2 SEARCH FOR ACTIVE INEQUALITIES

In [116] it is explored that optimality in the lower level implies that a minimum number of inequality constraints in problem (3.1) has to be satisfied as equations. This can be formulated by the help of additional (logical) inequalities using artificial variables. A branch-and-bound algorithm for the search for a global optimistic optimal solution can be developed. Branching is done by fixing one (or some of the) artificial variable(s) to zero or one. Fixing one variable to the value one is equivalent to demanding equality in the corresponding inequality constraint. Setting an artificial variable to zero means that the related inequality is strict or (by linear programming duality) that the dual variable is equal to zero. Both conditions are then used to decrease the number of variables in the lower level problem and to derive further additional (logical) relations. In doing so, the lower level problem step-by-step reduces to one with uniquely determined feasible solution for each parameter value which then can be inserted into the upper level problem. For computing bounds linear optimization problems are solved arising if the lower level objective function is dropped in the current problems.

3.6.3 DESCENT ALGORITHMS

This approach can be considered as an implementation of the necessary optimality conditions in Theorem 3.8.

Using the optimistic position, the bilevel programming problem (3.7), (3.18) can be transformed in that of minimizing the upper level objective function

$$\langle d^1, x \rangle + \langle d^2, y \rangle \to \min_{x,y}$$

subject to

$$x = (x_B^{\mathbf{B}}(y), 0)^{\top}, x_B^{\mathbf{B}}(y) = \mathbf{B}^{-1}(a - A^2 y^2), y \in \mathcal{R}(\mathbf{B}),$$
$$\mathbf{B} \text{ is a basic matrix }, A^3 y = b, \ y \geq 0.$$

This problem possesses its decomposition into a hierarchical problem

$$\Phi(\mathbf{B}) \to \min$$
$$\mathbf{B} \text{ is a basic matrix,} \tag{3.25}$$

where

$$\Phi(\mathbf{B}) = \min\{\langle d_B^1, \mathbf{B}^{-1}(a - A^2 y^2)\rangle + \langle d^2, y\rangle : y \in \mathcal{R}(\mathbf{B}), \ A^3 y = b, \ y \geq 0\}.$$

By

$$\mathcal{R}(\mathbf{B}) = \{y : \mathbf{B}^{-1}(a - A^2 y^2) \geq 0, \ y_B^{1\top} \mathbf{B}^{-1} A^1 - y^{1\top} \leq 0$$

evaluating $\Phi(\mathbf{B})$ requires solving a linear optimization problem. The algorithm starts with a first optimal basic matrix for the lower level problem (3.18) which can be obtained e.g. by first computing an optimal vertex solution (x^*, y^*) of the problem

$$\langle d^1, x\rangle + \langle d^2, y\rangle \to \min_{x,y}$$
$$A^1 x + A^2 y^2 = a, \ x \geq 0,$$
$$A^3 y = b, \ y \geq 0$$

followed by computing a vertex solution $\bar{x} \in \Psi_L(y^*)$. Then, $(\hat{x}, \hat{y}) = (\bar{x}, y^*)$ is a first feasible solution of the bilevel programming problem (3.7), (3.18) and the optimal basic matrix $\mathbf{B}$ corresponding to $\bar{x}$ can be used to start the process. At $(\hat{x}, \hat{y})$ satisfiability of the necessary optimality condition in Theorem 3.8 is tested:

- First, compute $\tilde{x} \in \underset{x}{\text{Argmin}} \ \{\langle d^1, x\rangle + \langle d^2, \hat{y}\rangle : x \in \Psi_L(\hat{y})\}$. This

  means that an optimistic solution corresponding to $\hat{y}$ is computed.

- Second, test by enumerating all the optimal basic matrices in the lower level problem for $\tilde{x}$ if there exists a descent direction. This part reduces to evaluating $\Phi(\mathbf{B})$.

After both steps have been performed a new better point $(\hat{x}, \hat{y}) = (\tilde{x}, \tilde{y})$ is obtained and the steps are repeated until the process stagnates.

In most of the cases there seems to be some hope that the number of different basic matrices for $\tilde{x}$ is small. In this case, the second step is not too expensive. If there does not exist a descent direction, then $(\tilde{x}, \hat{y})$ is a local optimal solution by Theorem 3.8. On the other hand, if a descent direction can be found than stepping forward along this direction a new parameter value $\tilde{y}$ is obtained for which a new basic matrix $\tilde{\mathbf{B}}$ of $\hat{x}$ appears. An algorithm of this type has first been developed in [66]. Note that this idea can also be used to compute a solution in the pessimistic case (3.9) and that the restriction to equations in the lower level problem is no loss of generality. Details are left to the interested reader.

3.6.4   PENALTY FUNCTION METHODS

Consider the bilevel programming problem in its optimistic version (3.1), (3.7). As already mentioned in (3.22) a feasible solution $x(\bar{y})$ to

(3.1) is optimal for $y = \overline{y}$ if and only if there exists a vector $\overline{\lambda}$ such that $(\overline{x}, \overline{y}, \overline{\lambda})$ satisfies the system

$$\begin{aligned}
&A^1 x + A^2 y \leq a, \ x \geq 0, \\
&A^{1\mathsf{T}} \lambda + c \geq 0, \ \lambda \geq 0, \\
&\langle c, x \rangle = \langle a - A^2 y, \lambda \rangle.
\end{aligned} \tag{3.26}$$

This can be used to formulate a penalty function approach for solving the problem (3.1), (3.7) [292]:

$$\begin{aligned}
&\langle d^1, x \rangle + \langle d^2, y \rangle + K[\langle c, x \rangle - \langle a - A^2 y, \lambda \rangle] \to \min_{x, y, \lambda} \\
&A^1 x + A^2 y \leq a, \ x \geq 0, \\
&A^{1\mathsf{T}} \lambda + c \geq 0, \ \lambda \geq 0, \\
&A^3 y = a, \ y \geq 0
\end{aligned} \tag{3.27}$$

with $K > 0$ sufficiently large. We use a decomposition approach to solve this problem for a fixed K:

$$\begin{aligned}
&\Phi_K(\lambda) \to \min_{\lambda} \\
&A^{1\mathsf{T}} \lambda + c \geq 0, \ \lambda \geq 0,
\end{aligned} \tag{3.28}$$

where $\Phi_K(\lambda)$ denotes the optimal value of the following parameterized problem

$$\begin{aligned}
&\langle d^1, x \rangle + \langle d^2, y \rangle + K[\langle c, x \rangle - \langle a - A^2 y, \lambda \rangle] \to \min_{x, y} \\
&A^1 x + A^2 y \leq a, \ x \geq 0, \\
&A^3 y = a, \ y \geq 0
\end{aligned} \tag{3.29}$$

for fixed λ. Then, the function $\Phi_K(\cdot)$ is concave [241] and the minimum in (3.28) is attained at a vertex of the feasible set

$$\{\lambda \geq 0 : A^{1\mathsf{T}} \lambda + c \geq 0\}$$

of problem (3.28) provided that the problem (3.27) has a solution. The following Theorem is an obvious implication of finiteness of the number of vertices of problem (3.28):

THEOREM 3.13 ([292]) *Let the set*

$$\{(x, y, \lambda) : A^1 x + A^2 y \leq a, \ A^{1\mathsf{T}} \lambda + c \geq 0, \ A^3 y = a, \ x, y, \lambda \geq 0\}$$

be nonempty and bounded. Then, there exists a finite value K^* such that, for each $K \geq K^*$ the global optimal solution $\overline{\lambda}$ of the problem (3.28) together with an optimal solution $(\overline{x}, \overline{y})$ of the problem (3.29) is also a global optimal solution of the problem (3.1), (3.7).

The difficulty with this approach is that for each value of the penalty parameter the outer nonconvex problem (3.28) has to be solved globally. But Theorem 3.13 extends also to the case when the upper level problem (3.28) is solved locally but the linear lower level problem (3.29) is solved globally (see also [187, 289]):

THEOREM 3.14 ([292]) *Let the set*

$$\{(x, y, \lambda) : A^1 x + A^2 y \le a, \; A^{1\top}\lambda + c \ge 0, \; A^3 y = a, \; x, y, \lambda \ge 0\}$$

be nonempty and bounded. Then, there exists a finite value K^* such that, for each $K \ge K^*$ the local optimal solution $\overline{\lambda}$ of the problem (3.28) together with an optimal solution $(\overline{x}, \overline{y})$ of the problem (3.29) is also a local optimal solution of the problem (3.1), (3.7).

In [292] some ideas for estimating the value K^* and an algorithm for computing an optimal solution for the optimistic bilevel linear programming problem based on these ideas are given. If we replace the lower level problem (3.1) by the more general (3.4) then problem (3.29) is no longer a linear optimization problem but a nonconvex quadratic one. This makes the investigation much more difficult. We will come back to the penalty function approach to bilevel programming in Section 6.3.

3.6.5   GLOBAL OPTIMIZATION

Consider the linear bilevel programming problem with inequality constraints using the optimistic position

$$\min_{x,y}\{\langle d^1, x\rangle + \langle d^2, y\rangle : A^3 y \le b, \; y \ge 0, \; x \in \Psi_L(y)\}, \qquad (3.30)$$

where $\Psi_L(y)$ again denotes the set of optimal solutions to the lower level problem (3.1). Let

$$\varphi_L(y) = \min_x\{\langle c, x\rangle : A^1 x + A^2 y \le a, \; x \ge 0\}$$

denote the optimal objective function value of problem (3.1).

THEOREM 3.15 ([33]) *The optimal value function $\varphi_L(\cdot)$ is convex and piecewise affine-linear on $\mathrm{dom}\varphi_L = \{y : \varphi_L(y) < \infty\}$. Moreover, either we have $\varphi_L(\cdot) \equiv -\infty$ or $\varphi_L(y) > -\infty$ for all y.*

The function $\varphi_L(\cdot)$ can be used to formulate an equivalent problem to (3.30):

$$\begin{aligned}
\langle d^1, x\rangle + \langle d^2, y\rangle &\to \min_{x,y} \\
A^1 x + A^2 y &\le a, \; x \ge 0 \\
A^3 y &\le b, \; y \ge 0, \langle c, x\rangle \le \varphi_L(y).
\end{aligned} \qquad (3.31)$$

This is an optimization problem with a reverse convex constraint. Such problems have been extensively investigated e.g. in [129]. In [276] a branch-and-bound algorithm is proposed for solving problem (3.31). First the problem is transformed into a quasiconcave optimization problem in the following way.

Let $(\bar{x}, \bar{y})$ be an optimal basic solution of the linear relaxed problem

$$\begin{aligned}
\langle d^1, x \rangle + \langle d^2, y \rangle &\to \min_{x,y} \\
A^1 x + A^2 y &\le a, \ x \ge 0 \\
A^3 y &\le b, \ y \ge 0.
\end{aligned} \qquad (3.32)$$

If this solution is feasible for (3.31) then, of course it is also an optimal solution for the bilevel problem (3.30). Hence, assume in the following $\langle c, \bar{x} \rangle > \varphi_L(\bar{y})$. Let

$$D = \{(x, y) : A^1 x + A^2 y \le a, \ A^3 y \le b, \ x, y \ge 0\} - \{(\bar{x}, \bar{y})\}$$

and

$$C = \{(x, y) : \langle c, x \rangle \ge \varphi_L(y)\} - \{(\bar{x}, \bar{y})\}.$$

Here $A - B$ denotes the Minkowski difference of the sets A, B:

$$A - B = \{z : \exists a \in A, \ \exists b \in B \text{ with } z = a - b\}.$$

The sets C, D have been constructed such that $0 \in (\text{int } C) \cap D$ and D is a convex polyhedron. Both sets C, D can be assumed to be nonempty since otherwise problem (3.31) has no solution. C is a convex set having a polar set

$$C^o = \{(\lambda, v) : \langle \lambda, x \rangle + \langle v, y \rangle \le 0 \ \forall (x, y) \in C\}.$$

Consider a point

$$(\bar{r}, \bar{s}) \in K := \{(r, s) : A^2 s \le 0, \ \langle c, r \rangle \ge 0\}.$$

Then, $A^1 \bar{x} + A^2 (\bar{y} + t\bar{s}) \le a, \ \langle c, \bar{x} + t\bar{r} \rangle \ge \langle c, \bar{x} \rangle$ for all $t \ge 0$. Hence, the convex cone K is a subset of C, $K \subseteq C$. Indeed, $\bar{x}$ is feasible for problem (3.1) with $y = \bar{y} + t\bar{s}$ which implies:

$$\langle c, \bar{x} + t\bar{r} \rangle \ge \langle c, \bar{x} \rangle \ge \varphi_L(\bar{y} + t\bar{s}) \ \forall \ t \ge 0.$$

Thus,

$$C^o \subseteq K^o = \text{cone } \{a^0, a^1, \ldots, a^p\}, \qquad (3.33)$$

where $a^0 = (-c \ 0)^\top, a^k = (0 \ A_k^2)^\top$ and A_k^2 denotes the k-th row of A^2 [241].

Using these sets problem (3.31) can equivalently be stated as

$$\min_{x,y}\{\langle d^1, x\rangle + \langle d^2, y\rangle : (x,y) \in D \setminus \text{int } C\}. \tag{3.34}$$

Let $(x,y) \notin C$. Then there exists $(\lambda, v) \in C^o$ such that $\langle \lambda, x\rangle + \langle v, y\rangle > 0$. Since C^o is a convex cone we can regularize this inequality such that the left-hand side is not smaller than 1 for suitable $(\lambda, v) \in C^o$. Thus, problem (3.34) possesses a decomposition into an hierarchical problem

$$\min_{(\lambda,v)}\{f(\lambda, v) : (\lambda, v) \in C^o\}, \tag{3.35}$$

where

$$f(\lambda, v) = \inf_{x,y}\{\langle c, x\rangle : (x,y) \in D, \ \langle \lambda, x\rangle + \langle v, y\rangle \geq 1\}.$$

THEOREM 3.16 ([276]) *The function $f(\cdot, \cdot)$ is quasiconcave and lower semicontinuous. The problems (3.30) and (3.35) are equivalent in the following sense: The global optimal values of both problems are equal and if $(\overline{\lambda}, \overline{v})$ solves (3.35), then any*

$$(\overline{x}, \overline{y}) \in \underset{x,y}{\text{Argmin}} \ \{\langle c, x\rangle : (x,y) \in D, \ \langle \lambda, x\rangle + \langle v, y\rangle \geq 1\}$$

solves (3.30).

For solving problem (3.30) it thus suffices to solve problem (3.35) which can be done by the application of the branch-and-bound approach. This starts with minimizing the function $f(\lambda, v)$ on some simplex T with $C^o \subset T$. By subdividing T into smaller and smaller simplices, an optimal solution of (3.35) is approached. The initial simplex is composed by the vectors a^j, $j = 0, \ldots, p$ and branching is essentially done by partitioning the cone K^o. For bounding, a linear program is solved whose optimal function value can be used to compute lower bounds for $f(\lambda, v)$. For a detailed description of the algorithm the reader is referred to [276]. A similar idea for globally solving bilevel linear programs is derived in [275].

3.7   PROOFS

PROOF OF THEOREM 3.1: By linear programming duality, $x \in \Psi_L(y)$ if and only if there exists $\lambda \in \mathbb{R}^p$ such that

$$\begin{aligned} A^1 x + A^2 y \leq a, \ x \geq 0, \ \lambda \geq 0, \ \langle \lambda, A^1 x + A^2 y - a\rangle = 0, \\ A^{1\top}\lambda + c \geq 0, \ \langle x, A^{1\top}\lambda + c\rangle = 0. \end{aligned} \tag{3.36}$$

For any sets $I \subseteq \{1, \ldots, n\}$, $J \subseteq \{1, \ldots, p\}$ consider the solution set $M(I, J)$ of the system of linear (in)equalities

$$(A^1 x + A^2 y - a)_i = 0, i \in J, \ (A^1 x + A^2 y - a)_i \leq 0, i \notin J,$$
$$x_j = 0, \ j \notin I, \ x_j \geq 0, \ j \in I, \ \lambda_i = 0, i \notin J, \ \lambda_i \geq 0, i \in J$$
$$(A^{1\top} \lambda + c)_j = 0, \ j \in I, \ (A^{1\top} \lambda + c)_j \geq 0, \ j \notin I.$$

Then, conditions (3.36) are satisfied. The set $M(I, J)$ is polyhedral and so is its projection on $\mathbb{R}^n \times \mathbb{R}^m$. Since the graph of $\Psi_L(\cdot)$ is equal to the union of the sets $M(I, J)$, the assertion follows. $\qquad \square$

It should be noted that if a set $M(I, J)$ is not empty then its projection on the space $\mathbb{R}^n \times \mathbb{R}^m$ is equal to the set of all solutions of the system

$$(A^1 x + A^2 y - a)_i = 0, i \in J, \ (A^1 x + A^2 y - a)_i \leq 0, i \notin J,$$
$$x_j = 0, \ j \notin I, \ x_j \geq 0, \ j \in I.$$

This set obviously determines a face of the convex polyhedral set $\{(x, y) : A^1 x + A^2 y \leq a, x \geq 0\}$. Moreover, if some inner point (x, y) of a face of the set $\{(x, y) : A^1 x + A^2 y \leq a, x \geq 0\}$ is feasible, i.e. $x \in \Psi_L(y)$, then the whole face has this property.

PROOF OF THEOREM 3.4: First, let $(\overline{x}, \overline{y})$ be feasible to (3.1), (3.2) but be not Pareto optimal for (3.10). Then, there exists $(\hat{x}, \hat{y})$ satisfying $A^1 \hat{x} + A^2 \hat{y} \leq a$, $A^3 \hat{y} = b$, $\hat{x}, \hat{y} \geq 0$ with $C(\hat{x}, \hat{y})^\top \leq C(\overline{x}, \overline{y})^\top$, $C(\hat{x}, \hat{y})^\top \neq C(\overline{x}, \overline{y})^\top$, where

$$C = \begin{pmatrix} 0 & A \\ 0 & -e^{m\top} A \\ c^\top & 0 \end{pmatrix}.$$

Thus, since $A\hat{y} \leq A\overline{y}$ and $-e^{m\top} A\hat{y} \leq -e^{m\top} A\overline{y}$ we have $A\hat{y} = A\overline{y}$. By rg $A = $ rg $(A^2, A^3, -E)^\top$, this leads to the equation $(A^2, A^3, -E)^\top \hat{y} = (A^2, A^3, -E)^\top \overline{y}$. Feasibility of $(\hat{x}, \hat{y})$ to (3.10) implies that $\hat{x}$ satisfies all constraints of the lower level problem (3.1) which by $C(\hat{x}, \hat{y})^\top \neq C(\overline{x}, \overline{y})^\top$ results in $\langle c, \hat{x} \rangle < \langle c, \overline{x} \rangle$ contradicting $\overline{x} \in \Psi_L(\overline{y})$.

Now, let $(\overline{x}, \overline{y})$ be Pareto optimal for the problem (3.10) but be not feasible for (3.1), (3.2). Of course, $\overline{x}$ is feasible for (3.1) for $y = \overline{y}$. Then, there exists $\hat{x} \geq 0$ satisfying $A^1 \hat{x} + A^2 \overline{y} \leq a$ such that $\langle c, \hat{x} \rangle < \langle c, \overline{x} \rangle$. But then, $C(\hat{x}, \overline{y})^\top \leq C(\overline{x}, \overline{y})^\top$, $C(\hat{x}, \overline{y})^\top \neq C(\overline{x}, \overline{y})^\top$ which contradicts the Pareto optimality of $(\overline{x}, \overline{y})$. $\qquad \square$

PROOF OF THEOREM 3.5: If $\overline{x}$ is a Pareto optimal solution of the problem (3.11) the optimal objective function value of the problem (3.12) for

$y = \overline{x}$ is equal to $e^{k^\top} C \overline{x}$ and $\overline{y} = \overline{x}$ is one optimal solution. This implies feasibility of $(\overline{x}, \overline{x})$ to (3.13). This shows that the first assertion in the Theorem implies the second one and the third one follows from the second assertion. Now, let $\overline{y}$ be such that $(\overline{x}, \overline{y})$ is a feasible problem to (3.13). Then, $\overline{x}$ is an optimal solution of (3.12) for $y = \overline{y}$ which implies that $\overline{x}$ is Pareto optimal for (3.11). $\square$

PROOF OF THEOREM 3.6: Let (x^0, y^0) be a local minimizer of problem (3.7). Then, by linear programming duality, there exists λ^0 such that (x^0, y^0, λ^0) is a local optimizer of the equivalent problem (3.16). By Theorem 6.1.1. in [61] there exists a nonvanishing vector $(\kappa_0, \kappa, \mu, \zeta, \xi)$ such that

$$0 \in \kappa_0 \begin{pmatrix} d^1 \\ d^2 \\ 0 \end{pmatrix} + \zeta^\top \mu + \begin{pmatrix} 0 \\ (A^3)^\top \kappa - \xi \\ 0 \end{pmatrix}$$

$$\zeta \in \partial \min\{F(x^0, y^0, \lambda^0), G(x^0, y^0, \lambda^0)\}$$
$$y^{0\top} \xi = 0, \ \xi \geq 0, \ \kappa_0 \geq 0,$$

where for a set M and a vector a the formula $M + a$ abbreviates the Minkowski sum $M + \{a\} = \{t + a : t \in M\}$. Now, if $G_i(x^0, y^0, \lambda^0) > 0$ set $\eta_i = \mu_i$, $\gamma_i = 0$, $\zeta_i = \nabla F_i(x^0, y^0, \lambda^0)$. But, if $F_i(x^0, y^0, \lambda^0) > 0$, then $\eta_i = 0$, $\gamma_i = \mu_i$, $\zeta_i = \nabla G_i(x^0, y^0, \lambda^0)$. In the third case, if $F_i(x^0, y^0, \lambda^0) = G_i(x^0, y^0, \lambda^0) = 0$, take $\eta_i = \mu_i \alpha$, $\gamma_i = \mu_i(1-\alpha)$ for some $\alpha \in [0,1]$ with $\zeta_i = \alpha \nabla F_i(x^0, y^0, \lambda^0) + (1 - \alpha) \nabla G_i(x^0, y^0, \lambda^0)$. Then, $\eta_i \gamma_i \geq 0$ for $F_i(x^0, y^0, \lambda^0) = G_i(x^0, y^0, \lambda^0) = 0$ and $F_i(x^0, y^0, \lambda^0) \eta_i = 0$, $G_i(x^0, y^0, \lambda^0) \gamma_i = 0$. Application of the special structure of the functions F and G gives the desired result. $\square$

REMARK 3.3 *It can be seen in the proof that all the inequalities* $\eta_i \gamma_i \geq 0$, $i = 1, \ldots, n + p$ *hold.*

PROOF OF THEOREM 3.7: To shorten the explanations we will use the functions F and G in the proof. The result will then follow if the structure of these functions is applied.

Let in the F. John conditions of Theorem 3.6 the multiplier $\kappa_0 = 0$. Note that by linearity of all the functions $F_i(x, y, \lambda)$ we have

$$\nabla F_i(x^0, y^0, \lambda^0)((x, y, \lambda)^\top - (x^0, y^0, \lambda^0)^\top) = F_i(x, y, \lambda) - F_i(x^0, y^0, \lambda^0)$$

and the same is true for the function $G_j(x, y, \lambda)$. Then using the points $z^0 = (x^0, y^0, \lambda^0)^\top$ and $\tilde{z} = (\tilde{x}, \tilde{y}, \tilde{\lambda})^\top$ from the generalized Slater's condition we get

$$\eta^\top \nabla F(x^0, y^0, \lambda^0)(\tilde{z} - z^0) + \gamma^\top \nabla G(x^0, y^0, \lambda^0)(\tilde{z} - z^0)$$
$$+ \ \kappa^\top A^3(\tilde{y} - y^0) - \xi^\top(\tilde{y} - y^0) = 0$$

which implies $\xi = 0$ since $\xi_i = 0$ for $y_i^0 > 0$ and $\tilde{y}_i - y_i^0 > 0$ otherwise and all other terms are zero by the properties of $(\eta, \kappa, \xi, \gamma)$ in Theorem 3.6 and $(\tilde{x}, \tilde{y}, \tilde{\lambda})$. Now, if $\xi = 0$ we get $(\eta, \kappa, \gamma) = 0$ by the generalized Slater's condition. This contradicts Theorem 3.6. Hence, $\kappa_0 > 0$ and without loss of generality, $\kappa_0 = 1$. $\square$

PROOF OF THEOREM 3.11: The proof is done in three steps:

- If $\varphi(h) = 0$ then the optimal solution $(\hat{x}, \hat{y}, \hat{\lambda})$ of problem (3.23) satisfies the conditions (3.22) as well as $A^3 y = b$, $y \geq 0$. Hence, by linear programming duality, the point $(\hat{x}, \hat{y})$ is feasible for the problem (3.7) and we have $\langle d^1, \hat{x} \rangle + \langle d^2, \hat{y} \rangle \leq h$.

- On the other hand, if $(\hat{x}, \hat{y})$ is feasible for the problem (3.7) then there exists a vector $\overline{\lambda} \in \mathbb{R}^p$ such that $(\hat{x}, \hat{y}, \overline{\lambda})$ is feasible for (3.23) with $h = \langle d^1, \hat{x} \rangle + \langle d^2, \hat{y} \rangle$ and the objective function value of this problem is equal to zero by linear programming duality. Since the objective function of this problem is bounded from below by zero this solution is then also an optimal one.

- Hence, $\varphi(h) > 0$ if and only if there is no feasible solution for the bilevel programming problem (3.7) having $\langle d^1, \hat{x} \rangle + \langle d^2, \hat{y} \rangle \leq h$. This implies the result.

                                                                                    $\square$

PROOF OF THEOREM 3.12: We adopt the proof in [81] and consider the problem 3-SAT: Let $X_1, X_2, \ldots, X_n$ be n Boolean variables, consider k clauses of disjunctions of maximal three of the Boolean variables and their negations. Does there exist an assignment of values to the n Boolean variables such that all the k clauses are at the same time true?

We will transform an arbitrary instance of 3-SAT into a special instance of (3.1), (3.2). Let, for each of the Boolean variables X_i two real upper level variables $y_i, \overline{y}_i$ be considered with the constraint $y_i + \overline{y}_i = 1$. Then, for each of the clauses construct an inequality $s_i + s_j + s_k \geq z$ where the variables s_t stand for one of the variables y_t or $\overline{y}_t$ if X_t resp. $\neg X_t$ appears in the clause. z is an additional upper level variable. Add n lower level variables x_i and the lower level constraints $0 \leq x_i \leq y_i$, $0 \leq x_i \leq \overline{y}_i$. Then, the lower level problem reads as

$$
\begin{aligned}
- \sum_{i=1}^{n} x_i \;\; &\to \min \\
0 \leq x_i \leq y_i, \quad &i = 1, \ldots, n \\
0 \leq x_i \leq \overline{y}_i, \quad &i = 1, \ldots, n.
\end{aligned}
\tag{3.37}
$$

Clearly, given y this problem has the optimal solution $x_i = \min\{y_i, \overline{y}_i\}$ for all i. To form the upper level problem, consider

$$
\begin{aligned}
\sum_{i=1}^{n} x_i - z \;\; &\to \min \\
s_i + s_j + s_k \geq z \;\; &\text{for each clause} \\
1 \geq z \geq 0, \quad 1 \geq y_i \geq 0, \;\; &i = 1, \ldots, n \\
1 \geq \overline{y}_i \geq 0, \quad y_i + \overline{y}_i = 1, \;\; &i = 1, \ldots, n \\
x \quad \text{solves (3.37)}, &
\end{aligned}
\tag{3.38}
$$

where the variables s_i are dummies for y_t or $\overline{y}_t$ as pointed out above. The decision version of the bilevel problem asks for a feasible solution of problem (3.38) having at most some predefined objective function value. Clearly, the instance of the problem 3-SAT has the answer yes if and only if the optimal objective function value of (3.38) is minus one. This shows that the problem 3-SAT is a subproblem of the bilevel linear programming problem. Since 3-SAT is $\mathcal{NP}$–complete [104], the linear bilevel programming problem is $\mathcal{NP}$–hard.

Now we show that the computation of an approximate solution to the linear bilevel programming problem is of the same complexity as computing an optimal solution. This is verified by transforming the approximate solution into a feasible solution with no worse objective function value in the upper level objective function. Let $(x^*, y^*, \overline{y}^*, z^*)$ be an ε–optimal solution, i.e. $\sum_{i=1}^{n} x_i^* - z^* \leq \varepsilon f^*$ for the optimal objective function value f^* and $0 < \varepsilon < 1$. If $y_i^* \in (0, 1/2)$ then $x_i^* = y_i^*$ in the lower level and we set $y_i = 0$, $\overline{y}_i = 1$, $x_i = 0$. Also, set z to a value between z^* and $z^* - x_i^*$ which is eventually necessary to retain feasibility (actually, z is decreased up to the next inequality constraint is satisfied as equation). Doing this, the upper level objective function value cannot increase. This can be performed one-by-one thus decreasing the number of non-integral variables but not increasing the objective function value.

In an analogous manner the case when $y_i^* \in [1/2, 1)$ can be treated: $y_i = 1$, $\overline{y}_i = 0$, $x_i = 0$. Hence, all variables x_i get integral values in this way, and the value of z will reach either zero or one and the objective function value will decrease to $f^* \in \{0, -1\}$. Clearly, this procedure is a polynomial one showing that it is as difficult to compute an ε–optimal solution as computing an optimal one. $\qquad\square$

Chapter 4

PARAMETRIC OPTIMIZATION

In this chapter we formulate some results from parametric optimization which will be used in the sequel. We start with some necessary definitions and (Lipschitz) continuity results of point-to-set maps. Upper resp. lower semicontinuity of the solution set mapping $\Psi(\cdot)$ will prove to be the essential assumption needed for guaranteeing the existence of optimal solutions to bilevel programming problems in Chapter 5. Uniqueness and continuity of the optimal solution is needed for converting the bilevel programming problem into a one-level one by means of the implicit function approach already mentioned in the Introduction. We will investigate this approach in Chapters 5 and 6. Some algorithms for solving the bilevel problem presuppose more than continuity of the optimal solution of the lower level problem. Namely, directional differentiability respectively Lipschitz continuity of the solution function of parametric optimization problems are needed. Below we will derive necessary conditions guaranteeing that this function is a PC^1-function, thus having these properties.

At many places we will not use the weakest possible assumptions for our results. This is done to keep the number of different suppositions small. For the investigation of bilevel programming problems we will need rather strong assumptions for the lower level problem making it not necessary that the auxiliary results are given by use of the most general assumptions. Proofs of many theorems in this Chapter are given in Section 4.6. In about two proofs validity of the desired result is verified under stronger assumptions than stated in the respective theorem. This is done since it was our aim to describe the main tools for the proof but to avoid lengthy additional investigations needed for the full proof.

Parametric optimization is a quickly developing field of research. Many deep results can be found in the monographs [17, 18, 46, 96, 113, 171].

4.1   OPTIMALITY CONDITIONS

We start with a short summary of results in smooth programming and an introduction of some notions used in the sequel.

Let the parametric optimization problem in the lower level be stated without loss of generality as

$$\min_{x}\{f(x,y) : g(x,y) \leq 0, \ h(x,y) = 0\}. \tag{4.1}$$

Here, $f : \mathbb{R}^n \times \mathbb{R}^m \to \mathbb{R}$, $g : \mathbb{R}^n \times \mathbb{R}^m \to \mathbb{R}^p$, $h : \mathbb{R}^n \times \mathbb{R}^m \to \mathbb{R}^q$, $g(x,y) = (g_1(x,y), \ldots, g_p(x,y))^\top$, $h(x,y) = (h_1(x,y), \ldots, h_q(x,y))^\top$ are sufficiently smooth (vector-valued) functions. For the investigation of continuity properties for parametric optimization problems the reader is referred e.g. to the papers [12, 120, 121, 152, 154, 155]. Quantitative and qualitative properties of the optimal solution and of the optimal objective function value of this problem with respect to parameter changes are of the main interest in this chapter.

Subsequently the gradient $\nabla w(x)$ of a function $w : \mathbb{R}^k \to \mathbb{R}$ is a row vector, and $\nabla_x g(x,y)$ denotes the gradient of the function g for fixed y with respect to x only. Problem (4.1) is a *convex parametric optimization problem* if all functions $f(\cdot, y)$, $g_i(\cdot, y)$, $i = 1, \ldots, p$, are convex and the functions $h_j(\cdot, y)$, $j = 1, \ldots, q$, are affine-linear on $\mathbb{R}^n$ for each fixed $y \in \mathbb{R}^m$. It is a *linear parametric optimization problem* whenever all functions $f(\cdot, y)$, $g_i(\cdot, y)$, $i = 1, \ldots, p$, $h_j(\cdot, y)$, $j = 1, \ldots, q$, are affine-linear on $\mathbb{R}^n$ for each fixed $y \in \mathbb{R}^m$. Let $\varphi : \mathbb{R}^m \to \mathbb{R}$ be the so-called *optimal value function* defined by the optimal objective function values of problem (4.1):

$$\varphi(y) := \min_{x}\{f(x,y) : g(x,y) \leq 0, \ h(x,y) = 0\} \tag{4.2}$$

and define the *solution set mapping* $\Psi : \mathbb{R}^m \to 2^{\mathbb{R}^n}$ by

$$\Psi(y) := \text{Argmin}_{x} \ \{f(x,y) : g(x,y) \leq 0, \ h(x,y) = 0\}. \tag{4.3}$$

Here, $2^{\mathbb{R}^n}$ is the *power set* of $\mathbb{R}^n$, i.e. the family of all subsets of $\mathbb{R}^n$. By definition, Ψ is a point-to-set mapping which maps $y \in \mathbb{R}^m$ to the set of global optimal solutions of problem (4.1). For convex parametric optimization problems, $\Psi(y)$ is a closed and convex, but possibly empty subset of $\mathbb{R}^n$. Denote by $M : \mathbb{R}^m \to 2^{\mathbb{R}^n}$ with

$$M(y) := \{x \in \mathbb{R}^n : g(x,y) \leq 0, \ h(x,y) = 0\} \tag{4.4}$$

another point-to-set mapping, the *feasible set mapping*. The set of local minimizers of problem (4.1) is denoted by

$$\Psi_{loc}(y) := \{x \in M(y) : \exists \varepsilon > 0 \text{ with } f(x,y) \le f(z,y) \forall z \in M(y) \cap V_\varepsilon(x)\},$$

where

$$V_\varepsilon(x) := \{z \in \mathbb{R}^n : \|x - z\| < \varepsilon\} \tag{4.5}$$

is an open neighborhood of x. We will need some regularity assumptions. Let $y^0 \in \mathbb{R}^m$, $x^0 \in M(y^0)$.

(MFCQ) We say that (MFCQ) is satisfied at (x^0, y^0) if there exists a direction $d \in \mathbb{R}^n$ satisfying

$$\begin{aligned}
\nabla_x g_i(x^0, y^0)d &< 0, \quad \text{for each } i \in I(x^0, y^0) := \{j : g_j(x^0, y^0) = 0\}, \\
\nabla_x h_j(x^0, y^0)d &= 0, \quad \text{for each } j = 1, \dots, q
\end{aligned} \tag{4.6}$$

and the gradients $\{\nabla_x h_j(x^0, y^0) : j = 1, \dots, q\}$ are linearly independent.

We will use the convention in what follows that, if (MFCQ) is supposed without reference to a point than we assume that it is valid for all feasible points $x \in M(y^0)$. (MFCQ) is the well-known Mangasarian-Fromowitz constraint qualification [195]. Denote by

$$L(x, y, \lambda, \mu) := f(x, y) + \lambda^\top g(x, y) + \mu^\top h(x, y)$$

the *Lagrangian function* of problem (4.1) and consider the set

$$\Lambda(x, y) := \{(\lambda, \mu) \in \mathbb{R}^p \times \mathbb{R}^q : \lambda \ge 0, \ \lambda^\top g(x, y) = 0, \ \nabla_x L(x, y, \lambda, \mu) = 0\}$$

of *Lagrange multipliers* corresponding to the point $(x, y) \in \mathbb{R}^n \times \mathbb{R}^m$. Then, we have the following result:

THEOREM 4.1 ([105]) *Consider problem (4.1) at $(x^0, y^0) \in \mathbb{R}^n \times \mathbb{R}^m$ with $x^0 \in \Psi_{loc}(y^0)$. Then, (MFCQ) is satisfied at (x^0, y^0) if and only if $\Lambda(x^0, y^0)$ is a nonempty, convex and compact polyhedron.*

The set

$$SP(y) := \{x \in M(y) : \Lambda(x, y) \ne \emptyset\}$$

is called the *set of stationary solutions of problem* (4.1). Clearly, if (MFCQ) is satisfied at each point (x, y) with $x \in \Psi_{loc}(y)$, then

$$\Psi(y) \subseteq \Psi_{loc}(y) \subseteq SP(y) \tag{4.7}$$

and we have $SP(y) = \Psi(y)$ if (4.1) is a convex parametric optimization problem.

It is a direct consequence of the definition that $x \in SP(y)$ if and only if there are vectors $\lambda \in \mathbb{R}^p$ and $\mu \in \mathbb{R}^q$ such that the triple (x, λ, μ) solves the *system of Karush-Kuhn-Tucker conditions* (KKT conditions for short) :

$$\nabla_x f(x, y) + \lambda^\top \nabla_x g(x, y) + \mu^\top \nabla_x h(x, y) = 0,$$
$$g(x, y) \leq 0, \ h(x, y) = 0, \ \lambda \geq 0, \lambda^\top g(x, y) = 0. \tag{4.8}$$

4.2   UPPER SEMICONTINUITY OF THE SOLUTION SET MAPPING

By the structure of bilevel optimization problems as stated in the Introduction, the investigation of the properties of parametric optimization problems is crucial for our considerations. We will only give a short explanation of the results used in the following. Since the optimal value function of parametric optimization problems is often not continuous and since the continuity properties of point-to-set mappings are of a different nature than those for functions, relaxed continuity properties of functions and point-to-set mappings will now be introduced:

DEFINITION 4.1 *A function* $w : \mathbb{R}^k \to \mathbb{R}$ *is lower semicontinuous at a point* $z \in \mathbb{R}^k$ *if for each sequence* $\{z^t\}_{t=1}^\infty \subset \mathbb{R}^k$ *with* $\lim\limits_{t \to \infty} z^t = z$ *we have*

$$\liminf_{t \to \infty} w(z^t) \geq w(z)$$

and upper semicontinuous at $z \in \mathbb{R}^k$ *provided that*

$$\limsup_{t \to \infty} w(z^t) \leq w(z)$$

for each sequence $\{z^t\}_{t=1}^\infty \subset \mathbb{R}^k$ *converging to* z. *The function is continuous at* $z \in \mathbb{R}^k$ *if it is lower as well as upper semicontinuous there.*

DEFINITION 4.2 *A point-to-set mapping* $\Gamma : \mathbb{R}^k \to 2^{\mathbb{R}^l}$ *is called upper semicontinuous at a point* $z \in \mathbb{R}^k$ *if, for each open set* Z *with* $\Gamma(z) \subset Z$, *there exists an open neighborhood* $U_\delta(z)$ *of* z *with* $\Gamma(z') \subset Z$ *for each* $z' \in U_\delta(z)$. Γ *is lower semicontinuous at* $z \in \mathbb{R}^k$ *if, for each open set* Z *with* $\Gamma(z) \cap Z \neq \emptyset$, *there is an open neighborhood* $U_\delta(z)$ *such that* $\Gamma(z') \cap Z \neq \emptyset$ *for all* $z' \in U_\delta(z)$.

For a nonempty set $A \subseteq \mathbb{R}^p$ and a point $x \in \mathbb{R}^p$ we define the *distance* of x from A as

$$\varrho(x, A) := \inf\{\|x - z\| : z \in A\}. \tag{4.9}$$

DEFINITION 4.3 *A point-to-set mapping* $\Gamma : \mathbb{R}^k \to 2^{\mathbb{R}^l}$ *with nonempty and compact image sets* $\Gamma(z)$ *is upper Lipschitz continuous at a point* $z \in \mathbb{R}^k$ *if there are an open neighborhood* $U_\delta(z)$ *and a constant* $L < \infty$ *such that for all* $z' \in U_\delta(z)$ *the inclusion* $\Gamma(z') \subseteq \Gamma(z) + L\|z - z'\| \mathcal{B}^l$ *holds, where* $\mathcal{B}^l$ *denotes the unit sphere in* $\mathbb{R}^l$. *It is locally Lipschitz continuous at* $z \in \mathbb{R}^k$ *if an open neighborhood* $U_\delta(z)$ *and a constant* $L < \infty$ *exist such that for all* $z', z'' \in U_\delta(z)$ *the inclusion* $\Gamma(z') \subseteq \Gamma(z'') + L\|z'' - z'\| \mathcal{B}^l$ *is satisfied.*

The following compactness assumption will make our considerations more easy. This assumption can be replaced by the inf-compactness assumption [290].

(C) The set $\{(x, y) \in \mathbb{R}^n \times \mathbb{R}^m : g(x, y) \leq 0, \ h(x, y) = 0\}$ is non-empty and compact.

Assumption (C) guarantees that the feasible set of problem (4.1) is non-empty and compact for each $y \in \{z : M(z) \neq \emptyset\}$, too. This implies that $\Psi(y) \neq \emptyset$ as well as $\Psi_{loc}(y) \neq \emptyset$ are compact sets for each $y \in \{z : M(z) \neq \emptyset\}$.

THEOREM 4.2 ([240]) *Consider problem (4.1) at* $y^0 \in \mathbb{R}^m$ *and let the assumptions (C) and (MFCQ) be satisfied for each* $x^0 \in SP(y^0)$. *Then, the point-to-set mappings* $\Lambda(\cdot, \cdot)$ *and* $SP(\cdot)$ *are upper semicontinuous at* y^0.

The local solution set mapping $\Psi_{loc}(\cdot)$ is in general not upper semicontinuous for non-convex parametric optimization problems. This can easily be verified when considering the unconstrained optimization problem of minimizing the function $f(x, y) = \frac{1}{3}x^3 - yx$ which has a local optimal solution at $x = \sqrt{y}$ only for positive y. With respect to the global solution set mapping we can maintain the upper semicontinuity:

THEOREM 4.3 ([17]) *If for the problem (4.1) the assumptions (C) as well as (MFCQ) are satisfied then, the global solution set mapping* $\Psi(\cdot)$ *is upper semicontinuous and the optimal value function* $\varphi(\cdot)$ *is continuous at* y^0.

The solution set mapping of the parametric programming problem (4.1) is generally not continuous and there is in general no continuous selection function $\psi(\cdot)$ for that point-to-set mapping. Here, a selection function $\psi(\cdot)$ of $\Psi(\cdot)$ is a function with

$$\psi(y) \in \Psi(y), \ \forall \ y \in \{z : \Psi(z) \neq \emptyset\}.$$

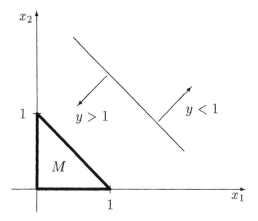

Figure 4.1.   Feasible set $M = M(1)$ and maximization direction in the Example on page 66

This can be shown by use of the following very simple example:

Example: Consider the problem:

$$(1-y)(x_1+x_2) \quad \rightarrow \quad \max_x$$
$$x_1 + x_2 \quad \leq \quad 1,$$
$$x_1, x_2 \quad \geq \quad 0.$$

Then,

$$\Psi(y) = \begin{cases} (0,0)^\top & \text{if } y > 1, \\ M & \text{if } y = 1, \\ \{(x_1, x_2)^\top \geq 0 : x_1 + x_2 = 1\} & \text{if } y < 1. \end{cases}$$

$\square$

Continuity of the solution set mapping can be guaranteed only under stronger assumptions (cf. e.g. [17]). Let $y^0 \in \mathbb{R}^m$, $x^0 \in SP(y^0)$.

(SSOC) The *strong sufficient optimality condition of second order* (SSOC) is said to be satisfied at (x^0, y^0) if, for each $(\lambda, \mu) \in \Lambda(x^0, y^0)$ and for each $d \neq 0$ satisfying

$$\nabla_x g_i(x^0, y^0)d = 0, \quad \text{for each } i \in J(\lambda) := \{j : \lambda_j > 0\},$$
$$\nabla_x h_j(x^0, y^0)d = 0, \quad j = 1, \ldots, q \tag{4.10}$$

we have

$$d^\top \nabla^2_{xx} L(x^0, y^0, \lambda, \mu)d > 0.$$

The local optimal solution $x^0 \in \Psi_{loc}(y^0)$ is called *strongly stable* if there exist open neighborhoods $U_\delta(y^0), \delta > 0$, of y^0 and $V_\varepsilon(x^0), \varepsilon > 0$, of x^0 and a uniquely determined continuous vector-valued function $x :$ $U_\delta(y^0) \to V_\varepsilon(x^0)$ such that $x(y)$ is the unique local optimal solution of problem (4.1) in $V_\varepsilon(x^0)$ for all $y \in U_\delta(y^0)$.

THEOREM 4.4 ([157]) *Let $y^0 \in \mathbb{R}^m$ and let the assumptions (MFCQ) and (SSOC) be satisfied for problem (4.1) at $(x,y) = (x^0, y^0)$ with $x^0 \in \Psi_{loc}(y^0)$. Then, the local optimal solution x^0 is strongly stable.*

REMARK 4.1 *The results in [157] are somewhat stronger than presented in Theorem 4.4. It is shown there that under (LICQ) a KKT point (x^0, λ^0, μ^0) is strongly stable if and only if a certain regularity condition is satisfied for the Hessian matrix of the Lagrangian function with respect to x hold (which is weaker than the (SSOC)). If the linear independence constraint qualification (LICQ) is violated but (MFCQ) holds, an optimal point x^0 is strongly stable if and only if (SSOC) is satisfied. The interested reader is also referred to the interesting paper [154] on strong stability.*

The following example borrowed from [107] shows that it is not possible to weaken the assumed second order condition.

Example: Consider the problem

$$\min_x \{-x_2 : -x_1^2 + x_2 \le y_1, \; x_1^2 + x_2 \le y_2\}.$$

The feasible set of this problem is illustrated in Fig. 4.2 for $y = (0,0)^\top$ and $y = (0,1)^\top$. Then, for $y^0 = (0,0)^\top$, the unique optimal solution is $x^0 = (0,0)^\top$ and

$$\Lambda(x^0, y^0) = \{\lambda \ge 0 : \lambda_1 + \lambda_2 = 1\}.$$

The (MFCQ) is satisfied and we have $I(x^0, y^0) = \{1, 2\}$,

$$\{d : \nabla_x g_i(x^0, y^0)d < 0, \text{ for } \lambda_i > 0\} = \{d : d_2 < 0\}.$$

Also,

$$d^\top \nabla_{xx}^2 L(x^0, y^0, \lambda)d = 2(\lambda_2 - \lambda_1)d_1^2$$

is positive for all $d \in \{d' : d_2' = 0\}$ if $\lambda_2 > \lambda_1$ but not if $\lambda_2 \le \lambda_1$. Hence, (SSOC) is not satisfied.

Now consider the optimal solution for $y(t) = t(0,2)^\top$. Then,

$$\Psi(y(t)) = \left\{ \left(-\sqrt{t}, t\right)^\top, \left(\sqrt{t}, t\right)^\top \right\}$$

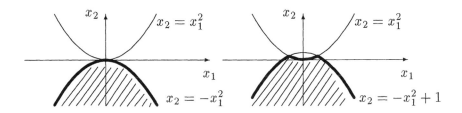

Figure 4.2.   The feasible set in the Example on page 67

which means that the optimal solution is not strongly stable.   □

In some cases strong stability of local optimal solutions is not enough, stronger properties are needed. The following theorem states upper Lipschitz continuity of the mapping $(\Psi_{\mathrm{loc}}(\cdot) \cap V_\varepsilon(x^0)) \times \Lambda(\cdot, \cdot)$ for some $\varepsilon > 0$. Upper Lipschitz continuity of the local solution set mapping $\Psi_{loc}(\cdot)$ alone has also been shown in in [108, 257]. Using the results in [126], this also implies upper Lipschitz continuity of the mapping $(\Psi_{\mathrm{loc}}(\cdot) \cap V_\varepsilon(x^0)) \times \Lambda(\cdot, \cdot)$.

THEOREM 4.5 ([240]) *Consider problem (4.1) at a point* $x^0 \in \Psi_{\mathrm{loc}}(y^0)$ *and let the assumptions (MFCQ) and (SSOC) be satisfied. Then, there exist open neighborhoods* $U_\delta(y^0), V_\varepsilon(x^0)$ *and a constant* $L < \infty$ *such that*

$$\varrho((x, \lambda, \mu), \{x^0\} \times \Lambda(x^0, y^0)) \le L\|y - y^0\|$$

for all $x \in \Psi_{loc}(y) \cap V_\varepsilon(x^0)$, $(\lambda, \mu) \in \Lambda(x, y)$, $y \in U_\delta(x^0)$.

Another quantitative stability property which is not touched here is Hölder continuity of the function $x(\cdot)$. This property has been investigated e.g. in the papers [46, 109, 209].

DEFINITION 4.4 *A function* $z : \mathbb{R}^p \to \mathbb{R}^q$ *is called locally Lipschitz continuous at a point* $x^0 \in \mathbb{R}^p$ *if there exist an open neighborhood* $V_\varepsilon(x^0)$ *of* x^0 *and a constant* $l < \infty$ *such that*

$$\|z(x) - z(x')\| \le l\|x - x'\| \ \forall x, x' \in V_\varepsilon(x^0).$$

It is locally upper Lipschitz continuous if for some open neighborhood $V_\varepsilon(x^0)$ *of* x^0 *and a certain constant* $l < \infty$ *the inequality*

$$\|z(x) - z(x^0)\| \le l\|x - x^0\| \ \forall x \in V_\varepsilon(x^0)$$

holds.

Note, that upper Lipschitz continuity is a weaker property than Lipschitz continuity in the sense that one of the points in the neighborhood $V_\varepsilon(x^0)$ is fixed to x^0. We remark that, to avoid a even larger number of different assumptions, we have used conditions in Theorem 4.5 which are more restrictive than those in the original paper. Under the original presumptions, the solution set mapping does not locally reduce to a function but it is locally upper Lipschitz continuous as a point-to-set mapping. Under the assumptions used, the optimal solution function is a locally upper Lipschitz continuous function due to strong stability, but it is generally not locally Lipschitz continuous as shown in the next example. Hence, for deriving Lipschitz continuity of local optimal solutions of problem (4.1) we need more restrictive assumptions than those established for guaranteeing strong stability of a local optimal solution.

Example: [257] Consider the problem

$$\frac{1}{2}(x_1 - 1)^2 + \frac{1}{2}x_2^2 \;\to\; \min$$
$$x_1 \leq 0$$
$$x_1 + y_1 x_2 + y_2 \leq 0 \tag{4.11}$$
$$-100 \leq x_i \leq 100, \qquad i = 1, 2$$
$$-100 \leq y_i \leq 100, \qquad i = 1, 2$$

at $y^0 = (0,0)^\top$. This is a convex parametric optimization problem satisfying the assumption (C). At the unique optimal solution $x^0 = (0,0)^\top$, the assumptions (MFCQ) and (SSOC) are also valid. Hence, the solution x^0 is strongly stable. It can easily be seen that the unique optimal solution $x(y)$ is given by

$$x(y) = \begin{cases} (0,0)^\top & \text{if } y_2 \leq 0, \\[2mm] \left(0, -\dfrac{y_2}{y_1}\right)^\top & \text{if } 0 < y_2 \leq y_1^2, \\[3mm] \left(\dfrac{y_1^2 - y_2}{1 + y_1^2}, -\dfrac{y_1 + y_1 y_2}{1 + y_1^2}\right)^\top, & \text{if } y_1^2 \leq y_2. \end{cases}$$

The components of the function $x(\cdot)$ are plotted in Figure 4.3 in a neighborhood of the point y^0. But, the function $x(\cdot)$ is not locally Lipschitz continuous at y^0. This can be seen by the following: Take $y_2 = y_1^2 - y_1^3 > 0$. Then

$$\frac{\|x(y_1, y_1^2 - y_1^3) - x(y_1, 0)\|}{\|(y_1, y_1^2 - y_1^3) - (y_1, 0)\|} = \frac{y_1 - y_1^2}{y_1^2 - y_1^3} \to \infty$$

for $y_1 \to 0$ from above. □

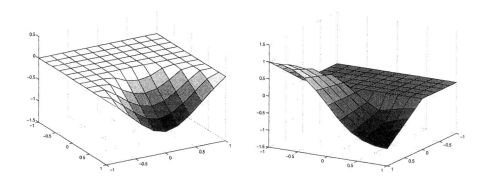

Figure 4.3. The components of the optimal solution in Example on page 69

A suitable approach to show local Lipschitz continuity of the optimal solution of a parametric optimization problem is the concept of a piecewise continuously differentiable function. Another approach to characterize optimization problems having locally Lipschitz continuous solutions is given in [172]. We will not follow this strong result here since our main aim is the computation of the directional derivative of the optimal solution and local Lipschitz continuity is rather a by-product in this way.

4.3   PIECEWISE CONTINUOUSLY DIFFERENTIABLE FUNCTIONS

Piecewise continuously differentiable functions are continuous selections of finitely many continuously differentiable functions. Examples are the pointwise maximum or minimum of finitely many smooth functions in $\mathbb{R}^n$. Such functions have been investigated in the last few years as important examples of nonsmooth functions. Optimization problems involving piecewise continuously differentiable functions are one interesting class within nonsmooth optimization. We will see that, under suitable assumptions, the solution function of a smooth optimization problem has this property, too. As an corollary of this property we derive that the optimal solution function is then also locally Lipschitzian and directionally differentiable. Hence, the bilevel programming problem proves to be equivalent to a Lipschitz optimization problem under the same assumptions. This will be useful in Chapters 5 and 6. A very comprehensive treatment of PC^1-functions can be found in [251].

DEFINITION 4.5 *A function $z : \mathbb{R}^p \to \mathbb{R}^q$ is called a piecewise continuously differentiable function (or PC^1-function for short) at x^0 if there exist an open neighborhood $V_\varepsilon(x^0)$ with $\varepsilon > 0$ and a finite number of continuously differentiable functions $z^i : V_\varepsilon(x^0) \to \mathbb{R}^q, i = 1, \ldots, t$, such that z is continuous on $V_\varepsilon(x^0)$ and*

$$z(w) \in \{z^1(w), \ldots, z^t(w)\} \; \forall \; w \in V_\varepsilon(x^0).$$

This means that the function z is a continuous selection of finitely many continuously differentiable functions. The index set of the active selection functions z^i at a point x^0 is denoted by $I_z(x^0)$:

$$I_z(x^0) = \{i \in \{1, \ldots, t\} : z^i(x^0) = z(x^0)\}. \qquad (4.12)$$

A selection function z^i is active on the set

$$\mathrm{Supp}(z, z^i) = \{x : z(x) = z^i(x)\}. \qquad (4.13)$$

By continuity, the sets $\mathrm{Supp}(z, z^i)$ are closed and $i \in I_z(x)$ for $x \in \mathrm{Supp}(z, z^i)$. In the following we will need also the definition of the contingent cone to a set:

DEFINITION 4.6 *Let $C \subset \mathbb{R}^q$ and $x^0 \in \mathrm{cl}\, C$. The set*

$$K_C(x^0) = \{v : \; \exists \{x^k\}_{k=1}^\infty \subseteq C, \exists \{t_k\}_{k=1}^\infty \subset \mathbb{R}_{++} \text{ with } \lim_{k\to\infty} x^k = x^0,$$
$$\lim_{k\to\infty} t_k = 0, \lim_{k\to\infty} (x^k - x^0)/t_k = v\}$$

is called the contingent cone to C at x^0.

The following example should illustrate the notion of a PC^1-function:

Example: Consider the problem

$$
\begin{aligned}
-x \;&\to\; \min_x \\
x \;&\le\; 1 \\
x^2 \;&\le\; 3 - y_1^2 - y_2^2 \\
(x - 1.5)^2 \;&\ge\; 0.75 - (y_1 - 0.5)^2 - (y_2 - 0.5)^2
\end{aligned}
$$

with two parameters y_1 and y_2. Then, x is a continuous selection of three continuously differentiable functions $x^1 = x^1(y), x^2 = x^2(y), x^3 = x^3(y)$ in an open neighborhood of the point $y^0 = (1, 1)^\top$:

$$
x(y) = \begin{cases}
x^1 = 1, & y \in \mathrm{Supp}(x, x^1), \\
x^2 = \sqrt{3 - y_1^2 - y_2^2}, & y \in \mathrm{Supp}(x, x^2), \\
x^3 = 1.5 - \sqrt{0.75 - (y_1 - 0.5)^2 - (y_2 - 0.5)^2}, & y \in \mathrm{Supp}(x, x^3),
\end{cases}
$$

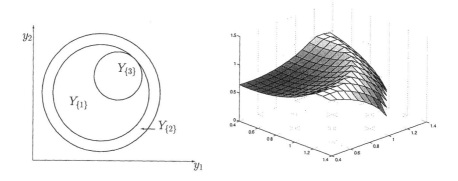

Figure 4.4. The sets $\mathrm{Supp}(x, x^i)$ and the optimal solution in the Example on page 71

where

$$\mathrm{Supp}(x, x^1) = \{y : y_1^2 + y_2^2 \leq 2, \ (y_1 - 0.5)^2 + (y_2 - 0.5)^2 \geq 0.5\},$$
$$\mathrm{Supp}(x, x^2) = \{y : 2 \leq y_1^2 + y_2^2 \leq 3\},$$
$$\mathrm{Supp}(x, x^3) = \{y : (y_1 - 0.5)^2 + (y_2 - 0.5)^2 \leq 0.5\}.$$

The sets $Y_{\{i\}} = \mathrm{Supp}(x, x^i)$ and the function x are illustrated in Figure 4.4. $\square$

PC^1-functions have a number of nice properties which are of great use in the following. We focus here on Lipschitz continuity and different differentiability properties as generalized differentiability in Clarke's sense, directional and pseudodifferentiability.

THEOREM 4.6 ([114]) *PC^1-functions are locally Lipschitz continuous. As a Lipschitz constant the largest Lipschitz constant of the selection functions can be used.*

For locally Lipschitz continuous functions the generalized gradient in the sense of Clarke can be defined. Since we deal only with finite dimensional spaces the following definition can be used for the generalized gradient [61]:

DEFINITION 4.7 *Let $z : \mathbb{R}^p \to \mathbb{R}$ be a locally Lipschitz continuous function. Then*

$$\partial z(x^0) = \mathrm{conv}\left\{ \lim_{k \to \infty} \nabla z(x^k) : \lim_{k \to \infty} x^k = x^0, \nabla z(x^k) \text{ exists } \forall k \right\}$$

is the generalized gradient of the function z at x^0.

For a vector valued differentiable function $z : \mathbb{R}^p \to \mathbb{R}^q$ we adopt the convention that the Jacobian of this function is also denoted by $\nabla z(x)$ and that the gradients of the component functions form the rows of the Jacobian. This will not lead to any confusion since in any case the kind of function considered will be clear. Then, the *generalized Jacobian* is analogously defined:

$$\partial z(x^0) = \text{conv} \left\{ \lim_{k \to \infty} \nabla z(x^k) : \lim_{k \to \infty} x^k = x^0, \nabla z(x^k) \text{ exists } \forall k \right\}.$$

THEOREM 4.7 ([163, 251]) *Let $z : \mathbb{R}^p \to \mathbb{R}$ be a PC^1-function. Then,*

$$\partial z(x^0) = \text{conv} \left\{ \nabla z^i(x^0) : x^0 \in \text{cl int Supp}(z, z^i) \right\}.$$

We will also apply this Theorem in an analogous manner to vector-valued functions. Then, it shows that the generalized Jacobian of the vector-valued function is equal to the convex hull of the Jacobians of all selection functions z^i for which $x^0 \in \text{cl int Supp}(z, z^i)$. We will call all these selection functions *strongly active* and use the abbreviation

$$I_z^0(x^0) = \{ i : x^0 \in \text{cl int Supp}(z, z^i) \}. \tag{4.14}$$

DEFINITION 4.8 *A locally Lipschitz continuous function $z : \mathbb{R}^p \to \mathbb{R}$ is called semismooth at a point x^0 if for each nonzero vector r the sequence $\{g^k r\}_{k=1}^\infty$ converges whenever $\{x^k\}_{k=1}^\infty$ is a sequence with $\lim_{k \to \infty} \dfrac{x^k - x^0}{\|x^k - x^0\|} = \dfrac{r}{\|r\|}$ and $g^k \in \partial z(x^k)$ for all k.*

THEOREM 4.8 ([58]) *PC^1-functions $z : \mathbb{R}^p \to \mathbb{R}$ are semismooth.*

Later on we will call a function $z : \mathbb{R}^p \to \mathbb{R}^q$ semismooth provided that each of its component functions $z_i : \mathbb{R}^p \to \mathbb{R}$, $i = 1, \ldots, q$, is semismooth in the sense of Definition 4.8. Then, Theorem 4.8 can analogously be applied to this case.

DEFINITION 4.9 *A function $z : \mathbb{R}^p \to \mathbb{R}$ is directionally differentiable at x^0 if for each direction $r \in \mathbb{R}^p$ the following one-sided limit exists:*

$$z'(x^0; r) := \lim_{t \to 0+} t^{-1} [z(x^0 + tr) - z(x^0)].$$

The value $z'(x^0; r)$ is the directional derivative of the function z at $x = x^0$ in direction r.

Having a closer look on the proof of Theorem 4.8 it can be seen that PC^1-functions are directionally differentiable. This is also a consequence of a result in [203]. Moreover, it is also obvious that for locally Lipschitz continuous, directionally differentiable functions z the function $r \mapsto z'(x^0; r)$ is Lipschitz continuous again (with the same Lipschitz constant). In [80] it is shown that the function $z : \mathbb{R}^p \to \mathbb{R}$ is uniformly directionally differentiable in the sense that for each $r \in \mathbb{R}^p$ and for each $\varepsilon > 0$ there exist numbers $\delta > 0$ and $\alpha_0 > 0$ such that

$$|z(x^0 + \alpha d) - z(x^0) - \alpha z'(x^0; d)| \leq \alpha \varepsilon$$

for all $\|d - r\| < \delta$ and $0 \leq \alpha \leq \alpha_0$ provided that the above assumptions are satisfied.

DEFINITION 4.10 *A directionally differentiable function $z : \mathbb{R}^p \to \mathbb{R}$ is called Bouligand differentiable (B-differentiable) at x^0 if the directional derivative gives a first-order approximation of z at x^0:*

$$\lim_{x \to x^0} \frac{z(x) - z(x^0) - z'(x^0; x - x^0)}{\|x - x^0\|} = 0.$$

A locally Lipschitz continuous, directionally differentiable function is also B-differentiable [251] and $z'(x^0; \cdot)$ is a continuous selection of linear functions:

$$z'(x^0; r) \in \{\nabla z^i(x^0) r : i \in I_z^0(x^0)\}. \tag{4.15}$$

This result can be improved to

$$z'(x^0; r) \in \{\nabla z^i(x^0) r : i \in \{j : \text{int } K_{\text{Supp}(z, z^j)}(x^0) \neq \emptyset\}\}$$

[280]. Since all these results follow more or less directly from the proof of Theorem 4.8 their proofs will be dropped.

DEFINITION 4.11 ([206]) *A function $z : \mathbb{R}^p \to \mathbb{R}$ is called pseudodifferentiable at x^0 if there exist an open neighborhood $V_\varepsilon(x^0)$ of x^0 and an upper semicontinuous point-to-set mapping $\Gamma_z : V_\varepsilon(x^0) \to 2^{\mathbb{R}^p}$ with nonempty, convex and compact values such that*

$$z(x) = z(x^0) + g(x - x^0) + o(x, x^0, g) \ \forall \ x \in V_\varepsilon(x^0),$$

where (the row vector) $g \in \Gamma_z(x)$ and

$$\lim_{k \to \infty} \frac{o(x^k, x^0, g^k)}{\|x^k - x^0\|} = 0 \tag{4.16}$$

for each sequences $\{x^k\}_{k=1}^\infty$, $\{g^k\}_{k=1}^\infty$ with $\lim_{k \to \infty} x^k = x^0$, $g^k \in \Gamma_z(x^k)$ for all k.

It has been shown in [206] that pseudodifferentiable functions are locally Lipschitz continuous and that locally Lipschitz continuous, semismooth functions are pseudodifferentiable. In the latter case, the generalized gradient in the sense of Clarke can be used as the *pseudodifferential* Γ_z. Since it is in general difficult to compute the Clarke generalized gradient but it is easy to compute a pseudodifferential with larger values for PC^1-functions we will use another proof to verify that a PC^1-function is pseudodifferentiable.

Note that a continuously differentiable function z is pseudodifferentiable with $\Gamma_z(x) = \{\nabla z(x)\}$.

THEOREM 4.9 ([206]) *Let $z : \mathbb{R}^p \to \mathbb{R}$ be a continuous function which is a selection of a finite number of pseudodifferentiable functions:*

$$z(x) \in \{z^i(x) : i = 1, \ldots, k\}$$

where $z^i : \mathbb{R}^p \to \mathbb{R}$ has the pseudodifferential $\Gamma_{z^i}(x), i = 1, \ldots, k < \infty$. Then z is pseudodifferentiable and, as pseudodifferential, we can take

$$\Gamma_z(y) = \mathrm{conv} \bigcup_{i \in I_z(y)} \Gamma_{z^i}(y).$$

It should be noticed that by simply repeating the proof of Theorem 4.8 it can be shown that semismoothness of a PC^1-functions is maintained if the Clarke generalized gradient is replaced by the pseudodifferential.

4.4 PROPERTIES OF OPTIMAL SOLUTIONS
4.4.1 DIRECTIONAL DIFFERENTIABILITY

In this section focus is on differential properties of the optimal solution of the parametric optimization problem (4.1):

$$\min_x \{f(x, y) : g(x, y) \le 0, \ h(x, y) = 0\}.$$

The example on page 69 has shown that without more restrictive assumptions than (MFCQ) and (SSOC) it is not possible to get Lipschitz continuity of the local solution function $x(\cdot)$. Let $x^0 \in \Psi_{loc}(y^0)$ and let $(\lambda^0, \mu^0) \in \Lambda(x^0, y^0)$. Then, using the Karush-Kuhn-Tucker conditions it is obvious that x^0 is also a stationary point of the enlarged problems

$$\min_x \{f(x, y^0) : g_i(x, y^0) \le 0, i \in J(\lambda^0), h(x, y^0) = 0\} \qquad (4.17)$$

and

$$\min_x \{f(x, y^0) : g_i(x, y^0) = 0, i \in J(\lambda^0), h(x, y^0) = 0\}. \qquad (4.18)$$

Moreover, if a sufficient second-order optimality condition is satisfied for (4.1) then x^0 is locally optimal for both problems, too. Now, (SSOC) at (x^0, y^0) for (4.1) guarantees strong stability of x^0 provided that a regularity condition is satisfied. (MFCQ) for (4.1) implies that this condition is also satisfied for (4.17) but not validity of a regularity condition for problem (4.18). The linear independence constraint qualification is satisfied for (4.18) if (MFCQ) holds for (4.1) and (λ^0, μ^0) is a vertex of $\Lambda(x^0, y^0)$. The following example shows that it is in general not possible to restrict the considerations to the vertices of $\Lambda(x^0, y^0)$ if the parametric properties of problem (4.1) are investigated and this problem is replaced by (4.18) to get a deeper insight into (4.1).

Example: Consider the following problem

$$
\begin{aligned}
(x_1 + 1)^2 + x_2^2 &\to \quad \min \\
(x_1 - 1)^2 + (x_2 + y_1)^2 - 1 &\leq \quad 0 \\
(x_1 - 1)^2 + (x_2 + y_2)^2 - 1 &\leq \quad 0 \\
-x_1 &\leq \quad 0
\end{aligned}
$$

at $y^0 = (0, 0)^\top$ with $r = (-q, 1)^\top$ for $0 < q < 1$. Define

$$
w(q, t) = \sqrt{1 - (q + 1)^2 t^2 / 4}, \; z(q) = \frac{1 - q}{1 + q}.
$$

Then,

$$
x(y^0 + tr) = (1 - w(q, t), -(1 - q)t/2)^\top, \; x^0 = (0, 0)^\top.
$$

Moreover,

$$
\Lambda(x^0, y^0) = \operatorname{conv} \{(1, 0, 0)^\top, (0, 1, 0)^\top, (0, 0, 2)^\top\}
$$

and

$$
\Lambda(x(tr), tr) = \left\{ \left(\frac{2 - w(q, t)}{2w(q, t)} - \frac{z(q)}{2}, \frac{2 - w(q, t)}{2w(q, t)} + \frac{z(q)}{2}, 0 \right) \right\}.
$$

Then, for $t \to 0+$ the Lagrange multiplier $\lambda(y^0 + tr)$ converges to

$$
\lambda^0 = \left\{ \left(\frac{1 - z(q)}{2}, \frac{1 + z(q)}{2}, 0 \right) \right\}
$$

for which both components are positive. But the gradients $\nabla_x g_1(x^0, y^0)$ and $\nabla_x g_2(x^0, y^0)$ are linearly dependent, λ^0 is not a vertex of $\Lambda(x^0, y^0)$.

□

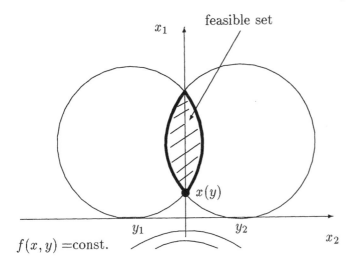

feasible set

x_1

$x(y)$

y_1 y_2

x_2

$f(x, y) = $const.

Figure 4.5. Feasible set and optimal solution in the Example on page 76

In order to ensure that each local optimal solution of (4.1) is also a local optimal solution of problem (4.18) for some vertex $(\lambda^0, \mu^0) \in \Lambda(x^0, y^0)$ (and hence the linear independence constraint qualification is satisfied) we use the following assumption:

(CRCQ) The *constant rank constraint qualification* is said to be satisfied for problem (4.1) at a point $(x, y) = (x^0, y^0)$ if there exists an open neighborhood $W_\varepsilon(x^0, y^0), \varepsilon > 0$, of (x^0, y^0) such that, for each subsets

$$I \subseteq I(x^0, y^0) := \{i : g_i(x^0, y^0) = 0\}, \quad J \subseteq \{1, \ldots, q\},$$

the family of gradient vectors $\{\nabla_x g_i(x, y) : i \in I\} \cup \{\nabla_x h_j(x, y) : j \in J\}$ has the same rank for all $(x, y) \in W_\varepsilon(x^0, y^0)$.

THEOREM 4.10 ([235]) *Consider the problem (4.1) at $y = y^0$ and let the assumptions (MFCQ), (SSOC) and (CRCQ) be satisfied at a stationary solution $x^0 \in SP(y^0)$. Then, the by Theorem 4.4 uniquely determined function $x(y)$ with $\{x(y)\} = \Psi_{loc}(y) \cap V_\varepsilon(x^0)$ is a PC^1-function.*

It has been shown in the proof of this theorem, that the local solution functions of the problems

$$\min_x \{f(x, y^0) : g_i(x, y^0) = 0, i \in I, h(x, y^0) = 0\}. \qquad (4.19)$$

can be used as selection functions for index sets I satisfying the following two conditions:

(C1) There is $(\lambda, \mu) \in E\Lambda(x^0, y^0)$ such that $J(\lambda) \subseteq I \subseteq I(x^0, y^0)$ and

(C2) The gradients $\{\nabla_x g_i(x^0, y^0) : i \in I\} \cup \{\nabla_x h_j(x^0, y^0) : j = 1, \ldots, q\}$ are linearly independent.

In the following we will denote the vertex set of $\Lambda(x^0, y^0)$ by $E\Lambda(x^0, y^0)$, the above selection functions of $x(\cdot)$ by $x^I(\cdot)$ and the family of all sets I satisfying both conditions (C1) and (C2) for a fixed vertex $(\lambda^0, \mu^0) \in \Lambda(x^0, y^0)$ by $\mathcal{I}(\lambda^0)$. As a corollary of the Theorems 4.6 to 4.10 we find

COROLLARY 4.1 *The local optimal solution function $x(\cdot)$ is*

- *locally Lipschitz continuous, the generalized Jacobian in the sense of Clarke is given by*

$$\partial x(y^0) = \text{conv } \{\nabla x^I(y^0) : I \in I^0_x(y^0)\},$$

- *directionally differentiable with*

$$x'(y^0; r) \in \{\nabla x^I(y^0) r : I \in I^0_x(y^0)\},$$

- *B-differentiable, and*

- *pseudodifferentiable with a pseudodifferential*

$$\Gamma_x(y^0) = \text{conv } \{\nabla x^I(y^0) : I \in I_x(y^0)\}.$$

At least if the lower level problem (1.1) is a parametric convex optimization problem, we can now replace the bilevel programming problem (1.1), (1.3) by the Lipschitz optimization problem

$$\min\{F(x(y), y) : G(x(y), y) \leq 0, H(x(x), y) = 0\}.$$

Then, solution algorithms for this problem could be based on ideas using the generalized Jacobian or the directional derivative of the function $x(y)$. Hence, we need effective formulae for the computation of at least one element of the generalized Jacobian and of the directional derivative of the function $x(\cdot)$. These formulae are given next. We start with the directional derivative.

THEOREM 4.11 ([235]) *Consider problem (4.1) at a point $y = y^0$ and let $x^0 \in \Psi_{loc}(y^0)$ be a local optimal solution where the assumptions (MFCQ), (SSOC), and (CRCQ) are satisfied. Then the directional derivative of the function $x(\cdot)$ at $y = y^0$ in direction r coincides with* *the unique optimal solution of the convex quadratic programming problem $QP(\lambda^0, \mu^0, r)$*

$$0.5d^{\top}\nabla^2_{xx}L(x^0, y^0, \lambda^0, \mu^0)d + d^{\top}\nabla^2_{yx}L(x^0, y^0, \lambda^0, \mu^0)r \to \min_{d}$$

$$\nabla_x g_i(x^0, y^0)d + \nabla_y g_i(x^0, y^0)r \begin{cases} = 0, & \text{if } i \in J(\lambda^0), \\ \leq 0, & \text{if } i \in I(x^0, y^0) \setminus J(\lambda^0), \end{cases}$$

$$\nabla_x h_j(x^0, y^0)d + \nabla_y h_j(x^0, y^0)r = 0 \text{ for all } j = 1, \ldots, q \qquad (4.20)$$

for arbitrary vectors $(\lambda^0, \mu^0) \in \Lambda(x^0, y^0)$ solving

$$\nabla_y L(x^0, y^0, \lambda, \mu)r \to \max_{(\lambda,\mu)\in\Lambda(x^0,y^0)} \qquad (4.21)$$

For later use, denote the solution set of problem (4.21) by

$$S(r) := \operatorname*{Argmax}_{(\lambda,\mu)} \{\nabla_y L(x^0, y^0, \lambda, \mu)r : (\lambda, \mu) \in \Lambda(x^0, y^0)\}. \qquad (4.22)$$

Summing up, for computing the directional derivative of the local optimal solution function $x(\cdot)$ at $y = y^0$ in direction r the following algorithm can be used:

Method for computing the directional derivative:
   Input: parametric optimization problem (4.1) at the point (x^0, y^0), where (MFCQ), (SSOC), and (CRCQ) hold and a direction r.
   Output: directional derivative of the function $x(\cdot)$ at $y = y^0$ in direction r.
   1. Compute an optimal solution (λ^0, μ^0) of the problem (4.21): $(\lambda^0, \mu^0) \in S(r)$,
   2. solve $QP(\lambda^0, \mu^0, r)$.

It should be noted that, for computing this directional derivative, the solution of two standard optimization problems is necessary which can be done in polynomial time [218].

The Example on page 69 can be used to illustrate the necessity of the assumptions in Theorem 4.10 especially that Lipschitz continuity cannot be achieved without (CRCQ). The local optimal solution function $x(\cdot)$ is directionally differentiable even without (CRCQ) and even under a weaker sufficient optimality condition of second order than (SSOC) [257]. Under these weaker assumptions the directional derivative proves to be the optimal solution of some minimax problem provided this problem has a unique optimal solution [257]. The paper [68] investigates directional differentiability of $x(\cdot)$ under (MFCQ) and (SSOC) without (CRCQ). It is shown that the directional derivative of $x(\cdot)$ can again be computed by solving $QP(\lambda, \mu, r)$ but now only for some suitably chosen vector $(\lambda, \mu) \in S(r)$. We will also mention that similar results under weaker directional versions of the (MFCQ) and the (SSOC) have been obtained e.g. in the papers [13, 107]. The Example on page 76 shows the difficulty in computing the directional derivative [68] in this case. The following Example shows that it is not possible to drop the strong differentiability assumptions while maintaining the directional differentiability of $x(\cdot)$.

Example: [176] Consider the problem

$$\min_x \{x_1^2 + x_2^2 - x_1 : A(y)x + a(y) \leq 0\},$$

where

$$A(y) = \begin{pmatrix} -1 & 0 \\ -1 & -y \end{pmatrix}, \quad a(y) = \begin{pmatrix} 1 \\ 1 + \frac{1}{2}y^2 \sin \sqrt[3]{\frac{1}{y}} \end{pmatrix}, \quad (\sin \frac{1}{0} := 0)$$

at $y = 0$. Then, the assumptions (C), (MFCQ) and (SSOC) are satisfied except of the smoothness assumption with respect to y: the function $a(\cdot)$ is only once continuously differentiable at zero. It is easy to see that

$$x(y) = (1, 0)^\top + \frac{1}{2}y \sin \sqrt[3]{\frac{1}{y}}(0, 1)^\top.$$

This function is continuous, but not directionally differentiable at zero in direction $r = 1$.   □

The Example on page 71 can be used to show that, even if the optimal solution function proves to be a PC^1-function, its directional derivative need not to be a continuous function of the reference point:

Example: Consider the Example on page 71 at $y^0 = (1, 1)^\top$. Let $r = (1, 1)^\top$. Then,

$$x(y) = \begin{cases} \sqrt{3 - y_1^2 - y_2^2}, & \text{if } y = y^0 + tr, t > 0, \\ 1.5 - \sqrt{0.75 - (y_1 - 0.5)^2 - (y_2 - 0.5)^2}, & \text{if } y = y^0 + tr, t < 0. \end{cases}$$

Hence,

$$x'(y; r) = \begin{cases} -\dfrac{y_1 + y_2}{\sqrt{3 - y_1^2 - y_2^2}}, & \text{if } y = y^0 + tr, \ t > 0, \\ \dfrac{y_1 + y_2 - 1}{\sqrt{0.75 - (y_1 - 0.5)^2 - (y_2 - 0.5)^2}}, & \text{if } y = y^0 + tr, \ t < 0. \end{cases}$$

Thus, for $t \to 0$, we obtain

$$\lim_{t \to +0} \varphi'(y^0 + tr; r) = -2, \quad \lim_{t \to -0} \varphi'(y^0 + tr; r) = 2.$$

□

4.4.2   THE GENERALIZED JACOBIAN

Now consider the problem of finding at least one element of the generalized Jacobian of the function $x(\cdot)$. By Theorems 4.7 and 4.9,

$$\partial x(y^0) = \text{conv} \bigcup_{I \in I_x^0(y^0)} \{\nabla x^I(y^0)\} \subseteq \Gamma_x(y^0) = \text{conv} \bigcup_{I \in I_x(y^0)} \{\nabla x^I(y^0)\}.$$

Here,

$$I_x(y^0) = \bigcup_{(\lambda,\mu)\in E\Lambda(x^0,y^0)} \mathcal{I}(\lambda).$$

In general it is not so easy to decide if

$$I \in I_x^0(y^0) = \{I : y^0 \in \text{cl int Supp}(x, x^I)\}$$

and the selection of $I \notin I_x^0(y^0)$ will falsify the generalized Jacobian of the function $x(\cdot)$. This can be seen in the following

Example: [78] The parametric optimization problem

$$(x_1 - y)^2 + x_2^2 \;\rightarrow\; \min$$

$$-x_1 + x_2 \;\leq\; 0$$

$$x_1 + x_2 \;\leq\; 0$$

has the solution function

$$x(y) = \begin{cases} \left(\dfrac{y}{2} \quad \dfrac{y}{2}\right)^{\mathsf{T}} & \text{if } y < 0, \\[2mm] (0 \quad 0)^{\mathsf{T}} & \text{if } y = 0, \\[2mm] \left(\dfrac{y}{2} \quad -\dfrac{y}{2}\right)^{\mathsf{T}} & \text{if } y > 0. \end{cases}$$

Let $y^0 = 0$ be the point of interest with optimal solution $x^0 = (0\ 0)^{\mathsf{T}}$. At this point, the solution function can be represented by a continuous selection of the continuously differentiable functions

$$x^{\{1\}}(y) = \begin{pmatrix} \dfrac{y}{2} \\[2mm] \dfrac{y}{2} \end{pmatrix} \qquad \text{and} \qquad x^{\{2\}}(y) = \begin{pmatrix} \dfrac{y}{2} \\[2mm] -\dfrac{y}{2} \end{pmatrix},$$

which both are strongly active at y^0. Hence, applying Theorem 4.7 immediately yields

$$\partial x(0) = \text{conv} \{\nabla x^i(0) \mid i = 1, 2\} = \text{conv} \left\{ \begin{pmatrix} \dfrac{1}{2} \\[2mm] \dfrac{1}{2} \end{pmatrix}, \begin{pmatrix} \dfrac{1}{2} \\[2mm] -\dfrac{1}{2} \end{pmatrix} \right\}.$$

At $(x^0, y^0) = \left(\begin{pmatrix} 0 \\ 0 \end{pmatrix}, 0 \right)$ it is $I(x^0, y^0) = \{1, 2\}$ and the conditions (MFCQ), (SSOC), and (CRCQ) hold with

$$L(x^0, y^0, \lambda) = (x_1 - y)^2 + x_2^2 + \lambda_1(x_1 + x_2) + \lambda_2(-x_1 + x_2).$$

One can easily show that $\lambda_1 = \lambda_2 = 0$ is the unique Lagrange multiplier. Therefore we have $\mathcal{I}(\lambda) = \{\emptyset, \{1\}, \{2\}, \{1, 2\}\}$ at the point (x^0, y^0). Now we consider the four possible choices for I:

- $I = \{1, 2\}$: Then $x^I(y) = (0\ 0)^\top$, $\mathrm{Supp}(x, x^I) = \{0\}$, and therefore cl int $\mathrm{Supp}(x, x^I) = \emptyset$. $\nabla_y x^{\{1,2\}}(0) = \begin{pmatrix} 0 \\ 0 \end{pmatrix}$ is not an element of the generalized Jacobian indeed.

- $I = \{2\}$:

$$x^I(y) \in \mathop{\mathrm{Argmin}}_{(x_1, x_2)} \left\{ (x_1 - y)^2 + x_2^2 \mid -x_1 + x_2 = 0 \right\},$$

  hence

$$x^{\{2\}}(y) = \left(\frac{y}{2}\ \ \frac{y}{2} \right)^\top \text{ and } \mathrm{Supp}(x, x^{\{2\}}) = (-\infty, 0].$$

  Here it is $0 \in$ cl int $\mathrm{Supp}(x, x^{\{2\}})$ and we get

$$\nabla_y x^{\{2\}}(0) = \begin{pmatrix} \frac{1}{2} \\ \frac{1}{2} \end{pmatrix} \in \partial x(0).$$

- Considering $I = \{1\}$ analogously to the latter case yields

$$\nabla_y x^{\{1\}}(0) = \begin{pmatrix} \frac{1}{2} \\ -\frac{1}{2} \end{pmatrix} \in \partial x(0).$$

- $I = \emptyset$ implies $x^\emptyset(y) = \begin{pmatrix} y \\ 0 \end{pmatrix}$ and $\mathrm{Supp}(x, x^\emptyset) = \{0\}$ with cl int $Y^\emptyset = \emptyset$. Again we have

$$\nabla_y x^\emptyset(0) = \begin{pmatrix} 1 \\ 0 \end{pmatrix} \notin \partial x(0).$$

$\square$

We recognize from this example that it is in general not easy to find one element of the generalized Jacobian of the function without additional assumptions. Such assumptions are:

(LIN) The gradients with respect to both x and y, $\{\nabla g_i(x^0, y^0) : i \in I(x^0, y^0)\} \cup \{\nabla h_j(x^0, y^0) : j = 1, \ldots, q\}$, are linearly independent.

In the following at some places we will use the abbreviations $z^0 = (x^0, y^0)$, $\nu^0 = (\lambda^0, \mu^0)$, $I_0 = I(x^0, y^0)$.

(FRR) For each vertex $\nu^0 \in \Lambda(z^0)$ the matrix

$$M := \begin{pmatrix} \nabla_{xx}^2 L(z^0, \nu^0) & \nabla_x^\top g_{J(\lambda^0)}(z^0) & \nabla_x^\top h(z^0) & \nabla_{yx}^2 L(z^0, \nu^0) \\ \nabla_x g_{I_0}(z^0) & 0 & 0 & \nabla_y g_{I_0}(z^0) \\ \nabla_x h(z^0) & 0 & 0 & \nabla_y h(z^0) \end{pmatrix}$$

has full row rank $n + |I_0| + q$.

(SCR) For a vertex $(\lambda^0, \mu^0) \in \Lambda(x^0, y^0)$, a set $I \in \mathcal{I}(\lambda^0)$ and a direction r strict complementarity slackness is satisfied in problem $QP(\lambda^0, \mu^0, r)$.

The condition (LIN) is much weaker than the usual linear independence constraint qualification

(LICQ) The gradients

$$\{\nabla_x g_i(x^0, y^0), i \in I(x^0, y^0)\} \cup \{\nabla_x h_j(x^0, y^0), j = 1, \ldots, q\}$$

are linearly independent

since it does not imply that the Lagrange multiplier of problem (4.1) is unique. It is trivially satisfied for any instance with right-hand side perturbed constraints. This assumption guarantees that for each vertex (λ^0, μ^0) of the set $\Lambda(x^0, y^0)$ there is some direction r such that (λ^0, μ^0) is the unique optimal solution of the problem (4.21) [68]:

$$\{(\lambda^0, \mu^0)\} = S(r) := \underset{(\lambda, \mu)}{\text{Argmax}} \{\nabla_y L(x^0, y^0, \lambda, \mu)r : (\lambda, \mu) \in \Lambda(x^0, y^0)\}$$

and thus, in some sense, that no basic representations of the vertices of the unperturbed Lagrange multiplier set are superfluous. Assumption (FRR) is a certain nondegeneracy assumption. If (λ^0, μ^0) is a vertex of $\Lambda(x^0, y^0)$ and (MFCQ) holds, the quadratic matrix

$$M := \begin{pmatrix} \nabla_{xx}^2 L(z^0, \nu^0) & \nabla_x^\top g_{J(\lambda^0)}(z^0) & \nabla_x^\top h(z^0) \\ \nabla_x g_{J(\lambda^0)}(z^0) & 0 & 0 \\ \nabla_x h(z^0) & 0 & 0 \end{pmatrix}$$

has full rank $n+|J(\lambda^0)|+q$ [157]. Hence, (FRR) means that the degeneracy generated by adding the rows $(\nabla_x g_i(x^0, y^0)\ 0\ 0)$ with $i \in I_0 \setminus J(\lambda^0)$ is compensated by adding the columns related to the gradients with respect to y.

Note that the assumption (SCR) makes sense only in the case when the Lagrange multiplier (γ^0, η^0) of the problem $QP(\lambda^0, \mu^0, r)$ is uniquely determined and the gradients $\{\nabla_x g_i(x^0, y^0) : i \in J(\lambda^0) \cup J(\gamma^0)\} \cup \{\nabla_x h_j(x^0, y^0) : j = 1, \dots, q\}$ are linearly independent.

THEOREM 4.12 *Consider the problem (4.1) at a point (x^0, y^0) and let the assumptions (MFCQ), (SSOC), and (CRCQ) be satisfied there.*

- *[71] Take any vertex $(\lambda^0, \mu^0) \in \Lambda(x^0, y^0)$ and a set $I \in \mathcal{I}(\lambda^0)$. Let r^0 be a direction such that condition (SCR) is satisfied. Then $\nabla x^I(y^0) \in \partial x(y^0)$.*

- *[73] If condition (FRR) is valid then*

$$\partial x(y^0) = \text{conv} \bigcup_{(\lambda^0, \mu^0) \in E\Lambda(x^0, y^0)} \bigcup_{I \in \mathcal{I}(\lambda^0)} \nabla x^I(y^0).$$

The assumption in the first assertion is difficult to be verified since we have to answer two questions together: First the question if the set of inner points of $K_{\text{Supp}(x, x^I)}(y^0)$ is not empty and then the question of how to find an element in this set. This is an even more restrictive task than to decide if $y^0 \in \text{cl int Supp}(x, x^I)$. In the second assertion the assumption guarantees that $y^0 \in \text{cl int Supp}(x, x^I) \neq \emptyset$ for all possible index sets. This is a rather unnecessarily restrictive assumption since we are in general interested in only one element of the generalized Jacobian of $x(\cdot)$ and in this case we compute the whole generalized Jacobian. The proof of this Theorem shows that it is sufficient to suppose that the matrix M has full row rank for just one vertex $(\lambda^0, \mu^0) \in \Lambda(x^0, y^0)$. Then we get an element of $\partial x(y^0)$, too. But this assumption again is difficult to be verified and how to find this vertex and the corresponding index set $I \in \mathcal{I}(\lambda^0)$?

REMARK 4.2 *It should be mentioned that in [190, 225] conditions have been formulated which guarantee the computability of generalized Jacobians of the function $x(\cdot)$. In both papers, the linear independence constraint qualification (LICQ) is used which implies uniqueness of the Lagrange multiplier. The condition in [225] can then be obtained from (FRR) by use of the Motzkin Theorem on the alternative. The result in [190] applies to problems with parameter-independent constraints only. In that case, (FRR) is implied by it. The assumption (SCR) together* *with the linear independence constraint qualification (LICQ) has been used in [94] to compute an element of the generalized Jacobian of $x(\cdot)$.*

The next simple example shows that the assumptions of the first assertion in Theorem 4.12 are possibly never satisfied.

Example: [71] Consider the problem

$$\min_{x}\{(x-y)^2 : x \leq y\}.$$

Then, $x(y) \equiv y$, assumptions (MFCQ), (SSOC), (CRCQ) are satisfied for each y, r. Moreover, $x'(y; r) = r$, hence $\{i : \nabla_x g_i(x, y)x'(y; r) + \nabla_y g_i(x, y)r = 0\} = \{1\}$. But $\lambda(y) \equiv 0$, i.e. assumption (SCR) cannot be satisfied.   □

Theorem 4.12 shows that it is possible to compute elements of the generalized Jacobian of a local solution function provided that some additional assumptions are satisfied. But this is not natural. Under (MFCQ), (SSOC), and (CRCQ) the function $x(\cdot)$ is locally Lipschitz continuous. Hence it is almost everywhere differentiable in which case the generalized Jacobian consists of just one element. Also, the directional derivative is piecewise linear, i.e. the parameter space decomposes into a finite number of convex cones and the directional derivative is linear on each of these cones. Take one direction pointing into the interior of one of these cones. Then, the local solution function behaves almost as a differentiable function on points in that direction. This is exploited to derive the next result. Formally, start with the Karush-Kuhn-Tucker conditions of problem $QP(\lambda, \mu, r)$ given as

$$\nabla_{xx}^2 L(z^0, \lambda, \mu)d + \nabla_{yx}^2 L(z^0, \lambda, \mu)r + \nabla_x^\top g_{I_0}(z^0)\gamma + \nabla_x^\top h(z^0)\eta = 0$$

$$\nabla_x g_i(z^0)d + \nabla_y g_i(z^0)r = 0, \ i \in J(\lambda),$$

$$\nabla_x g_i(z^0)d + \nabla_y g_i(z^0)r \leq 0, \ i \in I(z^0) \setminus J(\lambda),$$

$$\nabla_x h_j(z^0)d + \nabla_y h_j(z^0)r = 0, \ j = 1, \ldots, q,$$

$$\gamma_i \geq 0, \ (\nabla_x g_i(z^0)d + \nabla_y g_i(z^0)r)\gamma_i = 0, \ i \in I(z^0) \setminus J(\lambda).$$

If these conditions are satisfied, then there is at least one set I_a with

$$J(\lambda) \subseteq I_a \subseteq I(z^0)$$

such that

$$\nabla_x g_i(z^0)d + \nabla_y g_i(z^0)r = 0, \ \forall \ i \in I_a$$

and

$$\gamma_i = 0, \ \forall \ i \notin I_a,$$

i.e., such that the following system $KKT(r, I_a, \lambda)$ has a solution:

$$\nabla^2_{xx} L(z^0, \lambda, \mu)d + \nabla^2_{yx} L(z^0, \lambda, \mu)r + \nabla^T_x g_{I_a}(z^0)\gamma + \nabla^T_x h(z^0)\eta = 0$$

$$\nabla_x g_i(z^0)d + \nabla_y g_i(z^0)r = 0, \; i \in I_a,$$

$$\nabla_x g_i(z^0)d + \nabla_y g_i(z^0)r \leq 0, \; i \in I(z^0) \setminus I_a,$$

$$\nabla_x h_j(z^0)d + \nabla_y h_j(z^0)r = 0, \; j = 1, \ldots, q,$$

$$\gamma_i \geq 0, \; i \in I_a \setminus J(\lambda), \; \gamma_i = 0, \; i \notin I_a.$$

Clearly, the set I_a is in general not uniquely determined. For all index sets

$$I \in \bigcup_{(\lambda, \mu) \in E\Lambda(x^0, y^0)} \mathcal{I}(\lambda)$$

define the sets

$$R(I, \lambda) := \{r \in \mathbb{R}^m : \text{system } KKT(r, I, \lambda) \text{ has a solution}\}$$

i.e. the set of all r for which I can be used instead of I_a in the conditions $KKT(r, I, \lambda)$. The sets $R(I, \lambda)$ are generally not disjoint. But, due to Theorem 4.11 we have

$$\bigcup_{(\lambda, \mu) \in E\Lambda(z^0)} \left(\bigcup_{I \in \mathcal{I}(\lambda)} R(I, \lambda) \right) = \mathbb{R}^m \qquad (4.23)$$

for all problems (4.1) that meet the conditions (MFCQ), (SSOC), and (CRCQ) at z^0. As the solution set of a system of linear equations and inequalities, the set of all points (d, r, γ, η) satisfying $KKT(\cdot, I, \lambda)$ is a convex polyhedron and so is also the set $R(I, \lambda)$ (since the directional derivative $x'(y^0; \cdot)$ is positively homogeneous, the sets $R(I, \lambda)$ are even convex polyhedral cones). Hence, the boundary of $R(I, \lambda)$ is composed by a finite number of lower dimensional faces. Since the number of different sets $R(I, \lambda)$ (for vertices $(\lambda, \mu) \in E\Lambda(z^0)$ and sets $I \in \mathcal{I}(\lambda)$) is finite, the union of their boundaries is also of lower dimension. Assume that a vector r is chosen according to a uniform distribution over the unit sphere $\mathcal{B}^m$ in $\mathbb{R}^m$. Then, the vector $r \in \mathcal{B}^m \subset \mathbb{R}^m$ does not belong to any of these boundaries with probability one: More formally, let $Z(r)$ denote the event that, for fixed r, there exist a vertex $(\lambda, \mu) \in \Lambda(x^0, y^0)$ and an index set $I \in \mathcal{I}(\lambda)$ with $r \in \text{int } R(I, \lambda)$. Then,

$$P(Z(r)) = 1.$$

The following algorithm uses this probability result. It first selects a direction randomly and checks thereafter if it does belong to the interior of some of the sets $R(I, \lambda)$. This is performed by verifying the necessary conditions in

LEMMA 4.1 ([78]) *Let the problem (4.1) satisfy (MFCQ), (SSOC), (CRCQ), and (LIN) at the point z^0. Assume that there exists $r \in \mathbb{R}^m \setminus \{0\}$ such that $r \in \text{int } R(I, \lambda^0)$ for some index set I with $J(\lambda^0) \subseteq I$ for $(\lambda^0, \mu^0) \in \Lambda(x^0, y^0)$. Then,*

- *the optimal solution of problem (4.21) is unique and coincides with (λ^0, μ^0) and*

- *the gradients with respect to x*

$$\{\nabla_x g_i(z^0) : i \in I\} \cup \{\nabla_x h_j(z^0) : j = 1, \ldots, q\}$$

are linearly independent.

COROLLARY 4.2 *If the assumptions of the preceeding Lemma are satisfied, the optimal solution as well as the corresponding Lagrange multipliers of the problem $QP(\lambda^0, \mu^0, r)$ are unique and depend linearly on $r \in \text{int } R(I, \lambda)$.*

If we are luckily able to select a direction $r \in \text{int } R(I, \lambda^0)$ then it is possible to compute an element of the generalized Jacobian of the local solution function $x(\cdot)$:

THEOREM 4.13 ([78]) *Consider problem (4.1) for $y = y^0$ at a local optimal solution x^0 and assume that (MFCQ), (SSOC), and (CRCQ) are satisfied there. Let $r \in \mathbb{R}^m$ be chosen such that for the corresponding $(\lambda^0, \mu^0) \in S(r)$ there exists an I with $r \in \text{int } R(I, \lambda^0)$. Then there exists a set $\hat{I}$ with $J(\lambda^0) \subseteq \hat{I} \subseteq I(x^0, y^0)$ such that the continuously differentiable function $x^{\hat{I}}(y)$ is a strongly active selection function. Moreover*

$$\nabla x^I(y^0) = \nabla x^{\hat{I}}(y^0) \in \partial x(y^0).$$

REMARK 4.3 *Using $KKT(r, I, \lambda)$ it is not very difficult to see that $\nabla x^I(y^0) \in \partial x(y^0)$ for $I = J(\lambda^0) \cup J(\gamma^0)$ for a vertex (γ^0, η^0) of the Lagrange multiplier set of problem $QP(\lambda^0, \mu^0, r)$.*

Summing up we obtain the following algorithm computing an element of the generalized Jacobian of the function $x(\cdot)$ with probability one:

Method of active constraints:

   Input: parametric optimization problem (4.1) at the point (x^0, y^0), where (MFCQ), (SSOC), and (CRCQ) hold;

Output: a matrix, which is an element of the generalized Jacobian of the solution function with probability 1;

1. choose $r \in \mathcal{B}^m \setminus \{0\}$ randomly;
2. find a $(\lambda, \mu) \in S(r)$, which is furthermore a vertex of $\Lambda(x^0, y^0)$;
3. solve $QP(\lambda, \mu, r)$ and determine the set of active constraints I and Karush-Kuhn-Tucker multipliers of this problem;

   if strong complementarity holds in $QP(\lambda, \mu, r)$

   then calculate $\nabla x^I(y^0)$ according to [97]

   else calculate a vertex (γ, η) of the set of Lagrange multipliers of the problem $QP(\lambda, \mu, r)$ and delete all unnecessary inequality constraints. Calculate $\nabla x^I(y^0)$ according to [97].

It should be noted that this algorithm has a polynomial running time provided that the chosen direction belongs to the interior of some set $R(I, \lambda^0)$. Indeed, the linear optimization problem has then a unique optimal solution by Lemma 4.1 which can be computed in polynomial time e.g. by interior point algorithms [144]. The same class of algorithms can be used to solve quadratic, strictly convex optimization problems again in polynomial time [218]. Standard algorithms for solving systems of linear equations with quadratic regular coefficient matrix are also of polynomial time.

4.5 PROPERTIES OF THE OPTIMAL VALUE FUNCTION

Now we turn over to the investigation of the optimal value function

$$\varphi(y) = \min_x \{f(x, x) : g(x, y) \leq 0, \ h(x, y) = 0\}$$

of the problem (4.1) and investigate its continuity and differentiability conditions. Properties of this function can be found in the papers [111, 234, 242, 243]. The function $\varphi(\cdot)$ is locally Lipschitz continuous under presumably weak assumptions:

THEOREM 4.14 ([153]) *Consider problem (4.1) at $y = y^0$ and let the assumptions (C) and (MFCQ) be satisfied at all points (x, y) with $y = y^0$, $x \in \Psi(y^0)$. Then, the optimal value function φ is locally Lipschitz continuous at y^0.*

Let

$$\Psi_\varepsilon(y) := \{x \in M(y) : f(x, y) \leq \varphi(y) + \varepsilon\} \qquad (4.24)$$

denote the set of ε-optimal solutions of problem (4.1) for $\varepsilon \geq 0$. As a corollary of this Theorem we immadiately get

COROLLARY 4.3 ([153]) *Let the assumptions (C), (MFCQ) be satisfied for the convex optimization problem (4.1) at $y = y^0$. Then, there exists $\varepsilon^0 > 0$ such that $\Psi_\varepsilon(\cdot)$ is locally Lipschitz continuous at y^0 for each $0 < \varepsilon < \varepsilon^0$.*

The next Theorem gives some quantitative bounds on the local behavior of the function $\varphi(\cdot)$. For this we need the *upper and lower Dini directional derivatives* of a function $v : \mathbb{R}^m \to \mathbb{R}$. Let $r^0 \in \mathbb{R}^m$ be a fixed direction and $x^0 \in \mathbb{R}^m$ a fixed point. Then,

$$D^+ v(x^0; r^0) = \limsup_{t \to 0+} t^{-1} [v(x^0 + tr^0) - v(x^0)] \qquad (4.25)$$

denotes the upper and

$$D_+ v(x^0; r^0) = \liminf_{t \to 0+} t^{-1} [v(x^0 + tr^0) - v(x^0)] \qquad (4.26)$$

the lower Dini directional derivatives at x^0 in direction r^0. Clearly, if both directional derivatives coincide then the function v is directionally differentiable at the point x^0 in direction r^0.

THEOREM 4.15 ([106]) *Consider the problem (4.1) at a parameter value $y = y^0$ and let assumptions (C) and (MFCQ) be satisfied at all points $x \in \Psi(y^0)$. Then, for any direction r^0,*

$$D_+ \varphi(y^0; r^0) \geq \min_{x \in \Psi(y^0)} \min_{(\lambda, \mu) \in \Lambda(x, y^0)} \nabla_y L(x, y^0, \lambda, \mu) r^0$$

and

$$D^+ \varphi(y^0; r^0) \leq \inf_{x \in \Psi(y^0)} \max_{(\lambda, \mu) \in \Lambda(x, y^0)} \nabla_y L(x, y^0, \lambda, \mu) r^0.$$

COROLLARY 4.4 ([106]) *If in addition to the assumptions in Theorem 4.15 also the linear independence constraint qualification (LICQ) is satisfied at all points (x, y^0), $x \in \Psi(y^0)$, then the optimal value function $\varphi(\cdot)$ is directionally differentiable at $y = y^0$ and we have*

$$\varphi'(y^0; r) = \min\{\nabla_y L(x, y^0, \lambda^0, \mu^0) r : x \in \Psi(y^0)\} \; \forall r \in \mathbb{R}^m,$$

where for each $x \in \Psi(y^0)$ the set $\Lambda(x, y^0)$ reduces to a singleton (λ^0, μ^0) continuously depending on x.

For a similar result under weaker assumptions the interested reader is referred to the paper [108]. Investigations of methods for computing the directional derivative of the function $\varphi(\cdot)$ can be found e.g., in the papers [210, 211]. For existence results and properties of the optimal value function using second order multipliers we refer to the paper [243].

The next theorem shows that directional differentiability of the optimal value function can also be shown for convex optimization problems by use of the then necessary and sufficient saddle point condition:

THEOREM 4.16 ([127]) *Consider the convex parametric optimization problem (4.1) and assume that the conditions (C) and (MFCQ) are satisfied at some point $x \in M(y^0)$. Then, the optimal value function $\varphi(\cdot)$ is directionally differentiable at $y = y^0$ and we have*

$$\varphi'(y^0; r) = \max_{(\lambda,\mu)\in\Lambda(x,y^0)} \min_{x\in\Psi(y^0)} \nabla_y L(x, y^0, \lambda, \mu) r \ \forall r.$$

Note that in the differentiable case, (MFCQ) is satisfied at one feasible point of a convex optimization problem if and only if the so-called Slater's condition holds. Hence, (MFCQ) is satisfied at one feasible point if and only if it is valid at all feasible points. Next we will consider the generalized gradient of the locally Lipschitz continuous optimal value function $\varphi(\cdot)$.

THEOREM 4.17 ([106]) *Consider problem (4.1) and let the assumptions (C) and (MFCQ) at all points $x \in \Psi(y^0)$ be satisfied. Then, the optimal value function $\varphi(\cdot)$ is locally Lipschitz continuous by Theorem 4.14 and the generalized gradient $\partial\varphi(y^0)$ satisfies*

$$\partial\varphi(y^0) \subseteq \operatorname{conv} \bigcup_{x\in\Psi(y^0)} \bigcup_{(\lambda,\mu)\in\Lambda(x,y^0)} \nabla_y L(x, y^0, \lambda, \mu). \qquad (4.27)$$

In the next three corollaries we will see that, under more restrictive assumptions, we even get equality in Theorem 4.17. To achieve that result in the first case we need a further definition.

DEFINITION 4.12 *Let $v : \mathbb{R}^m \to \mathbb{R}$ be a locally Lipschitz continuous function with generalized gradient $\partial v(y^0)$ at $y^0 \in \mathbb{R}^m$. Then, the Clarke directional derivative $v^\circ(y^0; r)$ at $y^0 \in \mathbb{R}^m$ in direction $r \in \mathbb{R}^m$ is defined as*

$$v^\circ(y^0; r) = \limsup_{y\to y^0, t\to 0+} t^{-1}[v(y + tr) - v(y)].$$

The function v is called Clarke regular provided that the ordinary directional derivative $v'(y^0; r)$ exists and $v'(y^0; r) = v^\circ(y^0; r)$ for all $r \in \mathbb{R}^m$.

It is easy to see that $v^\circ(y^0; r) = \max_{\zeta\in\partial v(y^0)} \zeta r$ for all r [61].

COROLLARY 4.5 ([106]) *If the assumptions of Theorem 4.17 are satisfied and (LICQ) is satisfied at every $x \in \Psi(y^0)$, then*

$$\partial\varphi(y^0) = \operatorname{conv} \bigcup_{x\in\Psi(y^0)} \nabla_y L(x, y^0, \lambda, \mu),$$

where $\{(\lambda, \mu)\} = \Lambda(x, y^0)$. *Moreover, the function* φ *is Clarke regular in this case.*

The situation is better in the convex case as it is shown in

COROLLARY 4.6 ([261]) *Consider the convex parametric optimization problem (4.1) at the point* (x^0, y^0) *and let the assumptions (C) and (MFCQ) be satisfied there. If* $\Psi(y^0)$ *reduces to a singleton, then*

$$\partial\varphi(y^0) = \bigcup_{(\lambda,\mu)\in\Lambda(x^0,y^0)} \nabla_y L(x^0, y^0, \lambda, \mu).$$

If the problem functions are convex both in x and y we can even drop the uniqueness assumption for the set of global optimal solutions:

COROLLARY 4.7 ([261, 272]) *Let, under the assumptions of Corollary 4.6 all functions* f, g_i, $i = 1, \ldots, p$, *be convex in both* (x, y) *and all functions* h_j, $j = 1, \ldots, q$, *be affine linear in both variables. Then*

$$\partial\varphi(y^0) = \bigcup_{(\lambda,\mu)\in\Lambda(x^0,y^0)} \nabla_y L(x^0, y^0, \lambda, \mu).$$

We will end this Chapter with the remark that interesting additional theorems on properties of optimal value functions can be found in the monograph [46].

4.6 PROOFS

PROOF OF THEOREM 4.2: First it is shown that the point-to-set mapping $\Lambda(\cdot, \cdot)$ is *locally bounded* at (x^0, y^0), which means that there are an open neighborhood $W_\varepsilon(x^0, y^0)$, $\varepsilon > 0$ of (x^0, y^0), and a nonempty compact set $K \subseteq \mathbb{R}^p \times \mathbb{R}^q$ such that

$$\Lambda(x, y) \subset K \ \forall (x, y) \in W_\varepsilon(x^0, y^0).$$

Arguing from contradiction, let there be sequences $\{(x^k, y^k)\}_{k=1}^\infty$ converging to (x^0, y^0) and $\{(\lambda^k, \mu^k)\}_{k=1}^\infty$ with $\lim_{k\to\infty} \|(\lambda^k, \mu^k)\| = \infty$ and $(\lambda^k, \mu^k) \in \Lambda(x^k, y^k)$ for all k. Without loss of generality we can assume that the sequence $\left\{\dfrac{(\lambda^k, \mu^k)}{\|(\lambda^k, \mu^k)\|}\right\}_{k=1}^\infty$ converges to some vector (γ, η), with $\|(\gamma, \eta)\| = 1$. Then we have from (4.8),

$$\frac{\lambda_i^k}{\|(\lambda^k, \mu^k)\|} \geq 0, \frac{\lambda_i^k g_i(x^k, y^k)}{\|(\lambda^k, \mu^k)\|} \equiv 0$$

and hence $\gamma \geq 0$, $\gamma_i g_i(x^0, y^0) = 0$. Again from (4.8),

$$\frac{\nabla_x f(x^k, y^k)}{\|(\lambda^k, \mu^k)\|} + \sum_{i=1}^{p} \frac{\lambda_i^k}{\|(\lambda^k, \mu^k)\|} \nabla_x g_i(x^k, y^k)$$
$$+ \sum_{i=1}^{q} \frac{\mu_i^k}{\|(\lambda^k, \mu^k)\|} \nabla_x h_i(x^k, y^k) \equiv 0, \qquad (4.28)$$

which implies that

$$\gamma^\top \nabla_x g(x^0, y^0) + \eta^\top \nabla_x h(x^0, y^0) = 0.$$

Let d be a point satisfying the (MFCQ). Then there is some $\varepsilon > 0$ such that

$$g_i(x^0, y^0) + \nabla_x g_i(x^0, y^0) d + \varepsilon \gamma_i < 0 \ \forall i = 1, \ldots, p,$$
$$h_i(x^0, y^0) + \nabla_x h_i(x^0, y^0) d = 0 \ \forall i = 1, \ldots, q.$$

This implies

$$0 = \sum_{i=1}^{p} \gamma_i (g_i(x^0, y^0) + \nabla_x g_i(x^0, y^0) d) +$$
$$+ \sum_{i=1}^{q} \eta_i (h_i(x^0, y^0) + \nabla_x h_i(x^0, y^0) d) \leq -\varepsilon \sum_{i=1}^{p} \gamma_i^2 \leq 0.$$

Thus $\gamma = 0$. Hence, by (4.28)

$$\sum_{i=1}^{q} \eta_i \nabla_x h_i(x^0, y^0) = 0$$

with $\eta \neq 0$, which contradicts (MFCQ). This yields local boundedness of $\Lambda(\cdot, \cdot)$. Together with assumption (C) we obtain the existence of a compact set K' such that $SP(y) \times \Lambda(\cdot, y) \subseteq K'$ for all y sufficiently close to y^0.

Now, consider any convergent sequence $\{(x^k, y^k, \lambda^k, \mu^k)\}_{k=1}^{\infty}$ satisfying the conditions (4.8). Then, for the limit point of this sequence these conditions are also valid. This implies closedness of both point-to-set mappings $SP(\cdot)$ and $\Lambda(\cdot, \cdot)$. Together with local boundedness, this implies upper semicontinuity [17]. □

PROOF OF THEOREM 4.3: The proof is done in several steps. First we prove that the feasible set mapping $M(\cdot)$ is continuous.

- Continuity of the functions g_i, h_j together with assumption (C) imply upper semicontinuity of the mapping $M(\cdot)$.

- Let $\bar{x} \in M(y^0)$. Then, the (MFCQ) implies the existence of a sequence $\{x^k\}_{k=1}^{\infty}$ converging to $\bar{x}$ with

$$x^k \in M^0(y^0) := \{x : g_i(x, y^0) < 0 \; \forall i, \; h_j(x, y^0) = 0 \; \forall j\}.$$

Hence, $M(y^0) \subseteq \operatorname{cl} M^0(y^0)$. Now, for each $\bar{x} \in M^0(y^0)$ there exist constants $\varepsilon > 0, \delta > 0$ such that

$$g(x, y) < 0 \; \forall (x, y) : \|x - \bar{x}\| < \varepsilon, \|y - y^0\| < \delta.$$

Using the implicit function theorem, the linear independence of the gradients $\nabla_x h_j(x^0, y^0), j = 1, \ldots, q$ (cf. (MFCQ)) implies the existence of at least one continuous function $x(\cdot)$ defined on an open neighborhood $V_{\delta'}(y^0)$, $\delta' > 0$, with $h(x(y), y) \equiv 0, x(y^0) = \bar{x}$. But then, there is $0 < \gamma < \min\{\delta, \delta'\}$ with $\|x(y) - \bar{x}\| < \varepsilon$ for all $\|y - y^0\| < \gamma$. This guarantees that $x(y) \in M^0(y) \subseteq M(y)$ for all $\|y - y^0\| < \gamma$. Hence, $M(\cdot)$ is lower semicontinuous.

- Now, we show continuity of the optimal value function $\varphi(\cdot)$. Let $\{y^k\}_{k=1}^{\infty}$ be a sequence converging to y^0. First, let $\{x^k\}_{k=1}^{\infty}$ be defined by $x^k \in \Psi(y^k)$ for all k. Then, by (C), the sequence $\{x^k\}_{k=1}^{\infty}$ has at least one accumulation point $\bar{x}$ and $\bar{x} \in M(y^0)$ by upper semicontinuity of $M(\cdot)$. This implies $\liminf_{k \to \infty} \varphi(y^k) = \liminf_{k \to \infty} f(x^k, y^k) \geq \varphi(y^0)$.

- Now let $x^0 \in \Psi(y^0)$. Then, by lower semicontinuity of $M(\cdot)$ there exists a sequence $\{x^k\}_{k=1}^{\infty}$ converging to x^0 with $x^k \in M(y^k)$ for all k. This implies that $\limsup_{k \to \infty} \varphi(y^k) \leq \limsup_{k \to \infty} f(x^k, y^k) = \varphi(y^0)$.

- Now, since the optimal value function $\varphi(\cdot)$ is continuous, each accumulation point $\bar{x}$ of a sequence $\{x^k\}_{k=1}^{\infty}$ with $x^k \in \Psi(y^k)$ satisfies $f(\bar{x}, y^0) = \varphi(y^0)$, i.e. $\bar{x} \in \Psi(y^0)$ which by assumption (C) implies upper semicontinuity of $\Psi(\cdot)$.

$\square$

PROOF OF THEOREM 4.4: The original proof in [157] uses degree theory (cf. Chapter 6 in [222]). We will give here a proof which is (in part) valid under an additional assumption not using this theory. It should also be mentioned that Theorem 4.4 consists of only some part of the original theorem in [157].

- First note that validity of a sufficient second order condition implies that a local optimal solution $\bar{x}$ is isolated, i.e. that there is some open neighborhood $V_\varepsilon(\bar{x}), \varepsilon > 0$, of $\bar{x}$ not containing another local optimal solution.

- Now we show the existence of local optimal solutions sufficiently close to x^0 for perturbed problems (4.1). Consider the perturbed problem

$$\min_x \{f(x, y) : g(x, y) \leq 0, \ h(x, y) = 0, \ x \in \text{cl } V_\varepsilon(x^0)\}. \qquad (4.29)$$

If cl $V_\varepsilon(x^0)$ is assumed (without loss of generality) to be compact, the problem (4.29) has an optimal solution for all y for which the feasible set is not empty. The (MFCQ) is satisfied for this problem at the point (x^0, y^0) which implies that the feasible set of problem (4.29) is not empty for all y in some open neighborhood $U_1(y^0)$ of y^0. Thus, the set $\Psi_l(\cdot)$ of global optimal solutions of problem (4.29) is not empty, $\Psi_l(\cdot) \neq \emptyset$, on $U_1(y^0)$. Then, by (SSOC) we have $\{x^0\} = \Psi_l(y^0)$ and $\Psi_l(\cdot)$ is upper semicontinuous at y^0 by Theorem 4.3. This implies the existence of an open set $U_2(y^0) \subseteq U_1(y^0)$ such that $\Psi_l(y) \subset V_\varepsilon(x^0)$ for all $y \in U_2(y^0)$. Then, the additional constraint in (4.29) is not active and all points in $\Psi_l(y)$ are local optimal solutions of (4.1).

- Next we show that (SSOC) persists in some open neighborhood of (x^0, y^0). Assuming the contrary, let $\{(x^k, y^k, \lambda^k, \mu^k, d^k)\}_{k=1}^\infty$ be a sequence of points with

$$\lim_{k \to \infty} x^k = x^0, \ \lim_{k \to \infty} y^k = y^0, (\lambda^k, \mu^k) \in \Lambda(x^k, y^k) \text{ for all } k$$

and

$$d^{k\top} \nabla_{xx}^2 L(x^k, y^k, \lambda^k, \mu^k) d^k \leq 0$$

for some d^k satisfying

$$\nabla_x g_i(x^k, y^k) d^k = 0, \ i \in J(\lambda^k), \ \nabla_x h(x^k, y^k) d^k = 0,$$

where $J(\lambda) = \{j : \lambda_j > 0\}$. Without loss of generality we can assume that $\|d^k\| = 1$ for all k and that $\lim_{k \to \infty} d^k = d$. By upper semicontinuity of $\Lambda(\cdot, \cdot)$ (cf. Theorem 4.2) the sequence $\{\lambda^k, \mu^k\}_{k=1}^\infty$ converges without loss of generality to $(\lambda^0, \mu^0) \in \Lambda(x^0, y^0)$. This implies that $J(\lambda^0) \subseteq J(\lambda^k)$ and

$$\nabla_x g_i(x^0, y^0) d = 0, \ i \in J(\lambda^0), \ \nabla_x h(x^0, y^0) d = 0$$

as well as

$$d^\top \nabla_{xx}^2 L(x^0, y^0, \lambda^0, \mu^0) d \leq 0.$$

But this contradicts (SSOC).

- This implies that for each y in some open neighborhood $U_3(y^0) \subseteq U_2(y^0)$ there exist only a finite number of isolated local optimal solutions in $V_\varepsilon(x^0)$. We will show uniqueness of the local optimal solution of perturbed problems using the stronger linear independence constraint qualification (LICQ).

For the more general proof the reader is referred to the original paper [157]. The linear independence constraint qualification implies the (MFCQ) and guarantees that $\Lambda(x^0, y^0)$ consists of exactly one point. For $i = 1, 2$, let $\{x_i^k, y^k, \lambda_i^k, \mu_i^k\}_{k=1}^\infty$ be two sequences with $\{x_1^k, x_2^k\} \subseteq \Psi_l(y^k)$, $(\lambda_i^k, \mu_i^k) \in \Lambda(x_i^k, y^k)$ for all k and $\lim_{k\to\infty} y^k = y^0$. Let without loss of generality $f(x_1^k, y^k) \geq f(x_2^k, y^k)$ for all k. Then, by upper semicontinuity (cf. Theorems 4.2 and 4.3) we have $\lim_{k\to\infty} x_i^k = x^0$ and $\lim_{k\to\infty} (\lambda_i^k, \mu_i^k) = (\lambda^0, \mu^0) \in \Lambda(x^0, y^0)$. By $\lambda_1^k \geq 0$, $g(x_2^k, y^k) \leq 0$, $h(x_2^k) = 0$, and stationarity,

$$f(x_1^k, y^k) = L(x_1^k, y^k, \lambda_1^k, \mu_1^k) \geq f(x_2^k, y^k) \geq L(x_2^k, y^k, \lambda_1^k, \mu_1^k) \; \forall k.$$

Using Taylor's expansion formula (or the existence of the second order derivative),

$$
\begin{aligned}
&L(x_1^k, y^k, \lambda_1^k, \mu_1^k) \geq L(x_2^k, y^k, \lambda_1^k, \mu_1^k) \\
=\; &L(x_1^k, y^k, \lambda_1^k, \mu_1^k) + \nabla_x L(x_1^k, y^k, \lambda_1^k, \mu_1^k)(x_2^k - x_1^k) \\
+\; &0.5(x_2^k - x_1^k)^\top \nabla_{xx}^2 L(x_1^k, y^k, \lambda_1^k, \mu_1^k)(x_2^k - x_1^k) + o(\|x_2^k - x_1^k\|^2)
\end{aligned}
$$

which implies

$$
\begin{aligned}
0 \geq \; &\frac{x_2^k - x_1^k}{\|x_2^k - x_1^k\|}^\top \nabla_{xx}^2 L(x_1^k, y^k, \lambda_1^k, \mu_1^k)\frac{x_2^k - x_1^k}{\|x_2^k - x_1^k\|} + \frac{o(\|x_2^k - x_1^k\|^2)}{\|x_2^k - x_1^k\|^2} \\
&\underset{k\to\infty}{\longrightarrow} d^\top \nabla_{xx}^2 L(x^0, y^0, \lambda_1^0, \mu_1^0)d
\end{aligned}
\tag{4.30}
$$

where d is without loss of generality assumed to be the limit point of $\{(x_2^k - x_1^k)/\|x_2^k - x_1^k\|\}_{k=1}^\infty$. By (LICQ) and upper semicontinuity of $\Lambda(\cdot, \cdot)$ we have that $\lambda_{ij}^k > 0$ for $i = 1, 2$ and all sufficiently large k and $j \in J(\lambda^0)$. This implies $g_j(x_i^k, y^k) = 0$ for all $j \in J(\lambda^0), i = 1, 2$, and all sufficiently large k. Hence,

$$
\begin{aligned}
0 \;=\; &g_j(x_1^k, y^k) = g_j(x_2^k, y^k) \\
=\; &g_j(x_1^k, y^k) + \nabla_x g_j(x_1^k, y^k)(x_2^k - x_1^k) + o(\|x_2^k - x_1^k\|)
\end{aligned}
$$

or

$$
0 = \nabla_x g_j(x_1^k, y^k)\frac{x_2^k - x_1^k}{\|x_2^k - x_1^k\|} + \frac{o(\|x_2^k - x_1^k\|)}{\|x_2^k - x_1^k\|}
$$

for $j \in J(\lambda^0)$. For the limit point d of $\{(x_2^k - x_1^k)/\|x_2^k - x_1^k\|\}_{k=1}^\infty$ this implies

$$0 = \nabla_x g_j(x_0, y^0)d, j \in J(\lambda^0). \tag{4.31}$$

Analogously

$$0 = \nabla_x h_j(x^0, y^0)d, j = 1, \ldots, q. \tag{4.32}$$

Conditions (4.30)–(4.32) contradict (SSOC) which concludes the proof.

□

In the following proof we need the definition of a pseudo Lipschitz continuous point-to-set mapping:

DEFINITION 4.13 *A point-to-set mapping* $\Gamma : \mathbb{R}^k \to 2^{\mathbb{R}^l}$ *is called pseudo Lipschitz continuous at a point* (w^0, z^0) *with* $w^0 \in \Gamma(z^0)$ *if there exist open neighborhoods* $U_\varepsilon(w^0)$ *and* $V_\varepsilon(z^0)$ *(*$\varepsilon > 0$, $\delta > 0$*) and a constant* $L < \infty$ *such that*

$$\Gamma(z_1) \cap U_\varepsilon(w^0) \subset \Gamma(z_2) + L\|z_1 - z_2\|\mathcal{B}^l \ \forall z_1, z_2 \in V_\varepsilon(z^0).$$

PROOF OF THEOREM 4.5: We will only show that the function $x(\cdot)$ is upper Lipschitz continuous at y^0. For a proof of the full theorem, upper Lipschitz continuity of the solution set mapping of systems of linear (in)equalities [126] can be used or the original proof in [240]. The ideas for our proof are borrowed from [257]. Without loss of generality assume that $(x^0, y^0) = (0,0)$, $f(x^0, y^0) = 0$. Suppose that the the conclusion of the theorem is not true, i.e. that there exists a sequence $\{y^k\}_{k=1}^\infty$ such that

$$\lim_{k \to \infty} \frac{\|x(y^k)\|}{\|y^k\|} = \infty. \tag{4.33}$$

From Theorem 4.4 we know that $\lim_{k \to \infty} x(y^k) = 0$. Let d be an accumulation point of the bounded sequence $\{x(y^k)/\|x(y^k)\|\}_{k=1}^\infty$ and without loss of generality assume

$$d = \lim_{k \to \infty} \frac{x(y^k)}{\|x(y^k)\|}. \tag{4.34}$$

By Theorem 4.1 there exist $(\lambda^k, \mu^k) \in \Lambda(x(y^k), y^k)$ and the sequence $\{\lambda^k, \mu^k\}_{k=1}^\infty$ has all accumulation points in the set $\Lambda(x^0, y^0)$ by Theorem 4.2. If without loss of generality convergence of this sequence is assumed, we have $J(\lambda^0) \subseteq J(\lambda^k)$ for some vertex $(\lambda^0, \mu^0) \in \Lambda(x^0, y^0)$ for sufficiently large k. This implies that

$$g_i(x(y^k), y^k) = 0, \ i \in J(\lambda^0), \ x(y^k) \in M(y^k)$$

for sufficiently large k. Now, for $i \in J(\lambda^0)$ we have

$$\begin{aligned} 0 &= g_i(x(y^k), y^k) - g_i(0, 0) \\ &= \nabla_x g_i(x^0, y^0)x(y^k) + \nabla_y g_i(x(y^k), y^k)y^k + o(\|x(y^k)\|) \end{aligned}$$

which results in

$$0 = \nabla_x g_i(x(y^k), y^k) \frac{x(y^k)}{\|x(y^k)\|} + \nabla_y g_i(x(y^k), y^k) \frac{y^k}{\|x(y^k)\|} + \frac{o(\|x(y^k)\|)}{\|x(y^k)\|}.$$

By (4.34) and the assumption, this leads to

$$\nabla_x g_i(0,0)d = 0 \ \forall i \in J(\lambda^0). \tag{4.35}$$

Analogously,

$$\nabla_x h_i(0,0)d = 0 \ \forall i = 1, \ldots, q. \tag{4.36}$$

By $(\lambda^k, \mu^k) \in \Lambda(x(y^k), y^k)$,

$$f(x(y^k), y^k) = L(x(y^k), y^k, \lambda^k, \mu^k) \geq L(x(y^k), y^k, \lambda^0, \mu^0)$$

for all $\lambda^0 \geq 0, \mu^0$ and by $0 = f(0,0) = L(0,0,\lambda^0,\mu^0)$ for $(\lambda^0, \mu^0) \in \Lambda(0,0)$

$$
\begin{aligned}
L(x(y^k), y^k, \lambda^0, \mu^0) &= L(0,0,\lambda^0,\mu^0) + \nabla_y L(0,0,\lambda^0,\mu^0)y^k \\
&+ \ 0.5 x(y^k)^\top \nabla_{xx}^2 L(0,0,\lambda^0,\mu^0)x(y^k) + x(y^k)^\top \nabla_{yx}^2 L(0,0,\lambda^0,\mu^0)y^k \\
&+ \ 0.5 y^{k\top} \nabla_{yy} L(0,0,\lambda^0,\mu^0)y^k + o(\|(x(y^k), y^k)\|^2) \\
&= \ \nabla_y L(0,0,\lambda^0,\mu^0)y^k + 0.5\|x(y^k)\|^2 \frac{x(y^k)^\top}{\|x(y^k)\|} \nabla_{xx}^2 L(0,0,\lambda^0,\mu^0) \frac{x(y^k)}{\|x(y^k)\|} \\
&+ \ o(\|x(y^k)\|^2)
\end{aligned}
$$

by (4.33), (4.34). By (4.35), (4.36) and (SSOC) there exist $\alpha > 0$ such that

$$0.5\|x(y^k)\|^2 \frac{x(y^k)^\top}{\|x(y^k)\|} \nabla_{xx}^2 L(0,0,\lambda^0,\mu^0) \frac{x(y^k)}{\|x(y^k)\|} \geq \alpha \|x(y^k)\|^2 \tag{4.37}$$

for each sufficiently large k.

Let $(\lambda^1, \mu^1) \in \Lambda(0,0)$ be such that $(\lambda^1, \mu^1) \neq (\lambda^0, \mu^0)$ and $J(\lambda^1) \setminus J(\lambda^0) = \{i_0\}$. Then, $\{\nabla_x g_i(0,0) : i \in J(\lambda^0) \cup \{i_0\}\} \cup \{\nabla_x h_j(0,0) : j = 1, \ldots, q\}$ are linearly dependent. Hence, (4.35), (4.36) imply that also $\nabla_x g_{i_0}(0,0)d = 0$ which by (SSOC) has the consequence that (4.37) is valid for all vertices $(\lambda^0, \mu^0) \in \Lambda(0,0)$. Thus,

$$
\begin{aligned}
f(x(y^k), y^k) &\geq \max\{\nabla_y^\top L(0,0,\lambda,\mu)y^k : (\lambda, \mu) \in \Lambda(0,0)\} \\
&+ \ \alpha\|x(y^k)\|^2 + o(\|x(y^k)\|^2)
\end{aligned} \tag{4.38}
$$

for sufficiently large k. Let

$$(\lambda^k, \mu^k) \in \operatorname*{Argmax}_{(\lambda,\mu)} \{\nabla_y L(0,0,\lambda,\mu)y^k : (\lambda, \mu) \in \Lambda(0,0)\}.$$

Then, by finiteness of the number of vertices of the convex polyhedron $\Lambda(0,0)$ and parametric linear programming, without loss of generality $(\lambda^k, \mu^k) \equiv (\lambda^0, \mu^0)$ for all k. By pseudo Lipschitz continuity of the feasible set under (MFCQ) [244] there exist a sequence $\{w^k\}_{k=1}^\infty$ such that $w^k \in M(y^k)$ and $g_i(w^k, y^k) = 0$, $i \in J(\lambda^0)$ for all k and a number $\delta > 0$ such that $\|w^k\| \le \delta \|y^k\|$. This implies

$$
\begin{aligned}
f(x(y^k), y^k) &\le f(w^k, y^k) = \nabla_y L(0, 0, \lambda^0, \mu^0) y^k \\
&+ \ 0.5 w^{k\top} \nabla_{xx}^2 L(0, 0, \lambda^0, \mu^0) w^k + w^{k\top} \nabla_{yx}^2 L(0, 0, \lambda^0, \mu^0) y^k \\
&+ \ 0.5 y^{k\top} \nabla_{yy}^2 L(0, 0, \lambda^0, \mu^0) y^k + o(\|(w^k, y^k)\|^2) \\
&= \ \nabla_y L(0, 0, \lambda^0, \mu^0) y^k + O(\|y^k\|^2) \\
&= \ \max\{\nabla_y L(0, 0, \lambda, \mu) y^k : (\lambda, \mu) \in \Lambda(0, 0)\} + o(\|x(y^k)\|^2)
\end{aligned}
$$

by (4.33). This contradicts (4.38) and proves the Theorem.   □

To verify Theorem 4.6 we need

LEMMA 4.2 ([251]) *Let $f : [0, 1] \to \mathbb{R}^p$ be a continuous function, and let closed nonempty sets $\{A_i\}_{i=1}^q \subseteq \mathbb{R}^m$ be given with*

$$
f([0, 1]) := \bigcup_{x \in [0,1]} \{f(x)\} \subseteq \bigcup_{i=1}^q A_i.
$$

Then, there exist numbers $\{t_i\}_{i=0}^{r+1}$ with $t_0 = 0$, $t_{r+1} = 1$, $t_i < t_{i+1} \ \forall i$ and $\{j_i\}_{i=0}^r$, $r \le q$ such that

$$
\{f(t_i), f(t_{i+1})\} \subset A_{j_i}, \quad i = 0, \dots, r.
$$

PROOF The proof is done by induction on the number r of different sets A_i for which there exist points t such that we indeed have $f(t) \in A_i$. If $r = 1$ then the claim is obviously true. Let the proof be shown for all cases in which the curve $f([0, 1])$ meets not more than r of the sets A_i. Now, let $f([0, 1])$ intersect $r + 1$ of the sets A_i. Then, there is an index j_0 with $f(0) \in A_{j_0}$. Let $t_1 = \sup_t\{t \in [0, 1] : f(t) \in A_{j_0}\}$. By continuity of f and closedness of A_{j_0} we have $f(t_1) \in A_{j_0}$ and $f(t) \notin A_{j_0}$ for all $t > t_1$. Moreover, there exists a further set A_{j_1} with $f(t_1) \in A_{j_1}$. Then, for the interval $[t_1, 1]$ there are at most r of the sets A_i intersecting with $f([t_1, 1])$. The proof now follows by use of the result supposed.   □

PROOF OF THEOREM 4.6: The following proof can be found in [251]. Let x, y be arbitrary points in the neighborhood $V_\varepsilon(x^0)$ used in the definition of a PC^1-function z and $w(t) = z(x + t(y - x))$ for $t \in [0, 1]$. Let the selection functions be denoted by z_i and define

$$
w_i(t) := w(t) \text{ if } z_i(x + t(y - x)) = z(x + t(y - x)),
$$

$$A_i = \{x \in V_\varepsilon(x^0) : z_i(x) = z(x)\}.$$

Then, A_i are closed due to the supposed continuity of $z(\cdot)$. By Lemma 4.2 there exist points $t_0 = 0 < t_1 < \ldots < t_r < t_{r+1} = 1$ such that $\{w(t_i), w(t_{i+1})\} \subset A_{j_i}, i = 0, \ldots, r$. Then,

$$\|z(y) - z(x)\| = \|w(1) - w(0)\| = \left\|\sum_{i=0}^{r}(w(t_{i+1}) - w(t_i))\right\| \leq$$

$$\sum_{i=0}^{r} \|w(t_{i+1}) - w(t_i)\| = \sum_{i=0}^{r} \|w_{j_i}(t_{i+1}) - w_{j_i}(t_i)\| \leq$$

$$\sum_{i=0}^{r} L_{j_i}|t_{i+1} - t_i| \, \|y - x\| \leq \max_j L_j \sum_{i=0}^{r}(t_{i+1} - t_i)\|y - x\| =$$

$$\max_j L_j\|y - x\|,$$

where L_j is the Lipschitz constant of the j-th selection function which exists since continuously differentiable functions are Lipschitz continuous on each bounded set. This implies that the Lipschitz constant $L = \max_j L_j$ for the function $z(\cdot)$ is independent of the points x, y which proves the theorem. $\square$

PROOF OF THEOREM 4.7: Let the set C denote the right-hand side of the equation under consideration:

$$C = \text{conv}\,\{\nabla z^i(x^0) : x^0 \in \text{cl int Supp}(z, z^i)\}.$$

Since z^i are continuously differentiable and $x^0 \in \text{cl int Supp}(z, z^i)$ the inclusion $C \subseteq \partial z(x^0)$ is obvious.

To show the opposite inclusion first assume that x^0 is such that z is differentiable at x^0. Since the number of selection functions is finite there exists at least one selection function z^i with $x^0 \in \text{cl int Supp}(z, z^i)$. Then, there exists a sequence $\{x^k\}_{k=1}^{\infty} \subset \text{int Supp}(z, z^i)$ converging to x^0. Thus, we get

$$\lim_{k \to \infty} \|x^k - x^0\|^{-1}[z(x^k) - z(x^0)] = \lim_{k \to \infty} \|x^k - x^0\|^{-1}[z^i(x^k) - z^i(x^0)].$$

Since both $z(\cdot)$ and $z^i(\cdot)$ are differentiable at x^0, $\nabla z(x^0) = \nabla z^i(x^0)$ in this case.

Now, let z be not necessarily differentiable at x^0 and let $w \in \partial z(x^0)$ be an extreme point of $\partial z(x^0)$. Then, by definition, there exists a sequence $\{x^k\}_{k=1}^{\infty}$ converging to x^0 such that $\nabla z(x^k)$ exists for all k and $w = \lim_{k \to \infty} \nabla z(x^k)$. Hence, by use of the previous part of the proof, for each k there exist an index $i(k)$ with int $\text{Supp}(z, z^{i(k)}) \neq \emptyset$ and a sequence $\{x^{kl}\}_{l=1}^{\infty} \subset$ int $\text{Supp}(z, z^{i(k)})$ converging to x^k. By continuity, $i(k) \in \{i : x^0 \in \text{cl int } \text{Supp}(z, z^i)\}$ for sufficiently large k. But then, $\nabla z(x^k) = \nabla z^{i(k)}(x^k)$. By finiteness of $I_z(x^0)$ there exists a subsequence of $\{x^k\}_{k=1}^{\infty}$ such that $i(k) \equiv i^0$ for all elements in this subsequence. Let without loss of generality $\{x^k\}_{k=1}^{\infty}$ itself have this property. Then, $w = \lim_{k \to \infty} \nabla z^{i^0}(x^k) \in C$ which verifies $\partial z(x^0) \subseteq C$. $\qquad \square$

PROOF OF THEOREM 4.8: Take any direction $r \in \mathbb{R}^p$ with $\|r\| = 1$. Define

$$J = \{i : r \in K_{\text{Supp}(z, z^i)}(x^0)\}.$$

■ Define a relation on J as follows: Two indices $i, j \in J$ are said to be related if there exist indices $\{i_1, i_2, \ldots, i_\tau\} \subseteq J$ with $\tau \geq 2$ and $i_1 = i$, $i_\tau = j$ and

$$r \in K_{\text{Supp}(z, z^{i_l}) \cap \text{Supp}(z, z^{i_{l+1}})}(x^0), \quad l = 1, \ldots, \tau - 1.$$

By definition this means that there exist sequences $\{x^k\}_{k=1}^{\infty} \subseteq \text{Supp}(z, z^{i_l}) \cap \text{Supp}(z, z^{i_{l+1}})$ and $\{t_k\}_{k=1}^{\infty} \subset \mathbb{R}_{++}$ converging to x^0 and zero, resp., such that $\{(x^k - x^0)/t_k\}_{k=1}^{\infty}$ converges to r. Obviously, this relation is an equivalence relation. We will show now that there is only one equivalence class determined by this relation.

Let M_1 be one (non-empty) equivalence class and assume that the set $M_2 = J \setminus M_1$ is also not empty. Let

$$\delta_{ij} := \varrho(r, K_{\text{Supp}(z, z^i) \cap \text{Supp}(z, z^j)}(x^0)) \qquad (4.39)$$

denote the distance of r from $K_{\text{Supp}(z, z^i) \cap \text{Supp}(z, z^j)}(x^0)$. Then, if $i \in M_1$ and $j \in M_2$, $\delta_{ij} > 0$. Analogously, for $i \in I_z(x^0) \setminus J$ let

$$\varepsilon_i := \varrho(r, K_{\text{Supp}(z, z^i)}(x^0)). \qquad (4.40)$$

Then, $\varepsilon_i > 0$. Now, let

$$0 < \delta \leq \min\{\min\{\delta_{ij} : i \in M_1, j \in M_2\}, \min\{\varepsilon_i : i \in I_z(x^0) \setminus J\}\}/2$$

and take $\alpha \in (0, 1)$ such that $\|u - r\| \leq \delta$ for all

$$u \in C \cap \{v \in \mathbb{R}^p : \|v\| = 1\}$$

with the convex cone $C = \{v \in \mathbb{R}^p : \langle v, r \rangle \geq \alpha \|v\|\}$. Then, $r \in \text{int } C$ and zero is a extreme point of C. By the definition of δ and (4.39),

$$C \cap K_{\text{Supp}(z, z^i) \cap \text{Supp}(z, z^j)}(x^0) = \{0\}$$

for $i \in M_1$, $j \in M_2$. Let

$$\theta_{ij} := \inf\{\|x - x^0\| : x \neq x^0, x \in (C + x^0) \cap \mathrm{Supp}(z, z^i) \cap \mathrm{Supp}(z, z^j)\}.$$

Analogously, define

$$\sigma_i := \inf\{\|x - x^0\| : x \neq x^0, x \in (C + x^0) \cap \mathrm{Supp}(z, z^i)\}$$

for $i \in I_z(x^0) \setminus J$. It is easy to see that $\theta_{ij} > 0$, $\sigma_i > 0$ by (4.39) and (4.40). Take

$$0 < \theta \leq \min\{\min\{\theta_{ij} : i \in M_1, j \in M_2\}, \min\{\sigma_i : i \in I_z(x^0) \setminus J\}\}/2$$

and form the sets,

$$Q := (x^0 + C) \cap \{x : \|x - x^0\| \leq \theta\} \setminus \{x^0\}, \quad S_k = \bigcup_{i \in M_k} \mathrm{Supp}(z, z^i)$$

for $k = 1, 2$. The set Q is convex since x^0 is an extreme point of $x^0 + C$. Hence, it is connected. The sets S_k are closed. Then, by definition both

$$Q \subseteq S_1 \cup S_2 \text{ and } Q \cap S_1 \cap S_2 = \emptyset.$$

Since Q is connected, this implies that either $Q \cap S_1 = \emptyset$ or $Q \cap S_2 = \emptyset$. But each of these relations is not possible due to the definitions of the sets $\mathrm{Supp}(z, z^i)$ and the equivalence classes M_i. This shows that there is only one equivalence class.

- Let $i \in J$ and take any sequence $\{x^k\}_{k=1}^{\infty} \subset \mathrm{Supp}(z, z^i)$ converging to x^0. Then, by continuous differentiability of z^i, the sequence $\{\nabla z^i(x^k)r\}_{k=1}^{\infty}$ converges to $\nabla z^i(x^0)r$.

Now, assume that $i, j \in J$ are different indices and let $r \in K_{\mathrm{Supp}(z,z^i) \cap \mathrm{Supp}(z,z^j)}(x^0)$. Then, there exist sequences $\{x^k\}_{k=1}^{\infty} \subseteq \mathrm{Supp}(z, z^i) \cap \mathrm{Supp}(z, z^j)$, $\{t_k\}_{k=1}^{\infty} \subset \mathbb{R}_{++}$ converging to x^0 and zero, respectively, with $\lim_{k \to \infty} \frac{x^k - x^0}{t_k} = r$. Thus, $z^i(x^k) = z^j(x^k) = z(x^k)$ for all k by continuity of z. Hence we get the first order approximations

$$\begin{aligned}
z(x^k) &= z(x^0) + t_k \nabla z^i(x^0) \frac{x^k - x^0}{t_k} + o_i(t_k) \\
&= z(x^0) + t_k \nabla z^j(x^0) \frac{x^k - x^0}{t_k} + o_j(t_k),
\end{aligned}$$

which imply that $\nabla z^i(x^0)r = \nabla z^j(x^0)r$. Using the above equivalence relation we can show that $\nabla z^i(x^0)r$ is independent on the choice of $i \in J$.

- Now take arbitrary $\{x^k\}_{k=1}^{\infty}$ and $\{t_k\}_{k=1}^{\infty} \subset \mathbb{R}_{++}$ converging to x^0 and zero, respectively, with $\lim_{k\to\infty} \frac{x^k - x^0}{t_k} = r$. Take $\{g^k\}_{k=1}^{\infty}$ such that $g^k \in \partial z(x^k)$ for all k. Then, by Theorem 4.7 for each k there are numbers $\alpha_{i,k} \geq 0$, $i \in I_z(x^k)$, $\sum_{i \in I_z(x^k)} \alpha_{ik} = 1$, with

$$g^k = \sum_{i \in I_z(x^k)} \alpha_{ik} \nabla z^i(x^k).$$

By Lipschitz continuity, $I_z(x^k) \subseteq I_z(x^0)$ for sufficiently large k and the sequence $\{g^k\}_{k=1}^{\infty}$ is bounded, thus having accumulation points. Also the sequence of vectors $\{\alpha_{\cdot k}\}_{k=1}^{\infty}$ is bounded. Take any convergent subsequences of both sequences and assume without loss of generality that both sequences itself converge to $g \in \partial z(x^0)$ and $\alpha \geq 0$ with $\sum_{i \in I_z(x^0)} \alpha_i = 1$. Then, by the second part of this proof,

$$
\lim_{k\to\infty} g^k \frac{x^k - x^0}{t_k} = \lim_{k\to\infty} \sum_{i \in I_z(x^k)} \alpha_{ik} \nabla z^i(x^k) \frac{x^k - x^0}{t_k}
$$

$$
= \sum_{i \in I_z(x^0)} \alpha_i \nabla z^i(x^0) r = \nabla z^i(x^0) r
$$

for each $i \in J$ independently of the chosen accumulation point.

$\square$

PROOF OF THEOREM 4.9: The arguments used here mainly parallel those used in the proof of Theorem 1.5 in [206]. We show the result for $k = 2$. The desired result then follows by induction. Clearly, the point-to-set mapping Γ_z has nonempty, convex and compact values. It is easy to see that Γ_z is a closed point-to-set mapping. Since each of the mappings Γ_{z^i} is locally bounded, the same can be said for Γ_z which in turn implies that Γ_z is upper semicontinuous [17].

We now show that (4.16) holds or, in other words, that for each $\varepsilon > 0$ there exists a $\delta = \delta(\varepsilon) > 0$ such that

$$
-\varepsilon \leq \frac{z(x) - z(x^0) - g(x - x^0)}{\|x - x^0\|} \leq \varepsilon \tag{4.41}
$$

for all points in the set $\{x : \|x - x^0\| \leq \delta, \ x \neq x^0\}$ and for all $g \in \Gamma_z(x)$. Since the functions z^i, $i = 1, 2$, are pseudodifferentiable there exist corresponding $\delta_i = \delta_i(\varepsilon) > 0$ such that for all $g^i \in \Gamma_{z^i}(x)$ and all points in $\{x : \|x - x^0\| \leq \delta_i, x \neq x^0\}$, we have

$$
-\varepsilon \leq \frac{z^i(x) - z^i(x^0) - g^i(x - x^0)}{\|x - x^0\|} \leq \varepsilon. \tag{4.42}
$$

To show that inequalities (4.42) imply (4.41) we will consider all the different cases of coincidence between the functions $z(\cdot)$, $z^1(\cdot)$ and $z^2(\cdot)$. If $z(x^0) = z^1(x^0) \neq z^2(x^0)$ then, by continuity, there is some open neighborhood $\{x : \|x - x^0\| \leq \delta\}$, $\delta > 0$ of x such that $z(x) = z^1(x) \neq z^2(x)$ for all x in that neighborhood. Then for each x in that neighborhood, the result follows from (4.42).

Now let the function values at $x = x^0$ coincide: $z(x^0) = z^1(x^0) = z^2(x^0)$ and put $\delta = \delta(\varepsilon) = \min\{\delta_1(\varepsilon), \delta_2(\varepsilon)\}$. Take a fixed point $x \in \{x : \|x - x^0\| \leq \delta, \ x \neq x^0\}$. If it happens that $z(x) = z^1(x) \neq z^2(x)$ then the result again follows from (4.42) since we have only to consider $z^1(x)$ for evaluating (4.41). Consider now the last case when $z(x) = z^1(x) = z^2(x)$. Then $g \in \Gamma_z(x)$ if there exists $\beta \in [0, 1]$ with $g = \beta g^1 + (1 - \beta)g^2$ for some $g^i \in \Gamma_{z^i}(x)$. Hence, if we multiply the relations (4.42) for $i = 1$ by β and for $i = 2$ by $1 - \beta$ and add the two, we obtain (4.41). $\qquad\square$

PROOF OF THEOREM 4.10: By continuity of the function $x(\cdot)$ and (MFCQ) there exist open neighborhoods $U_1(y^0) \subseteq U_\delta(y^0)$ and $V_1(x^0) \subseteq V_\varepsilon(x^0)$ such that $I(x(y), y) \subseteq I(x^0, y^0)$ on $V_1(x^0) \times U_1(y^0)$. Let $(x(y), y) \in V_1(x^0) \times U_1(y^0)$. Since the (MFCQ) persists on a sufficiently small neighborhood of (x^0, y^0) we can assume that it is satisfied in $(x(y), y)$ (else we shrink the neighborhoods $U_\delta(y^0)$, $V_\varepsilon(x^0)$ again). Then, there exists a vertex $(\lambda(y), \mu(y))$ of (the bounded, nonempty and polyhedral set) $\Lambda(x(y), y)$ such that

$$\nabla_x f(x(y), y) + \lambda(y)^\top \nabla_x g(x(y), y) + \mu(y)^\top \nabla_x h(x(y), y) = 0,$$
$$\lambda(y)^\top g(x(y), y) = 0.$$

Hence, $x(y)$ is also a stationary point of the problem

$$\min_x \{f(x, y) : g_i(x, y) = 0, i \in J(\lambda(y)), h(x, y^0) = 0\}. \tag{4.43}$$

By Theorem 4.2 and (CRCQ) all the accumulation points (λ^0, μ^0) of the functions $(\lambda(y), \mu(y))$ for y tending to y^0 are vertices of the set $\Lambda(x^0, y^0)$ and $J(\lambda^0) \subseteq J(\lambda(y))$ for all $y \in U_1(y^0)$. Moreover, by (CRCQ) the gradients $\{\nabla_x g_i(x^0, y^0) : i \in J(\lambda(y))\} \cup \{\nabla_x h_j(x^0, y^0) : j = 1, \ldots, q\}$ are linearly independent. Hence, (SSOC) is valid on a certain open neighborhood of (x^0, y^0) and the point $x(y)$ is also a local optimal solution of (4.43). Let $x^{J(\lambda(y))}(\cdot)$ denote the unique continuous function of local optimal solutions in $V_1(x^0)$ of problem (4.43) for $y \in U_1(y^0)$ (cf. Theorem 4.4). Then, by [97] the function $x^{J(\lambda(y))}(\cdot)$ is continuously differentiable on some open neighborhood of y^0. Let, without loss of generality, this neighborhood coincide with $U_1(y^0)$.

Now we have shown that, for each point $(x(y), y)$ sufficiently close to (x^0, y^0) there exists a continuously differentiable function $x^{J(\lambda(y))}(\cdot)$

with $x(y) = x^{J(\lambda(y))}(y)$. Since there are only a finite number of different sets $J(\lambda(y))$ the proof follows.   □

To prove Theorem 4.11 we need some auxiliary results and some notation. Let

$$S(r) := \operatorname*{Argmax}_{(\lambda,\mu)} \{\nabla_y L(x^0, y^0, \lambda, \mu)r : (\lambda, \mu) \in \Lambda(x^0, y^0)\}$$

denote the set of optimal solutions of problem (4.21) and $T_\lambda(r)$ be the set of all points d satisfying

$$\nabla_x g_i(x^0, y^0)d + \nabla_y g_i(x^0, y^0)r \begin{cases} = 0, & \text{if } i \in J(\lambda), \\ \leq 0, & \text{if } i \in I(x^0, y^0) \setminus J(\lambda), \end{cases}$$

$$\nabla_x h_j(x^0, y^0)d + \nabla_y h_j(x^0, y^0)r = 0, j = 1, \ldots, q.$$

Then, by use of the Karush-Kuhn-Tucker conditions for problem (4.21) it is easy to see that the set $T_\lambda(r)$ does not depend on (λ, μ) as long as $(\lambda, \mu) \in S(r)$ [68]. This set is empty whenever $(\lambda, \mu) \notin S(r)$. Moreover, the problem $QP(\lambda, \mu, r)$ has a feasible (and thus a uniquely determined optimal) solution if and only if $T_\lambda(r) \neq \emptyset$.

LEMMA 4.3 ([235]) *Consider the perturbed Karush-Kuhn-Tucker conditions for problem (4.1)*

$$\nabla_x f(x, y) + \lambda^\top \nabla_x g(x, y) + \mu^\top \nabla_x h(x, y) = 0$$
$$g_i(x, y) - g_i(x(y), y) \leq 0, \ \lambda_i(g_i(x, y) - g_i(x(y), y)) = 0, \ i \in J^0(r)$$
$$g_i(x, y) \leq 0, \ \lambda_i g_i(x, y) = 0, \ i \notin J^0(r) \qquad (4.44)$$
$$\lambda_i \geq 0, \ i = 1, \ldots, p$$
$$h(x, y) = 0$$

where $J^0(r) = \{j : \exists(\lambda, \mu) \in S(r) \text{ with } \lambda_j > 0\}$ and let $\Lambda^0(r, x, y)$ be the set of Lagrange multipliers of this problem. Then, for each direction r, for each sequence $\{y^k\}_{k=1}^\infty$ converging to y^0 with $\lim_{k\to\infty} y^k/\|y^k\| = r$ and each $(\lambda^0, \mu^0) \in S(r)$ we have

$$\lim_{k\to\infty} \varrho((\lambda^0, \mu^0), \Lambda^0(r, x(y^k), y^k)) = 0.$$

PROOF Fix a direction r, a vector $(\lambda^0, \mu^0) \in S(r)$ and a sequence $\{y^k\}_{k=1}^\infty$ converging to y^0. Since $S(r)$ is convex we can assume that $\lambda_i^0 > 0$ for all $i \in J^0(r)$. Let $(\lambda^k, \mu^k) \in \Lambda(x(y^k), y^k)$. Then, by Theorem 4.2 the sequence $\{(\lambda^k, \mu^k)\}$ is bounded and converges without loss of generality to some $(\lambda, \mu) \in \Lambda(x^0, y^0)$. As in [68] it is easy to see that $(\lambda, \mu) \in S(r)$. Hence $J(\lambda) \subseteq J^0(r)$.

Let $\Delta\lambda = \lambda^0 - \lambda$, $\Delta\mu = \mu^0 - \mu$. Then, $\Delta\lambda_i > 0$ if $i \in J^0(r) \setminus J(\lambda)$ and $\Delta\lambda_i = 0$ if $i \notin J^0(r)$. From the Karush-Kuhn-Tucker conditions of (4.1) we get

$$\Delta\lambda^\top \nabla_x g(x^0, y^0) + \Delta\mu^\top \nabla_x h(x^0, y^0) = 0.$$

Due to the (CRCQ) and [17] the point-to-set mapping

$$\mathcal{N}_I(x,y) = \{(\alpha, \beta) \in \mathbb{R}^{|I|} \times \mathbb{R}^q : \sum_{i \in I} \alpha_i \nabla_x g_i(x,y) + \sum_{j=1}^q \beta_j \nabla_x h_j(x,y) = 0\}$$

is lower semicontinuous for $I \subseteq I(x^0, y^0)$. Thus, taking $I = J^0(r)$ there exists a sequence $\{(\Delta\lambda^k, \Delta\mu^k)\}_{k=1}^\infty$ converging to $(\Delta\lambda, \Delta\mu)$ such that $\Delta\lambda_i^k = 0$ for $i \notin J^0(r)$ and

$$\Delta\lambda^{k\top} \nabla_x g(x(y^k), y^k) + \Delta\mu^{k\top} \nabla_x h(x(y^k), y^k) = 0 \qquad (4.45)$$

for sufficiently large k.

Now, we construct points $(\gamma^k, \eta^k) \in \Lambda^0(r, x(y^k), y^k)$ converging to (λ^0, μ^0). Let $i \in J(\lambda)$ and put

$$t_i^k = \max\{t \in [0,1] : \lambda_i^k + t\Delta\lambda_i^k \geq 0\}.$$

If $\lambda_i > 0$ then $\lambda_i^k > 0$ for sufficiently large k. Moreover,

$$\lim_{k \to \infty} (\lambda_i^k + \Delta\lambda_i^k) = \lambda_i + \Delta\lambda_i = \lambda_i^0 > 0, \ \forall i \in J^0(r).$$

Thus, we have that t_i^k converges to 1 for $k \to \infty$. Let $t^k := \min\{t_i^k : i \in J(\lambda)\}$ and

$$\gamma^k = \lambda^k + t^k \Delta\lambda^k, \ \eta^k = \mu^k + t^k \Delta\mu^k$$

for all k. Then, t^k again converges to 1 and (γ^k, η^k) converges to (λ^0, μ^0) for $k \to \infty$.

For proving the lemma we have to show that $(\gamma^k, \eta^k) \in \Lambda^0(r, x(y^k), y^k)$ for all large k. First, by construction, $\gamma_i^k \geq 0$ for $i \in J(\lambda)$, $\lambda_i^k \geq 0 = \Delta\lambda_i^k$ for $i \notin J^0(r)$ and $\Delta\lambda_i^k > 0$ for $i \in J^0(r) \setminus J(\lambda)$. This implies $\gamma_i^k \geq 0$ for all k. Considering analogously all these three cases it is easy to see that the complementarity condition $\gamma_i^k(g_i(x(y^k), y^k) - g_i(x(y^k), y^k)) = 0$ for $i \in J^0(r)$ and $\gamma_i^k g_i(x(y^k), y^k) = 0$ for $i \notin J^0(r)$ is satisfied.

To conclude the proof we mention that due to the definition of the mapping $\mathcal{N}_{J^0(r)}(x,y)$ the condition

$$\nabla_x f(x(y^k), y^k) + \gamma^{k\top} \nabla_x g(x(y^k), y^k) + \eta^{k\top} h(x(y^k), y^k) = 0$$

is satisfied for all k. □

PROOF OF THEOREM 4.11: The vector $d^0 \in \mathbb{R}^n$ solving the problem $QP(\lambda^0, \mu^0, r)$ is characterized as the unique solution of the following affine variational inequality describing first order necessary and sufficient optimality conditions for the problem $QP(\lambda^0, \mu^0, r)$

$$d^0 \in T_{\lambda^0}(r) \text{ and } \forall d \in T_{\lambda^0}(r),$$
$$\left\langle \nabla^2_{xx} L(x^0, y^0, \lambda^0, \mu^0) d^0 + \nabla^2_{yx} L(x^0, y^0, \lambda^0, \mu^0) r, d - d^0 \right\rangle \geq 0. \tag{4.46}$$

Let $d^0 = x'(y^0; r)$ which exists by Corollary 4.1 and take any sequence $\{t_k\}_{k=1}^{\infty} \subset \mathbb{R}_{++}$ converging to zero. Then,

$$\lim_{k \to \infty} t_k^{-1} [x(y^0 + t_k r) - x^0] = x'(y^0; r).$$

Let $\{(\lambda^k, \mu^k)\}_{k=1}^{\infty}$ be a sequence with $(\lambda^k, \mu^k) \in \Lambda(x(y^0 + t_k r), y^0 + t_k r)$ for all k. Then, without loss of generality, there exists

$$\lim_{k \to \infty} (\lambda^k, \mu^k) = (\overline{\lambda}, \overline{\mu}) \in \Lambda(x^0, y^0).$$

Considering the difference

$$0 \equiv [\nabla_x L(x(y^0 + t_k r), y^0 + t_k r, \lambda^k, \mu^k) - \nabla_x L(x^0, y^0, \overline{\lambda}, \overline{\mu})]/t_k$$

for $k \to \infty$ yields

$$0 = \nabla^2_{xx} L(x^0, y^0, \overline{\lambda}, \overline{\mu}) x'(y^0; r) + \nabla^2_{yx} L(x^0, y^0, \overline{\lambda}, \overline{\mu}) r$$
$$+ \lim_{k \to \infty} \sum_{i=1}^{p} \frac{\lambda_i^k - \overline{\lambda}_i}{t_k} \nabla_x^\top g_i(x^0, y^0) + \lim_{k \to \infty} \sum_{j=1}^{q} \frac{\mu_i^k - \overline{\mu}_i}{t_k} \nabla_x^\top h_j(x^0, y^0).$$
$$\tag{4.47}$$

Our aim is it to show that $x'(y^0; r)$ satisfies the above variational inequality posed using an arbitrary vector $(\lambda^0, \mu^0) \in S(r)$. First we show that $x'(y^0; r)$ belongs to $T_{\overline{\lambda}}(r)$. If $\overline{\lambda}_i > 0$ then $\lambda_i^k > 0$ for large k yielding $\nabla_x g_i(x^0, y^0) x'(y^0; r) + \nabla_y g_i(x^0, y^0) r = 0$. Similar considerations in the case $i \in I(x^0, y^0) \setminus J(\overline{\lambda})$ and for the equality constraints show that $x'(y^0; r) \in T_{\overline{\lambda}}(r)$. Since $T_{\lambda^0}(r) = T_{\overline{\lambda}}(r)$ we have $x'(y^0; r) \in T_{\lambda^0}(r)$.

Now take a sequence (γ^k, η^k) converging to (λ^0, μ^0) with $(\gamma^k, \eta^k) \in \Lambda^0(r, x^0, y^0)$ whose existence is guaranteed by Lemma 4.3. Then, similarly to (4.47) we get

$$0 = \nabla^2_{xx} L(x^0, y^0, \lambda^0, \mu^0) x'(y^0; r) + \nabla^2_{yx} L(x^0, y^0, \lambda^0, \mu^0) r$$
$$+ \lim_{k \to \infty} \sum_{i=1}^{p} \frac{\gamma_i^k - \lambda_i^0}{t_k} \nabla_x^\top g_i(x^0, y^0) + \lim_{k \to \infty} \sum_{j=1}^{q} \frac{\eta_i^k - \mu_i^0}{t_k} \nabla_x^\top h_j(x^0, y^0).$$
$$\tag{4.48}$$

Due to

$$\nabla_x g_i(x^0, y^0)(d - x'(y^0; r)) = 0, \ \nabla_x h_j(x^0, y^0)(d - x'(y^0; r)) = 0,$$

for all $d \in T_{\lambda^0}(r)$ and $i \in J(\lambda^0)$ and for all j we get

$$\langle \nabla_{xx}^2 L(x^0, y^0, \lambda^0, \mu^0) x'(y^0; r) + \nabla_{yx}^2 L(x^0, y^0, \lambda^0, \mu^0) r, d - x'(y^0; r) \rangle =$$

$$- \lim_{k \to \infty} \sum_{i \in I(x^0, y^0) \setminus J(\lambda^0)} \frac{\gamma_i^k}{t_k} \nabla_x^\top g_i(x^0, y^0)(d - x'(y^0; r)).$$

(4.49)

Let $i \in I(x^0, y^0) \setminus J(\lambda^0)$. If $\nabla_x g_i(x^0, y^0) x'(y^0; r) + \nabla_y g_i(x^0, y^0) r < 0$ then $g_i(x(y^0 + t_k r), y^0 + t_k r) < 0$ which implies $\gamma_i^k = 0$ for sufficiently large k. On the other hand, if $\nabla_x g_i(x^0, y^0) x'(y^0; r) + \nabla_y g_i(x^0, y^0) r = 0$ then $\nabla_x g_i(x^0, y^0)(x'(y^0; r) - d) \geq 0$ by $d \in T_{\lambda^0}(r)$ and $\gamma_i^k \geq 0$. This implies (4.46). □

PROOF OF THEOREM 4.12:

■ We start with the first assertion: If r^0 is a direction such that (SCR) is satisfied, then problem $QP(\lambda^0, \mu^0, r^0)$ is solvable and $x'(y^0; r^0)$ is equal to the unique optimal solution of this problem. By linear programming duality and non-emptiness of the feasible set of problem $QP(\lambda^0, \mu^0, r^0)$, we have $S(r^0) = \{(\lambda^0, \mu^0)\}$. Let (γ^0, η^0) denote the (unique) Karush-Kuhn-Tucker vector of problem $QP(\lambda^0, \mu^0, r^0)$. Directional differentiability of $x(\cdot)$, (SCR) and $\nabla_x g_i(x^0, y^0) x'(y^0; r) + \nabla_y g_i(x^0, y^0) r < 0$ for all $i \notin I := J(\lambda^0) \cup J(\gamma^0)$ imply that $g_i(x(y^0 + tr), y^0 + tr) < 0$ for all $t > 0$ sufficiently small. Hence, $x(y^0 + tr^0)$ is also the unique optimal solution of the enlarged problem

$$\min_x \{f(x, y) : g_i(x, y) \leq 0, \ i \in I, \ h(x, y) = 0\}$$

for $y = y^0 + tr^0$ and small $t > 0$. For this problem, the linear independence constraint qualification is satisfied at $y = y^0$ (by $I \in \mathcal{I}(\lambda^0)$). Hence, its Karush-Kuhn-Tucker multiplier vector is uniquely determined and is also directionally differentiable [137]. This multiplier $(\overline{\lambda}(y^0), \overline{\mu}(y^0))$ is equal to $((\lambda_i^0)_{i \in I}, \mu^0)$, its directional derivative is $((\gamma_j^0)_{j \in I}, \eta^0)$ at $y = y^0$. Hence, $(\lambda(y^0 + tr^0), \mu(y^0 + tr^0))$ with $\lambda_i(y^0 + tr^0) := \overline{\lambda}_i(y^0 + tr^0) = \lambda_i^0 + t\gamma_i^0 + o(t) > 0$, $i \in I$, $\lambda_j(y^0 + tr^0) := 0$, $j \notin I$, $\mu_j(y^0 + tr^0) := \overline{\mu}_j(y^0 + tr^0) = \mu_j^0 + t\eta_j^0 + o(t)$, is one Karush-Kuhn-Tucker multiplier vector for problem (4.1) for sufficiently small $t > 0$, where $\lim_{t \to 0} \frac{o(t)}{t} = 0$. Thus, $g_i(x(y^0 + tr^0), y^0 + tr^0) = 0$, $\forall i \in I$

and small $t > 0$. Consequently, $x(y^0+tr^0) = x^I(y^0+tr^0)$ for each sufficiently small $t > 0$. All these arguments remain true if the direction r^0 is slightly perturbed which implies that $r^0 \in \text{int } K_{\text{Supp}(x,x^I)}(y^0)$. Hence, $x(y)$ is differentiable at each point $y = y^0 + tr^0$ with $t > 0$ and sufficiently small, which leads to $\nabla x^I(y^0) \in \partial x(y^0)$.

- Now consider the second assertion: Let $I \in \mathcal{I}(\lambda^0)$ for some vertex $\nu^0 \in \Lambda(z^0)$. Then, the optimal solution function $x^I(\cdot)$ of the problem (4.19) is continuously differentiable [97] and $x^I(y^0) = x(y^0)$. Consider the necessary and sufficient optimality conditions of first order for problem $QP(\lambda^0, \mu^0, r)$

$$\nabla^2_{xx}L(z^0,\nu^0)d + \nabla^2_{yx}L(z^0,\nu^0)r + \nabla^T_x g_{I^0}(z^0)\gamma + \nabla^T_x h(z^0)\eta = 0,$$
$$\nabla_x g_i(z^0)d + \nabla_y g_i(z^0)r = 0, \ i \in J(\lambda^0),$$
$$\nabla_x g_i(z^0)d + \nabla_y g_i(z^0)r \leq 0, \ i \in I(z^0) \setminus J(\lambda^0),$$
$$\nabla_x h_j(z^0)d + \nabla_y h_j(z^0)r = 0, \ j = 1,\ldots,q,$$
$$\gamma_i(\nabla_x g_i(z^0)d + \nabla_y g_i(z^0)r) = 0, \ i \in I(z^0) \setminus J(\lambda^0),$$
$$\gamma_i \geq 0, i \in I(z^0) \setminus J(\lambda^0).$$

Let, without loss of generality,

$$J(\lambda^0) = \{1,\ldots,s\}, \ I = \{1,\ldots,u\}, \ I(z^0) = \{1,\ldots,v\}$$

for $s \leq u \leq v$. Then, by (FRR), the matrix M^0 defined by

$$\begin{pmatrix} \nabla^2_{xx}L(z^0,\nu^0) & \nabla^T_x g_{I^0}(z^0) & \nabla^T h(z^0) & \nabla^2_{xy}L(z^0,\nu^0) \\ \nabla_x g_{I^0}(z^0) & 0 & 0 & \nabla_y g_{I^0}(z^0) \\ \nabla_x h(z^0) & 0 & 0 & \nabla_y h(z^0) \\ 0 & e^T_{s+1} & 0 & 0 \\ 0 & \vdots & 0 & 0 \\ 0 & e^T_v & 0 & 0 \end{pmatrix}$$

has rank $n + |I(z^0)| + q + |I(z^0) \setminus J(\lambda^0)|$ (note that we have added $|I(z^0) \setminus J(\lambda^0)|$ columns and the same number of rows which contain a unit matrix of full dimension). Here e_i denotes the i-th unit vector. Hence, the system of linear equations $M^0(d,\gamma,\eta,r)^T = a$ has a solution for arbitrary right-hand side a. Take a right-hand side vector a which has the value $-\varepsilon < 0$ for each component corresponding to a left-hand side

$$\nabla_x g_i(z^0)d + \nabla_y g_i(z^0)r, \ i \in I(x^0,y^0) \setminus I,$$

the value ε in each component for left-hand side γ_i, $i \in I \setminus J(\lambda^0)$, and vanishes in all other components. Let $(d^0, \gamma^0, \eta^0, r^0)^\top$ be a solution of the resulting linear system. Then, $(d^0, \gamma^0, \eta^0)^\top$ satisfies the Karush-Kuhn-Tucker conditions for the problem $QP(\lambda^0, \mu^0, r^0)$. Moreover, strict complementary slackness is satisfied for this system. Together with the first part, this completes the proof.

$\square$

PROOF OF LEMMA 4.1:

- Since the problem $QP(\lambda, \mu, \widehat{r})$ has a feasible solution if and only if $(\lambda, \mu) \in S(\widehat{r})$, for each $\widehat{r} \in R(I, \lambda^0)$ we have $(\lambda^0, \mu^0) \in S(\widehat{r})$. Let there exist a vector $(\widehat{\lambda}, \widehat{\mu}) \neq (\lambda^0, \mu^0)$ with $(\widehat{\lambda}, \widehat{\mu}) \in S(\widehat{r})$. Then,

$$\nabla_y L(x^0, y^0, \widehat{\lambda}, \widehat{\mu})\widehat{r} = \nabla_y L(x^0, y^0, \lambda^0, \mu^0)\widehat{r} \geq \nabla_y L(x^0, y^0, \lambda', \mu')\widehat{r}$$

for all $\widehat{r} \in \mathrm{int}\, R(I, \lambda^0)$ and all $(\lambda', \mu') \in \Lambda(x^0, y^0)$. Then, by direct calculation we get

$$\sum_{i \in I(x^0, y^0)} (\widehat{\lambda}_i - \lambda_i^0) \nabla_y g_i(x^0, y^0)\widehat{r} + \sum_{j=1}^{q} (\widehat{\mu}_j - \mu_j^0) \nabla_y h_j(x^0, y^0)\widehat{r} = 0$$

for all $\widehat{r} \in \mathrm{int}\, R(I, \lambda^0)$, i.e. in an open neighborhood of some $r \in R(I, \lambda^0)$. This implies

$$\sum_{i \in I(x^0, y^0)} (\widehat{\lambda}_i - \lambda_i^0) \nabla_y g_i(x^0, y^0) + \sum_{j=1}^{q} (\widehat{\mu}_j - \mu_j^0) \nabla_y h_j(x^0, y^0) = 0.$$

Since $\{(\widehat{\lambda}, \widehat{\mu}), (\lambda^0, \mu^0)\} \subset \Lambda(x^0, y^0)$ we also have

$$\sum_{i \in I(x^0, y^0)} (\widehat{\lambda}_i - \lambda_i^0) \nabla_x g_i(x^0, y^0) + \sum_{j=1}^{q} (\widehat{\mu}_j - \mu_j^0) \nabla_x h_j(x^0, y^0) = 0.$$

But this contradicts assumption (LIN). Hence, the first assertion is true.

- Assume that there exist numbers δ_i, $i \in I$, ν_j, $j = 1, \ldots, q$, such that $\sum_{i \in I} \delta_i^2 + \sum_{j=1}^{q} \nu_j^2 > 0$ and

$$\sum_{i \in I} \delta_i \nabla_x g_i(x^0, y^0) + \sum_{j=1}^{q} \nu_j \nabla_x h_j(x^0, y^0) = 0. \tag{4.50}$$

Since the relations $\nabla_x g_i(x^0, y^0) x'(y^0, \widehat{r}) + \nabla_y g_i(x^0, y^0)\widehat{r} = 0 \; \forall \; i \in I$ and $\nabla_x h_j(x^0, y^0) x'(y^0, \widehat{r}) + \nabla_y h_j(x^0, y^0)\widehat{r} = 0 \; \forall \; j = 1,\ldots,q$, are satisfied, this implies that also

$$\left(\sum_{i \in I} \delta_i \nabla_y g_i(x^0, y^0) \right) \widehat{r} + \left(\sum_{j=1}^q \nu_i \nabla_y h_j(x^0, y^0) \right) \widehat{r} = 0$$

for all $\widehat{r} \in \text{int } R(I, \lambda^0)$ which is an open neighbourhood of r. Thus,

$$\sum_{i \in I} \delta_i \nabla_y g_i(x^0, y^0) + \sum_{j=1}^q \nu_i \nabla_y h_j(x^0, y^0) = 0.$$

Together with (4.50), this contradicts the assumption (LIN). Hence, also the second assertion is valid.

$\square$

PROOF OF COROLLARY 4.2: The matrix

$$B_I = \begin{pmatrix} \nabla^2_{xx} L(x^0, y^0, \lambda^0, \mu^0) & \nabla^{\mathsf{T}}_x g_I(x^0, y^0) & \nabla^{\mathsf{T}}_x h(x^0, y^0) \\ \nabla_x g_I(x^0, y^0) & 0 & 0 \\ \nabla_x h(x^0, y^0) & 0 & 0 \end{pmatrix}$$

is quadratic and regular [157]. This implies uniqueness and linearity of the solution of $KKT(\cdot, I, \lambda^0)$. Since (d, γ, η) is a solution of $KKT(r, I, \lambda^0)$ if and only if it satisfies also the additional constraints $\nabla_x g_i(x^0, y^0) d + \nabla_y g_i(x^0, y^0) r \leq 0$, $i \in I(x^0, y^0) \setminus I$ and $\gamma_i \geq 0$, $i \in I \setminus J(\lambda^0)$, existence of this solution is guaranteed if and only if $r \in R(I, \lambda^0)$. The corollary follows now since the Karush-Kuhn-Tucker conditions are necessary and sufficient for optimality for $QP(\lambda^0, \mu^0, r)$. $\square$

PROOF OF THEOREM 4.13: The proof is done in two steps: First, using the assumptions, we derive that the directional derivative $x'(y^0; r)$ is equal to $\nabla x(y^0) r$ for all $r \in R(I, \lambda^0)$ and some active selection function $x^I(\cdot)$ for the PC^1-function $x(\cdot)$. Since it is not obvious that this implies that the active selection function $x^I(\cdot)$ is also strongly active, we select a suitable strongly active selection function in the second step. Using that there are only a finite number of different selection functions we will finally derive the desired result.

- Take $I \in \mathcal{I}(\lambda^0)$, $(\lambda^0, \mu^0) \in S(r)$ as given in the assumptions. Consider the differentiable selection function $x^I(\cdot)$ which is active at $y = y^0$ due to our assumptions. Let the matrices Γ, H denote the gradients of the Lagrange multiplier functions of problem (4.19) which exist by [97]. Then, inserting $(\nabla x^I(y^0)r, r, \Gamma r, Hr)$ into $KKT(r, I, \lambda^0)$, we see that these conditions are satisfied. Uniqueness of the optimal solution of problem $QP(\lambda, \mu, r)$ then yields

$$\nabla x^I(y^0)r = x'(y^0; r) \ \forall \ r \in R(I, \lambda^0).$$

- By Corollary 4.1 there exists a strongly active selection function $x^{\widehat{I}}(y^0)$ with $x'(y^0; r) = \nabla_y x^{\widehat{I}}(y^0)r$. Because the number of elements in the set $I_x^0(y^0)$ is finite and because the directional derivative $x(y^0; \cdot)$ is continuous with respect to changes of the direction, we can assume without loss of generality that int $K_{\mathrm{Supp}(x, x^{\widehat{I}})}(y^0) \neq \emptyset$. Moreover, it is easy to see that $x'(y^0; r) = \nabla x^{\widehat{I}}(y^0)r$ for all $r \in$ int $K_{\mathrm{Supp}(x, x^{\widehat{I}})}(y^0)$. But then $S := $ int $(K_{\mathrm{Supp}(x, x^{\widehat{I}})}(y^0) \cap R(I, \lambda^0)) \neq \emptyset$ and we have $x'(y^0; r) = \nabla_y x^{\widehat{I}}(y^0)r = \nabla_y x^I(y^0)r$ for each $r \in S$. This implies $\nabla_y x^I(y^0) = \nabla_y x^{\widehat{I}}(y^0)$.

$\square$

PROOF OF THEOREM 4.14: The main ideas of the following proof date back to [64]. The differentiability assumptions imply that the objective function $f(\cdot, \cdot)$ is locally Lipschitz continuous with, say, a Lipschitz modulus $L_f < \infty$ on an open set $W_\gamma(x^0, y^0)$, $\gamma > 0$. (MFCQ) also implies pseudo Lipschitz continuity of the feasible set mapping $M(\cdot)$ [245] with a Lipschitz modulus $L_M < \infty$ on the set $U_\delta(y^0)$, $\delta > 0$. Let without loss of generality $\delta \leq \gamma$. Take any points $y^1, y^2 \in U_\delta(y^0)$ and let $x^1 \in \Psi(y^1)$. By upper semicontinuity of the solution set mapping $\Psi(\cdot)$ (cf. Theorem 4.3) we can assume that δ is small enough to guarantee that $x^1 \in V_\varepsilon(x^0), \varepsilon > 0$, where $V_\varepsilon(x^0)$ comes from the definition of pseudo Lipschitz continuity for some $x^0 \in \Psi(y^0)$ and $\varepsilon \leq \gamma$. Then there exists $x^2 \in M(y^2)$ such that $\|x^1 - x^2\| \leq L_M\|y^1 - y^2\|$. By $\varphi(y^2) \leq f(x^2, y^2)$ and $\varphi(y^1) = f(x^1, y^1)$ this implies

$$\begin{aligned}
\varphi(y^2) - \varphi(y^1) &\leq f(x^2, y^2) - f(x^1, y^1) \leq |f(x^2, y^2) - f(x^1, y^1)| \\
&\leq L_f(\|x^1 - x^2\| + \|y^1 - y^2\|) \leq L_f(L_M\|y^1 - y^2\| + \|y^1 - y^2\|) \\
&= L_f(L_M + 1)\|y^1 - y^2\|.
\end{aligned}$$

Changeing the places of y^1, y^2 shows the Theorem. $\square$

PROOF OF THEOREM 4.15: First we prove the upper bound on the upper Dini directional derivative. Let $x^0 \in \Psi(y^0)$ be some optimal solution and consider the linearized problem

$$\lambda^\top \nabla_y g(x^0, y^0) r + \mu^\top \nabla_y h(x^0, y^0) r \to \max_{(\lambda, \mu)}$$

$$(\lambda, \mu) \in \Lambda(x^0, y^0). \tag{4.51}$$

for a fixed direction $r = r^0$ which coincides with problem (4.21). By (MFCQ) the feasible set of this problem is nonempty and compact [105] (cf. Theorem 4.1). Hence, its dual is also feasible. Let d^0 be an optimal solution of the dual to (4.51):

$$\nabla_x f(x^0, y^0) d \to \min_d$$

$$\nabla_x g_i(x^0, y^0) d + \nabla_y g_i(x^0, y^0) r^0 \leq 0, \ i \in I(x^0, y^0),$$

$$\nabla_x h_j(x^0, y^0) d + \nabla_y h_j(x^0, y^0) r^0 = 0, \ j = 1, \ldots, q. \tag{4.52}$$

Let $\bar{d}$ be a direction satisfying the (MFCQ) and consider the direction $d(\alpha) = \alpha \bar{d} + d^0$. Then, for all $\alpha > 0$ we have

$$\nabla_x f(x^0, y^0) d(\alpha) = \alpha \nabla_x f(x^0, y^0) \bar{d}$$
$$+ \max_{(\lambda, \mu) \in \Lambda(x^0, y^0)} \{\lambda^\top \nabla_y g(x^0, y^0) r^0 + \mu^\top \nabla_y h(x^0, y^0) r^0\}$$

$$\nabla_x g_i(x^0, y^0) d(\alpha) + \nabla_y g_i(x^0, y^0) r^0 < 0, \ i \in I(x^0, y^0),$$

$$\nabla_x h_j(x^0, y^0) d(\alpha) + \nabla_y h_j(x^0, y^0) r^0 = 0, \ j = 1, \ldots, q.$$

By the last two relations together with the (MFCQ) and an implicit function theorem (see e.g. Lemma 2.3 in [106]), for each $\alpha > 0$ there exists a function $\beta(t, d(\alpha))$ such that the function

$$X(t, d(\alpha)) = x^0 + \beta(t, d(\alpha)) + td(\alpha)$$

is feasible for sufficiently small $t > 0$:

$$g_i(X(t, d(\alpha)), y^0 + tr^0) < 0, \ i = 1, \ldots, p,$$

$$h_j(X(t, d(\alpha)), y^0 + tr^0) = 0, \ j = 1, \ldots, q.$$

Moreover, the function $X(t, d(\alpha))$ is continuously differentiable on an open neighborhood of zero with $\frac{d}{dt} X(0, d(\alpha)) = d(\alpha)$. Hence,

$$f(X(t, d(\alpha)), y^0 + tr^0) \geq \varphi(y^0 + tr^0)$$

which implies

$$
\begin{aligned}
D^+ \varphi(y^0; r^0) &= \limsup_{t \to 0+} t^{-1}[\varphi(y^0 + tr^0) - \varphi(y^0)] \\
&\leq \limsup_{t \to 0+} t^{-1}[f(X(t, d(\alpha)), y^0 + tr^0) - f(x^0, y^0)] \\
&= \nabla_x f(x^0, y^0) d(\alpha) + \nabla_y f(x^0, y^0) r^0 \\
&= \alpha \nabla_x f(x^0, y^0) \bar{d} + \max_{(\lambda, \mu) \in \Lambda(x^0, y^0)} \nabla_y L(x^0, y^0, \lambda, \mu) r^0.
\end{aligned}
$$

The result follows now by $\alpha \to 0$ and since $x^0 \in \Psi(y^0)$ has been arbitrarily chosen.

Now we show the lower bound for the lower Dini directional derivative. Fix a direction r^0 and take any positive sequence $\{t_k\}_{k=1}^{\infty}$ converging to zero such that

$$D_+\varphi(y^0; r^0) = \liminf_{t \to 0+} \frac{\varphi(y^0 + tr^0) - \varphi(y^0)}{t} = \lim_{k \to \infty} \frac{\varphi(y^0 + t_k r^0) - \varphi(y^0)}{t_k}.$$

By (C) and (MFCQ) the sets $M(y^0 + t_k r^0)$ are not empty which is also valid for the solution sets $\Psi(y^0 + t_k r^0)$ for sufficiently large k (cf. Theorem 4.3). Set $y^k = y^0 + t_k r^0$ and take $\{x^k\}_{k=1}^{\infty}$ with $x^k \in \Psi(y^k)$ for all k. Again by (C) this sequence has accumulation points $x^0 \in \Psi(y^0)$ by Theorem 4.3. Let without loss of generality $\lim_{k \to \infty} x^k = x^0$.

Let $\bar{d}$ be a direction satisfying (MFCQ) at (x^0, y^0). By persistence of (MFCQ) in an open neighborhood of (x^0, y^0) there exists a sequence $\{\bar{d}^k\}_{k=1}^{\infty}$ such that $\bar{d}^k$ satisfies (MFCQ) at (x^k, y^k) for all k and $\lim_{k \to \infty} \bar{d}^k = \bar{d}$ [17]. Let d^0 be an optimal solution of (4.52) for the negative direction $r = -r^0$ and let $\{d^k\}_{k=1}^{\infty}$ be a sequence converging to d^0 such that

$$\nabla_x h(x^k, y^k) d^k - \nabla_y h(x^k, y^k) r^0 = 0.$$

Note that this sequence exists by (MFCQ) and [17]. Take $\alpha > 0$ and consider

$$\widehat{d}^k = \alpha \bar{d}^k + d^k.$$

Let $\widehat{d} = \lim_{k \to \infty} \widehat{d}^k = \alpha \bar{d} + d^0$. Again there exists a continuous function $X(t, x, y, d, r)$ defined on a certain neighborhood $(-\bar{t}, \bar{t}) \times W(x^0, y^0) \times V(\widehat{d}, r^0)$ such that

$$h(X(t, x, y, d, r), y - tr) = h(x, y) + t\{\nabla_x h(x, y) d - \nabla_y h(x, y) r\}$$

where $\frac{d}{dt} X(t, x, y, d, r)$ exists with $\lim_{t \to 0+} \frac{d}{dt} X(t, x, y, d, r) = d$ and we have $X(0, x, y, d, r) = x$. Now let

$$X^k(t) = X(t, x^k, y^k, \widehat{d}^k, r^0).$$

Then, it can be shown that $X^k(t_k) \in M(y^0)$ for sufficiently large k and, thus, $f(X^k(t_k), y^0) \geq \varphi(y^0)$. This implies

$$D_+\varphi(y^0; r^0) \geq \lim_{k \to \infty} t_k^{-1}[f(x^k, y^k) - f(X^k(t_k), y^0)] = -\lim_{k \to \infty} \frac{d}{dt} f^k(\tau_k t_k)$$

with $\tau_k \in (0,1)$ by the mean value theorem, where

$$f^k(t) = f(X^k(t), y^k - tr^0).$$

By differentiability,

$$
\begin{aligned}
\frac{d}{dt}f^k(\tau_k t_k) &= \nabla_x f(X^k(\tau_k t_k), y^k - \tau_k t_k r^0)\frac{d}{dt}X^k(\tau_k t_k) - \\
&\quad \nabla_y f(X^k(\tau_k t_k), y^k - \tau_k t_k r^0)r^0.
\end{aligned}
$$

But then,

$$
\begin{aligned}
\lim_{k\to\infty}\frac{d}{dt}f^k(\tau_k t_k) &= \nabla_x f(x^0, y^0)\hat{d} - \nabla_y f(x^0, y^0)r^0 = \\
&\quad \nabla_x f(x^0, y^0)(\alpha\overline{d} + d^0) - \nabla_y f(x^0, y^0)r^0.
\end{aligned}
$$

Now, letting $\alpha \to 0$ and using that d^0 is an optimal solution to (4.52) for $r = -r^0$ we get

$$
\begin{aligned}
D_+\varphi(y^0; r^0) &\geq - \max_{(\lambda,\mu)\in\Lambda(x^0,y^0)} \nabla_y L(x^0, y^0, \lambda, \mu)(-r^0) \\
&= \min_{(\lambda,\mu)\in\Lambda(x^0,y^0)} \nabla_y L(x^0, y^0, \lambda, \mu)r^0.
\end{aligned}
$$

To show the Theorem we prove that the infimum in

$$\inf_{x\in\Psi(y^0)}\min_{(\lambda,\mu)\in\Lambda(x^0,y^0)} \nabla_y L(x^0, y^0, \lambda, \mu)r^0$$

is attained. By Theorem 4.2 the mapping $\Lambda(\cdot, \cdot)$ is upper semicontinuous. Hence, by continuity of the function $\nabla_y L(\cdot, \cdot, \cdot, \cdot)$ the function $\kappa(x) = \min_{(\lambda,\mu)\in\Lambda(x,y^0)} \nabla_y L(x, y^0, \lambda, \mu)r^0$ is lower semicontinuous [17] and the infimum is attained. $\square$

PROOF OF THEOREM 4.16: By convexity and regularity, $\overline{x} \in \Psi(\overline{y})$ if and only if the saddle point inequality

$$L(\overline{x}, \overline{y}, \lambda, \mu) \leq L(\overline{x}, \overline{y}, \overline{\lambda}, \overline{\mu}) \leq L(x, \overline{y}, \overline{\lambda}, \overline{\mu}) \ \forall x \in \mathbb{R}^n, \lambda \geq 0, \mu \in \mathbb{R}^q$$

for each $(\overline{\lambda}, \overline{\mu}) \in \Lambda(\overline{x}, \overline{y})$. Moreover, $\Lambda(x, \overline{y})$ is independent of $x \in \Psi(y^0)$ (see e.g. [30]). Let $r^0 \in \mathbb{R}^m$ be fixed, $\{t_k\}_{k=1}^{\infty}$ be any sequence converging to zero from above and set $y^k = y^0 + t_k r^0$ for all k. Let $x^0 \in \Psi(y^0)$, $(\lambda^0, \mu^0) \in \Lambda(x^0, y^0)$, $x^k \in \Psi(y^k)$, and $(\lambda^k, \mu^k) \in \Lambda(x^k, y^k)$ for all k which exist by conditions (C) and (MFCQ). Using the saddle point inequalities at the points $(\overline{x}, \overline{y}, \overline{\lambda}, \overline{\mu}) = (x^0, y^0, \lambda^0, \mu^0)$ and $(\overline{x}, \overline{y}, \overline{\lambda}, \overline{\mu}) = (x^k, y^k, \lambda^k, \mu^k)$ we derive

$$
\begin{aligned}
\varphi(y^k) - \varphi(y^0) &\leq L(x^0, y^k, \lambda^k, \mu^k) - L(x^0, y^0, \lambda^k, \mu^k) \\
&= \nabla_y L(x^0, y^0, \lambda^k, \mu^k)(y^k - y^0) + o(\|y^k - y^0\|),
\end{aligned}
$$

where $\lim_{k\to\infty} \frac{o(\|y^k-y^0\|)}{\|y^k-y^0\|} = 0$. Due to upper semicontinuity of the Lagrange multiplier set mapping (Theorem 4.2) this implies that we have

$$\lim_{k\to\infty} (t_k^{-1})[\varphi(y^k) - \varphi(y^0)] \leq \max_{(\lambda,\mu)\in\Lambda(x^0,y^0)} \{\nabla_y L(x^0, y^0, \lambda, \mu)r^0\}$$

for each $x^0 \in \Psi(y^0)$ or

$$\lim_{k\to\infty} (t_k^{-1})[\varphi(y^k) - \varphi(y^0)] \leq \min_{x^0\in\Psi(y^0)} \max_{(\lambda,\mu)\in\Lambda(x^0,y^0)} \{\nabla_y L(x^0, y^0, \lambda, \mu)r^0\}.$$

In the same way we derive the second inequality:

$$\begin{aligned} \varphi(y^k) - \varphi(y^0) &\geq L(x^k, y^k, \lambda^0, \mu^0) - L(x^k, y^0, \lambda^0, \mu^0) \\ &= \nabla_y L(x^k, y^0, \lambda^0, \mu^0)(y^k - y^0) + o(\|y^k - y^0\|) \end{aligned}$$

which by upper semicontinuity of the global optimal solution set mapping (see Theorem 4.3) implies

$$\lim_{k\to\infty} (t_k^{-1})[\varphi(y^k) - \varphi(y^0)] \geq \min_{x^0\in\Psi(y^0)} \{\nabla_y L(x^0, y^0, \lambda^0, \mu^0)r^0\}$$

for each $(\lambda^0, \mu^0) \in \Lambda(x^0, y^0)$. Hence,

$$\lim_{k\to\infty} (t_k^{-1})[\varphi(y^k) - \varphi(y^0)] \geq \max_{(\lambda,\mu)\in\Lambda(x^0,y^0)} \min_{x^0\in\Psi(y^0)} \{\nabla_y L(x^0, y^0, \lambda, \mu)r^0\}.$$

Since both $\Psi(y^0)$ and $\Lambda(x^0, y^0)$ are compact this implies that the directional derivative of $\varphi(\cdot)$ exists as well as the desired result. $\square$

PROOF OF THEOREM 4.17: Let $\zeta \in \partial\varphi(y^0)$ be such that there exists a sequence $\{y^k\}_{k=1}^\infty$ converging to y^0 such that $\nabla\varphi(y^k)$ exists for all k and $\lim_{k\to\infty} \nabla\varphi(y^k) = \zeta$. Note that by Definition 4.7 the generalized gradient $\partial\varphi(y^0)$ is equal to the convex hull of all such points ζ. Then, by Theorem 4.15, for each direction r we have

$$\nabla\varphi(y^k)r = D^+\varphi(y^k; r) \leq \nabla_y L(x^k, y^k, \lambda^k, \mu^k)r$$

for some $x^k \in \Psi(y^k)$, $(\lambda^k, \mu^k) \in \Lambda(x^k, y^k)$. By upper semicontinuity of $\Psi(\cdot)$ and $\Lambda(\cdot, \cdot)$ (cf. Theorems 4.2 and 4.3) there exist $x^0 \in \Psi(y^0)$ and $(\lambda^0, \mu^0) \in \Lambda(x^0, y^0)$ such that

$$\begin{aligned} \zeta r &\leq \nabla_y L(x^0, y^0, \lambda^0, \mu^0)r \\ &\leq \max\{\eta r : \eta \in \bigcup_{x\in\Psi(y^0)} \bigcup_{(\lambda,\mu)\in\Lambda(x,y^0)} \nabla_y L(x, y^0, \lambda, \mu)\}. \end{aligned}$$

It is easy to see that for arbitrary sets $A, B \subseteq \mathbb{R}^m$,

$$\max\{\alpha r : \alpha \in A\} \leq \max\{\beta r : \beta \in B\} \; \forall r \in \mathbb{R}^m$$

if and only if $A \subseteq \text{cl conv } B$. Hence

$$\zeta \in \text{cl conv} \bigcup_{x \in \Psi(y^0)} \bigcup_{(\lambda, \mu) \in \Lambda(x, y^0)} \nabla_y L(x, y^0, \lambda, \mu)$$

and

$$\partial \varphi(y^0) \subseteq \text{cl conv} \bigcup_{x \in \Psi(y^0)} \bigcup_{(\lambda, \mu) \in \Lambda(x, y^0)} \nabla_y L(x, y^0, \lambda, \mu)$$

by convexity of the right–hand side.

The set $\bigcup_{x \in \Psi(y^0)} \bigcup_{(\lambda, \mu) \in \Lambda(x, y^0)} \nabla_y L(x, y^0, \lambda, \mu)$ is compact [261]. Hence, its convex hull is closed and the theorem follows. □

PROOF OF COROLLARY 4.5: From Corollary 4.4 we have

$$\begin{aligned}
(-\varphi)'(y^0; -r) &= -\min\{-\nabla_y L(x, y^0, \lambda^0, \mu^0) r : x \in \Psi(y^0)\} \\
&\leq (-\varphi)^\circ(y^0; -r) = \max_{\zeta \in \partial(-\varphi)(y^0)} \{-\zeta r\} = \max_{\zeta \in \partial \varphi(y^0)} \{\zeta r\}.
\end{aligned}$$

This implies

$$\max\{\zeta r : \zeta \in \bigcup_{x \in \Psi(y^0)} \nabla_y L(x, y^0, \lambda^0, \mu^0)\} \leq \max\{\zeta r : \zeta \in \partial \varphi(y^0)\}.$$

Hence,

$$\bigcup_{x \in \Psi(y^0)} \nabla_y L(x, y^0, \lambda^0, \mu^0) \subseteq \text{cl conv } \partial \varphi(y^0).$$

Since $\partial \varphi(y^0)$ is a closed and convex set by definition we derive

$$\text{conv} \bigcup_{x \in \Psi(y^0)} \nabla_y L(x, y^0, \lambda^0, \mu^0) \subseteq \partial \varphi(y^0)$$

which together with Theorem 4.17 implies the desired result. □

PROOF OF COROLLARY 4.6: First, by Theorem 4.14 the optimal value function is locally Lipschitz continuous and we have the implication

$$\partial \varphi(y^0) \subseteq \text{conv} \{\nabla_y L(x^0, y^0, \lambda, \mu) : (\lambda, \mu) \in \Lambda(x^0, y^0)\}$$

by Theorem 4.17. Since the image of a convex set via a linear function is convex, we can drop the convex hull operation in this inclusion. By Theorem 4.16 the function $\varphi(\cdot)$ is directionally differentiable and we have

$$\varphi'(y^0; r) = \max_{(\lambda, \mu) \in \Lambda(x^0, y^0)} \nabla_y L(x^0, y^0, \lambda, \mu) r \leq \max_{v \in \partial \varphi(y^0)} vr = \varphi^\circ(y^0; r)$$

for all directions r by the definition of the Clarke directional derivative. By the properties of support functions [124] this implies that

$$\bigcup_{(\lambda,\mu)\in\Lambda(x^0,y^0)} \nabla_y L(x^0, y^0, \lambda, \mu) \subseteq \partial\varphi(y^0).$$

□

Before proving the next Corollary we need

LEMMA 4.4 ([127]) *Consider a convex optimization problem*

$$\min_x \{f(x, y) : x \in X\},$$

where the function f is jointly convex in both x, y and differentiable and X is a convex set. Then, $\nabla_y f(x^1, y) = \nabla_y f(x^2, y)$ for all optimal solutions x^1, x^2.

PROOF Let x^1, x^2 be different optimal solutions of the parametric optimization problem $\min_x \{f(x, y) : x \in X\}$. Let $\alpha \in (0, 1)$. Then, by convexity of X, the point $x^1 + \alpha(x^2 - x^1)$ is also an optimal solution. This implies

$$0 = \nabla_x f(x^1, y)(x^2 - x^1) = \nabla_x f(x^2, y)(x^2 - x^1). \qquad (4.53)$$

Let $\nabla_y f(x^1, y) \neq \nabla_y f(x^2, y)$. Then there is a vector r with $\|r\| = 1$ and

$$\nabla_y f(x^1, y)r < \nabla_y f(x^2, y)r.$$

Then, by first-order Taylor's expansion

$$f(x^1 + \alpha(x^2 - x^1), y + \alpha r)$$
$$= f(x^1, y) + \alpha \left(\nabla_x f(x^1, y)(x^2 - x^1) + \nabla_y f(x^1, y)r \right) + o^1(\alpha)$$

and

$$f(x^2 - \alpha(x^2 - x^1), y - \alpha r)$$
$$= f(x^2, y) - \alpha \left(\nabla_x f(x^2, y)(x^2 - x^1) + \nabla_y f(x^2, y)r \right) + o^2(\alpha)$$

with $\alpha^{-1} o^j(\alpha)$ tending to zero for α converging to zero. Now, using (4.53) we derive

$$f(x^1 + \alpha(x^2 - x^1), y + \alpha r) - f(x^1, y) + f(x^2 - \alpha(x^2 - x^1), y - \alpha r)$$
$$- f(x^2, y) = \alpha \left(\nabla_y f(x^1, y)r - \nabla_y f(x^2, y)r \right) + o^1(\alpha) + o^2(\alpha) < 0$$

for sufficiently small $\alpha > 0$. Hence,

$$0.5\left(f(x^1, y) + f(x^2, y)\right)$$
$$> \quad 0.5\left(f(x^1 + \alpha(x^2 - x^1), y + \alpha r) + f(x^2 - \alpha(x^2 - x^1), y - \alpha r)\right)$$
$$\geq \quad f(0.5(x^1 + x^2), y)$$

where convexity is used for the second inequality. But then, the first inequality contradicts optimality of the points x^1, x^2. □

PROOF OF COROLLARY 4.7: Let $(\lambda^1, \mu^1) \in \Lambda(x^1, y^0)$ and $(\lambda^2, \mu^2) \in \Lambda(x^2, y^0)$. Then,

$$L(x^1, y^0, \lambda^2, \mu^2) \leq L(x^1, y^0, \lambda^1, \mu^1) \leq L(x^2, y^0, \lambda^1, \mu^1)$$

and

$$L(x^2, y^0, \lambda^1, \mu^1) \leq L(x^2, y^0, \lambda^2, \mu^2) \leq L(x^1, y^0, \lambda^2, \mu^2)$$

by the saddle point inequalities for regular convex optimization problems. This implies that $\Lambda(x^1, y^0) = \Lambda(x^2, y^0)$ for all $x^1, x^2 \in \Psi(y^0)$. Hence the equation $L(x^1, y^0, \lambda^1, \mu^1) = L(x^2, y^0, \lambda^1, \mu^1)$ holds for all $x^1, x^2 \in \Psi(y^0)$ and $(\lambda^1, \mu^1) \in \Lambda(x^2, y^0)$. By Lemma 4.4 this implies that $\nabla_y L(x^1, y^0, \lambda^1, \mu^1) = \nabla_y L(x^2, y^0, \lambda^1, \mu^1)$ for all $(\lambda^1, \mu^1) \in \Lambda(x^1, y^0)$ and for all $x^1, x^2 \in \Psi(y^0)$. Since

$$\varphi'(y^0; r) = \min_{x \in \Psi(y^0)} \max_{(\lambda, \mu) \in \Lambda(x, y^0)} \nabla_y L(x, y^0, \lambda, \mu) r$$
$$= \max_{(\lambda, \mu) \in \Lambda(x, y^0)} \min_{x \in \Psi(y^0)} \nabla_y L(x, y^0, \lambda, \mu) r$$

this implies

$$\varphi'(y^0; r) = \max_{(\lambda, \mu) \in \Lambda(x, y^0)} \nabla_y L(x, y^0, \lambda, \mu) r.$$

As in the proof of Corollary 4.6 this implies the desired equation.   □

Chapter 5

OPTIMALITY CONDITIONS

In this chapter we start with the investigation of the bilevel programming problem in its general formulation. Our first topic will be the notion of an optimal solution. In general programming, a point is called a local optimal solution if there is no better feasible point with respect to the objective function in a certain sufficiently small neighborhood of this point. As pointed out in Section 3.3, in bilevel programming we have to distinguish at least between problems with unique lower level solutions for all parameter values and such problems not having this property. While in the first case the usual optimality notion can be used, the second case calls for the definition of an auxiliary problem. Two such problems will be considered here: one reflecting an optimistic position of the leader and the other modelling the leader's task to bound the damage resulting from an undesirable choice of the follower if the latter cannot be forced to cooperate with the leader.

After having defined the notion of an optimal solution, we describe several different necessary and sufficient optimality conditions using more or less restrictive assumptions. The first set of such conditions is based in the restrictive conditions guaranteeing that the lower level problem possesses a locally Lipschitz continuous optimal solution. The resulting conditions are of a combinatorial nature and use the (non-convex) contingent cone to the feasible set of the bilevel programming problem. Dualizing this cone leads to a generalized equation describing an equivalent necessary optimality condition.

After that we will reformulate the bilevel problem as a one-level one by the help of the Karush-Kuhn-Tucker conditions applied to the lower level problem. Note that both problems are equivalent in the case when the lower level problem is a convex regular problem, but only for the

119

optimistic case and only if optimistic optimality is defined via a problem similar to (3.7). The relations between the original bilevel problem and its reformulations will be discussed. We will start our investigations with a closer look on possible regularity conditions. It is shown that the (MFCQ) cannot be valid at any feasible point if a differentiable version of the KKT reformulation is used but the (LICQ) is a generic assumption if a nonsmooth version is applied. The resulting optimality conditions are of the F. John resp. the Karush-Kuhn-Tucker types.

Using a reformulation of the bilevel programming problem via the optimal value function of the lower level problem, the application of nonsmooth analysis enables us again to describe KKT type necessary optimality conditions. They also apply to the optimistic auxiliary problem.

The last approach via generalized PC^1-functions applies to both the optimistic and the pessimistic auxiliary problems but supposes that all optimal solutions in the lower level problem are strongly stable.

We will remark an approach to optimality conditions via the use of Mordukhovich's generalized gradient investigated e.g. in the papers [227, 295, 296, 301] which is not touched here. We will only consider first order optimality conditions here. Necessary and sufficient optimality conditions using second order information can e.g. be found in [94, 188, 249, 297].

This introduction is closed with a remark to a related optimization problem with set-valued objective functions. Let $M \subseteq \mathbb{R}^n$ be a nonempty set, K be a convex and pointed cone in $\mathbb{R}^m$ and consider the point-to-set mapping $F : M \to 2^{\mathbb{R}^m}$. Then, in [32, 132] the set-valued optimization problem

$$F(y) \to \text{``} \min_{y \in M} \text{''}$$

is considered. For this problem, a point $y \in M$ is called optimal, if it is a minimal point of the set $F(M) := \bigcup_{y \in M} F(y)$ with respect to the cone K, i.e. if

$$((F(M) - y) \cap (-K) = \{0\},$$

where $F(M) - y := \{z : z = x - y \text{ for some } x \in F(M)\}$ denotes the Minkowski sum. Let $m = 1$, then this optimality notion coincides with the optimistic approach to solving a bilevel programming problem with a mapping

$$F(y) = \bigcup_{x \in \Psi(y)} \{f(x, y)\}.$$

Necessary and sufficient optimality conditions for this problem have been described in [32, 132] by use of the contingent epiderivative for the mapping $F(\cdot)$. For most of these results, the point-to-set mapping $F(\cdot)$ needs to be K-convex, i.e. the epigraph of $F(\cdot)$ is convex in the sense

$$\lambda F(x^1) + (1 - \lambda)F(x^2) \subseteq F(\lambda x^1 + (1 - \lambda)x^2) + K.$$

Unfortunately, this is in general not valid for bilevel programming and, due to the inherent properties of bilevel programming is seems also to be difficult to find necessary conditions guaranteeing this property.

5.1   OPTIMALITY CONCEPTS

As in Chapter 4 let

$$\Psi(y) = \operatorname*{Argmin}_{x} \{f(x,y) : g(x,y) \le 0, \ h(x,y) = 0\}$$

denote the solution set mapping of a smooth parametric optimization problem

$$\min_{x}\{f(x,y) : g(x,y) \le 0, \ h(x,y) = 0\}. \tag{5.1}$$

This problem is called the *lower level problem* or the *follower's problem*. Using $\Psi(\cdot)$ the bilevel programming problem can be defined as

$$\text{``}\min_{y}\text{''}\{F(x,y) : x \in \Psi(y), \ y \in Y\}, \tag{5.2}$$

where $F : \mathbb{R}^n \times \mathbb{R}^m \to \mathbb{R}$, $Y \subseteq \mathbb{R}^m$, is a closed set. This problem is also called the *upper level problem* or the *leader's problem*. At many places the set Y is given by explicit inequality constraints: $Y = \{y : G(y) \le 0\}$, where $G : \mathbb{R}^m \to \mathbb{R}^l$. We will not use explicit equality constraints for simplicity of writing (such constraints can easily be added) and we will not consider problems with coupling upper level constraints. The reason for the latter can be seen in Section 3.2. Again, the quotation marks are used to express the uncertainity in the definition of the bilevel programming problem in the case of nonuniquely determined lower level optimal solutions. The following two examples are used to illustrate the difficulties arising in this case.

Example: [186] Consider the following convex parametric optimization problem

$$\Psi(y) = \operatorname*{Argmin}_{x} \{-xy : 0 \le x \le 1\}$$

and the bilevel programming problem

$$\text{``}\min_{y}\text{''}\{x^2 + y^2 : x \in \Psi(y), \ -1 \le y \le 1\}.$$

Then,

$$\Psi(y) = \begin{cases} \{0\}, & \text{if } y < 0, \\ \{1\}, & \text{if } y > 0, \\ [0,1], & \text{if } y = 0. \end{cases} \qquad F(x(y),y) \begin{cases} = y^2, & \text{if } y < 0, \\ = 1 + y^2, & \text{if } y > 0, \\ \in [0,1], & \text{if } y = 0. \end{cases}$$

The mapping $F(x(y), y)$ is plotted in fig. 5.1. In this figure it can be seen that the upper level objective function value is unclear unless the follower has announced his selection to the leader.

Solvability of the resulting problem of the leader depends on this reaction of the follower: the upper level problem is solvable only in the case when the follower selects $x(0) = 0 \in \Psi(0)$. Hence, the infimal function value 0 of the bilevel programming problem is not attained if the follower does not take $x(0) = 0$.

The notion of an optimal value is also not clear: There are choices for y leading to upper level objective function values sufficiently close to zero, but it is not clear whether the value zero can be attained. It is also not possible to overcome this difficulty if the follower is allowed to have a free choice and the leader only wants to be in a position where he is able to observe all the selections of the follower. In this case, if the follower takes $0.5 \in \Psi(0)$, then the problem in the above example does also not have a solution.   □

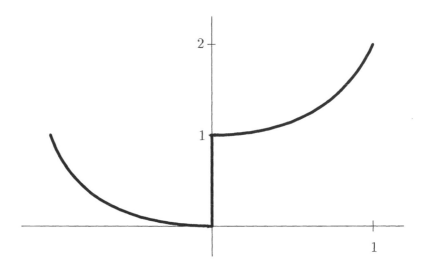

Figure 5.1.   A typical mapping to be minimized if the lower level has a non-unique optimal solution

To define different notions of optimal solutions we start with the easiest case when the lower level optimal solution is uniquely determined for all parameter values. Then, we can adopt the usual notions from mathematical optimization to bilevel programming:

DEFINITION 5.1 *Let $\Psi(y)$ consist of at most one point for all values of $y \in Y$. Then, a point $(x^*, y^*) \in \mathbb{R}^n \times \mathbb{R}^m$ is called locally optimal for problem (5.2) if $y^* \in Y$, $x^* \in \Psi(y^*)$ and there exists an open neighborhood $U_\delta(y^*), \delta > 0$, with $F(x, y) \geq F(x^*, y^*)$ for all (x, y) satisfying $y \in Y \cap U_\delta(y^*)$, $x \in \Psi(y)$. It is called a global optimal solution if $\delta = \infty$ can be selected.*

In this case, the quotation marks in (5.2) can be dropped. Definition 5.1 describes the case when the optimal solution $x(y)$ of the follower can be predicted by the leader. Hence, he has to solve the problem

$$\min_y \{F(x(y), y) : y \in Y\},$$

which is a continuous optimization problem provided that the lower level optimal solution function is continuous at all points $y \in Y$. Then we have the following

THEOREM 5.1 *Consider the bilevel programming problem (5.1), (5.2) and let the assumptions (C) and (MFCQ) at all points $(x, y) \in \mathbb{R}^n \times Y$ with $x \in M(y)$ be satisfied. Moreover, let (5.1) have a unique optimal solution for each $y \in Y$. Then, the bilevel programming problem has a global optimal solution provided it has a feasible solution.*

Consider now the case when the lower level optimal solution is not uniquely determined. In most of the publications in which the unique solvability of the lower level problem is not assumed, either an optimistic (cf. e.g. [24, 43, 66, 117]) or a pessimistic position (cf. e.g. [181, 186]) is applied. Both approaches lead to three–level problems.

If the leader is able to persuade the follower to select that global optimal solution which is the best one from the leader's point of view then he has to solve the problem

$$\min_y \{\varphi_o\{y\} : y \in Y\}, \tag{5.3}$$

where

$$\varphi_o(y) = \min_x \{F(x, y) : x \in \Psi(y)\}. \tag{5.4}$$

This motivates the notion of an *optimistic solution* which is a point $y^* \in \mathbb{R}$ solving the problem (5.3):

DEFINITION 5.2 *A point* $(x^*, y^*) \in \mathbb{R}^n \times \mathbb{R}^m$ *is called a local optimistic solution for problem (5.2) if* $y^* \in Y$, $x^* \in \Psi(y^*)$ *with*

$$F(x^*, y^*) \le F(x, y^*) \ \forall x \in \Psi(y^*)$$

and there exists an open neighborhood $U_\delta(y^*), \delta > 0$, *with*

$$\varphi_o(y^*) \le \varphi_o(y) \ \forall y \in Y \cap U_\delta(y^*).$$

It is called a global optimistic solution if $\delta = \infty$ *can be selected.*

Note that this definition is slightly different from the idea behind problem (3.7). From the leader's point of view it seems to be more appropriate to use this definition since the leader has only control over y. We will come back to this in Theorem 5.15.

As already pointed out in [43] while investigating linear bilevel programming problems resulting from certain economic situations, under suitable assumptions the optimistic position can be realized if e.g. the follower is participated in the profit of the leader. In the context of closed-loop Stackelberg games, the problem of finding a decision for the leader guaranteeing that the follower will agree him to use the optimistic approach has been considered also in [212].

If the assumptions in Theorem 4.3 are satisfied, then the set

$$\{(x, y) : x \in \Psi(y)\}$$

is closed and the intersection of this set with $\mathbb{R}^n \times Y$ is compact if Y is closed due to assumption (C). This implies that the minimum in the problem

$$\min_{x,y}\{F(x, y) : x \in \Psi(y), \ y \in Y\} \tag{5.5}$$

is attained. Thus, the global optimal values in (5.3) and (5.5) coincide [187]. This is in general not true for local optimal solutions. Since the lower level problem (5.1) is equivalent to its KKT conditions provided that it is a convex and regular one, this can be seen in Theorem 5.15 and the subsequent example. Nevertheless, equivalence of problems (5.3) and (5.5) with respect to global optimal solutions implies

THEOREM 5.2 *Let the assumptions (C) and (MFCQ) at all points* $(x, y) \in \mathbb{R}^n \times Y$ *with* $x \in M(y)$ *be satisfied. Then, a global optimistic solution of the bilevel programming problem (5.1), (5.3), (5.4) exists provided there is a feasible solution.*

The existence of optimistic optimal solutions has also been investigated in [117, 301] under somewhat slightly weaker assumptions. We will call problem (5.1), (5.3), (5.4) the *optimistic bilevel problem* in what follows.

The optimistic position seems not to be possible without any trouble at least in the cases when cooperation is not allowed (e.g. in the matter of legislation), when cooperation is not possible (e.g. in games of a human being against nature), or if the follower's seriousness of keeping the agreement is not granted. Then, when the leader is not able to influence the follower's choice, he has the way out of this unpleasant situation to bound the damage resulting from an undesirable selection of the follower. This leads to the problem

$$\min_{y}\{\varphi_p(y) : y \in Y\}, \tag{5.6}$$

where

$$\varphi_p(y) = \max_{x}\{F(x,y) : x \in \Psi(y)\}. \tag{5.7}$$

DEFINITION 5.3 *A point* $(x^*, y^*) \in \mathbb{R}^n \times \mathbb{R}^m$ *is called a local pessimistic solution for problem (5.2) if* $y^* \in Y$, $x^* \in \Psi(y^*)$ *with*

$$F(x^*, y^*) \geq F(x, y^*) \ \forall x \in \Psi(y^*)$$

and there exists an open neighborhood $U_\delta(y^*), \delta > 0$, *with*

$$\varphi_p(y^*) \leq \varphi_p(y) \ \forall y \in Y \cap U_\delta(y^*).$$

It is called a global pessimistic solution if $\delta = \infty$ *can be selected.*

Problem (5.1), (5.6), (5.7) is called the *pessimistic bilevel problem* in the sequel. The pessimistic approach for solving bilevel programming problems with non-unique lower level solutions has been intensively investigated by P. Loridan and J. Morgan and their co-authors. In most cases, they consider a slightly more general problem than we did in the sense that also the functions f, g, h, F, G, H describing the problem are perturbed and convergence to the unperturbed ones is investigated (see e.g. [179], [180], [181]). In [184] the concept of strict ε-optimal solutions in the lower level is used to regularize the pessimistic bilevel problem. A comparison of both the pessimistic and the optimistic approaches is given in [185].

THEOREM 5.3 *Consider the bilevel programming problem (5.1), (5.6) (5.7). Let the point-to-set mapping* $\Psi(\cdot)$ *be lower semicontinuous at all points* $y \in Y$ *and assume (C) to be satisfied. Then, a global pessimistic solution exists provided that problem (5.6) has a feasible solution.*

A discussion of this result especially with respect to the possibilities to satisfy the assumptions (which seem to be very restrictive) can be found in [250]. It should be noted that a pessimistic solution exists generically [147] which means that for almost all bounded functions F, f, g, h the problem (5.6) has a uniquely determined global optimal solution.

Example: With respect to the last example, the situation is as follows:

$$\varphi_p(y) = \left\{ \begin{array}{ll} y^2, & \text{if } y > 0, \\ 1 + y^2, & \text{if } y \leq 0, \end{array} \right. \qquad \varphi_o(y) = \left\{ \begin{array}{ll} y^2, & \text{if } y \geq 0, \\ 1 + y^2, & \text{if } y < 0, \end{array} \right.$$

Hence, the optimal solution of the upper level problem exists if the optimistic position is used while it does not exist in the pessimistic one.

$\square$

Being aware of the difficulties arising when global optimal solutions of non-convex optimization problems are to be computed, some authors [131, 182, 250] use optimality definitions which determine whole sets of optimal solutions. We can define the following set

$$S := \{(x, y) : y \in Y, x \in \Psi(y), F(x, y) \leq \inf_z \varphi_p(z)\} \qquad (5.8)$$

as a set of optimal solutions of the bilevel programming problem. We will call the points in this set *lower optimal solutions* since in the case of the Stackelberg game a similar set has been called "set of lower Stackelberg equilibrium points" [182]. Figure 5.2 can be used to illustrate the different notions of optimal solutions in the case of non-unique lower level solutions. In this figure, the pessimistic solution is denoted by y_p^*, while y_o^* is the optimistic solution. The set S of optimal solutions is given as the set of points (x, y) with $y \in \text{Proj } S$ and corresponding values for $x \in \Psi(y)$ such that $F(x, y) \leq \varphi_p(y_p^*)$.

The optimality definition in the sense S has the drawback that the upper bound for the values of $F(x, y)$ will in general not be attained unless the function $\varphi_p(\cdot)$ is lower semicontinuous. To overcome this difficulty we can try to replace this function by its lower envelope

$$\varphi_p^0(y) := \liminf_{z \to y} \varphi_p(z).$$

The function $\varphi_p^0(\cdot)$ is lower semicontinuous on its domain and coincides with $\varphi_p(\cdot)$ at all points where the second function is lower semicontinuous. The relations between both functions as well as some possibilities for the computation of at least an approximation of the values of $\varphi_p^0(\cdot)$ have been investigated in [250]. Clearly, optimal solutions in the sense S exist whenever optimistic solutions exist.

We will shortly touch another difficulty closely connected with non-uniqueness of the lower level solution in the case when there are also

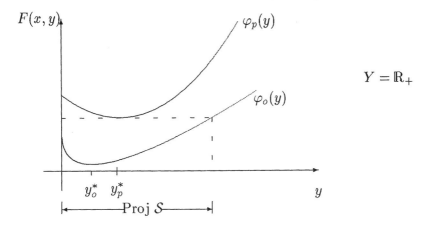

Figure 5.2. The different optimality definitions in bilevel programming

upper level constraints of the kind $G(x, y) \leq 0$ depending on the lower level optimal solution. Constraints of this type are some sort of an a-posteriori feasibility test for the leader's selection. If the lower level optimal solution is uniquely determined for all possible choices of the parameter, then feasibility of that choice can be verified easily by calculating the values of the left-hand side functions in the upper level constraints. But, if the optimal solution in the lower level is not uniquely determined then, feasibility of the parameter chosen depends on the concrete choice of an optimal solution in the lower level. Even if the leader knows the solution set mapping $\Psi(\cdot)$ he is not able to predict feasibility of his selection before the follower's selection is known. This is a very difficult situation which will not be investigated here.

But also in case of a uniquely determined lower level optimal solution for all parameter values, the appearance of connecting upper level constraints extremely complicates the problem. The following example shows that the position of connecting constraints in the two levels of hierarchy is important.

Example: [76] Consider the problem

$$\min_{y}\{x^2 + y : -x - y \leq 0, \ x \in \Psi(y)\},$$

where

$$\Psi(y) = \underset{x}{\text{Argmin}} \ \{x : x \geq 0\}.$$

Then, the optimal solution of this problem is equal to $x = y = 0$. But, if the upper level constraint $-x - y \leq 0$ is shifted into the lower level, we get the optimal solution $x = 0.5$, $y = -0.5$. On the other hand, if the lower level constraint $x \geq 0$ is shifted into the upper level, then the lower level problem has no optimal solution at all for arbitrary selection of y, i.e. the problem becomes unsolvable.   □

We will close this section with one further example showing some surprising property of bilevel programming problems: The optimal solution is in general not independent of inactive constraints (in the lower level problem).

Example: [189] Consider the following bilevel programming problem

$$\min_{y}\{(x - 1)^2 + (y - 1)^2 : x \in \Psi(y)\}$$

with

$$\Psi(y) = \operatorname*{Argmin}_{x}\ \{0.5x^2 + 500x - 50xy\}.$$

Then, this problem is equivalent to

$$\min_{x,y}\{(x - 1)^2 + (y - 1)^2 : x + 500 - 50y = 0\}.$$

The unique optimal solution of this problem is $(x^*, y^*) = (10.02, 0.82)$ with an optimal function value $F(x^*, y^*) = 81.33$.

Now add the simple lower level constraint $x \geq 0$ to the problem. Then, (x^*, y^*) is feasible for the resulting problem and $x^* > 0$. But, the optimal solution of the lower level problem is

$$x(y) = \begin{cases} 50y - 500 & \text{if } y \geq 10 \\ 0 & \text{if } y < 10. \end{cases}$$

Inserting this function into the upper level objective function yields

$$F(x(y), y) = \begin{cases} (50y - 501)^2 + (y - 1)^2 & \text{if } y \geq 10 \\ 1 + (y - 1)^2 & \text{if } y < 10. \end{cases}$$

The minimal value of this function is 1 which is attained for $(x^0, y^0) = (0, 1)$.   □

Necessary and sufficient conditions guaranteeing that a global (optimistic) optimal solution of a bilevel programming problem is independent of the addition or cancellation of inactive constraints can be found in [189].

5.2 OPTIMALITY CONDITIONS BASED ON STRONG STABILITY

If the lower level problem (5.1) has a unique optimal solution for all values of the parameter and this solution defines a continuous function $x(\cdot)$ over the feasible set Y of the upper level problem, then the bilevel problem (5.2) can be reformulated as

$$\min\{F(x(y), y) : G(y) \leq 0\}, \qquad (5.9)$$

if $Y = \{y : G(y) \leq 0\}, G : \mathbb{R}^m \to \mathbb{R}^l$. In what follows we will use an upper level regularity assumption:

(ULR) Given a feasible solution (x^0, y^0), the upper level regularity assumption is satisfied if the set

$$\{r : \nabla G_j(y^0)r < 0, \ j \in \{i : G_i(y^0) = 0\}\} \neq \emptyset.$$

Applying this idea the following necessary and sufficient optimality conditions using the directional derivative can be derived:

THEOREM 5.4 ([67]) *Let (x^0, y^0) be a local optimal solution of the bilevel programming problem (5.1), (5.2) and assume that the lower level problem (5.1) is a convex parametric optimization problem satisfying the conditions (MFCQ), (SSOC), and (CRCQ) at (x^0, y^0). Then the following optimization problem has a nonnegative optimal objective function value:*

$$
\begin{aligned}
\alpha &\to \min_{\alpha, r} \\
\nabla_x F(x^0, y^0)x'(y^0; r) + \nabla_y F(x^0, y^0)r &\leq \alpha \\
\nabla G_i(y^0)r &\leq \alpha, \forall \ i : G_i(y^0) = 0 \\
\|r\| &\leq 1.
\end{aligned}
\qquad (5.10)
$$

Moreover, if (ULR) is satisfied, problem (5.10) can be replaced by

$$
\begin{aligned}
\nabla_x F(x^0, y^0)x'(y^0; r) + \nabla_y F(x^0, y^0)r &\to \min_r \\
\nabla G_i(y^0)r &\leq 0, \ \forall \ i : G_i(y^0) = 0 \\
\|r\| &\leq 1.
\end{aligned}
\qquad (5.11)
$$

If the convexity assumption to the lower level problem is dropped the uniqueness of the global optimal solution of this problem is difficult to be guaranteed [141]. This can have tremendous implications on optimality conditions. We will come back to this in Sections 5.5 and 5.7. An analogous necessary and sufficient optimality condition for Mathematical Programs with Equilibrium Constraints can be found in [188]. Theorem 5.10 easily follows from directional differentiability of the lower level optimal solution (cf. Theorem 4.11) together with the formula for the directional derivative of composite functions. Hence, no explicit proof is necessary. Applying theorems on the alternative the first condition (5.10) can be posed as a kind of a Fritz John condition whereas (5.11) is equivalent to a Karush-Kuhn-Tucker type condition. But here the used multipliers depend on the direction as in [67] which is not convenient. Using the formulae for the computation of the directional derivative of the lower level solution function (cf. Theorem 4.11) we get a combinatorial optimization problem to be solved for evaluating the necessary optimality conditions:

$$\min\{v_\alpha(\lambda, \mu, K) : (\lambda, \mu) \in E\Lambda(x^0, y^0), J(\lambda) \subseteq K \subseteq I(x^0, y^0)\} \geq 0,$$
(5.12)

where $v_\alpha(\lambda, \mu, K)$ denotes the optimal objective function value of the following problem, where the abbreviation $z^0 = (x^0, y^0)$ is used:

$$\alpha \to \min_{\alpha, r, d, \gamma, \eta}$$

$$\nabla_x F(z^0)d + \nabla_y F(z^0)r \leq \alpha$$

$$\nabla G_i(y^0)r \leq \alpha, \forall\, i : G_i(y^0) = 0$$

$$\nabla^2_{xx}L(z^0, \lambda, \mu)d + \nabla^2_{yx}L(z^0, \lambda, \mu)r + \nabla^\top_x g(z^0)\gamma + \nabla^\top_x h(z^0)\eta = 0$$

$$\nabla_x g_i(z^0)d + \nabla_y g_i(z^0)r = 0,\ i \in K,$$

$$\nabla_x g_i(z^0)d + \nabla_y g_i(z^0)r \leq 0,\ i \in I(z^0) \setminus K,$$

$$\nabla_x h_j(z^0)d + \nabla_y h_j(z^0)r = 0,\ j = 1, \ldots, q,$$

$$\gamma_i \geq 0,\ i \in K \setminus J(\lambda),\ \gamma_i = 0,\ i \notin K,$$

$$\|r\| \leq 1.$$

Here again, the right-hand side of the second set of inequalities can be replaced by zero in the regular case and the family $\{K : J(\lambda) \subseteq K \subseteq I(x^0, y^0)\}$ is allowed to be reduced to $\mathcal{I}(\lambda)$.

The following simple example shows that the condition (5.12) is not a sufficient optimality condition for the bilevel problem:

Example: [67]: Consider the problem

$$\min_y\{(x + 4)^2 + (y - 3.5)^2 : x \in \Psi(y)\},$$
(5.13)

where

$$\Psi(y) = \operatorname*{Argmin}_x\ \{(x - 3)^2 : x^2 \leq y\}.$$
(5.14)

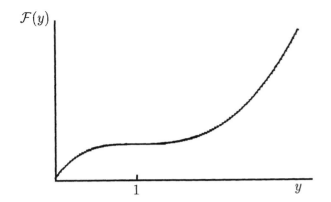

Figure 5.3. The function $\overline{F}(y)$.

Then, at $(x^0, y^0) = (1,1)$ we get $\Lambda(x^0, y^0) = \{2\}$, $K = \{1\}$ which implies

$$v_\alpha(2, 0, \{1\}) = \min\{10d - 5r : 6d + 2\gamma = 0, \ 2d - r = 0\} = 0.$$

This problem has the optimal solution

$$d = x'(1; r) = r/2, \ \gamma = -3r/2, \ \forall r.$$

On the other hand, the unique optimal solution of problem (5.14) is $x(y) = \sqrt{y}$ for $0 < y < 9$, which gives

$$\overline{F}(y) = F(x(y), y) = (\sqrt{y} + 4)^2 + (y - 3.5)^2.$$

This function is plotted in Fig. 5.3. Then,

$$\overline{F}'(1) = \overline{F}''(1) = 0, \overline{F}'''(1) = 3 \neq 0.$$

Hence, (x^0, y^0) is stationary but not locally optimal. In this example even the stronger (LICQ) is satisfied for the lower level problem. $\qquad\square$

It should be noted that the optimality conditions in [67] are obtained using the weaker assumption

(A) For each sequence $\{y^k\}_{k=1}^\infty$ converging to y^0 and each index set $K \subseteq I(x(y^k), y^k)$ for all k such that the gradients

$$\{\nabla_x g_i(x(y^k), y^k) : i \in K\} \cup \{\nabla_x h_j(x(y^k), y^k) \ \forall j\}$$

are linearly independent, also the gradients in

$$\{\nabla_x g_i(x(y^0), y^0) : i \in K\} \cup \{\nabla_x h_j(x(y^0), y^0) \ \forall j\}$$

are linearly independent.

This assumption together with (MFCQ) and (SSOC) guarantees that the directional derivative of the lower level problem (5.1) can be computed by means of the quadratic optimization problem $QP(\lambda, \mu, r)$ (cf. formula (4.20) in Theorem 4.11) for some optimal vertex (λ, μ) of problem (4.21). But in this case the local optimal solution function $x(\cdot)$ is in general not locally Lipschitz continuous and, consequently, its directional derivative is then not (Lipschitz) continuous with respect to variations of the direction. This implies that the cone

$$\{(d, r) : \nabla G_i(y^0) r \leq 0, \ \forall \ i : G_i(y^0) = 0, \ d = x'(y^0; r)\}$$

is in general not closed. Hence, some weaker necessary optimality conditions are obtained. This is illustrated in the following example:

Example: Consider the problem

$$x_2 + y_1^2 + y_2^2 \to \min$$

subject to

$$x \in \Psi(y), \ y_1^2 + (y_2 + 1)^2 - 1 \leq 0,$$

where

$$\Psi(y) = \underset{x}{\text{Argmin}} \ \left\{ \frac{1}{2}(x_1 - 1)^2 + \frac{1}{2}x_2^2 : x_1 \leq 0, \ x_1 + x_2 y_1 + y_2 \leq 0 \right\}$$

at $y^0 = (0, 0)^\top$, $x^0 = (0, 0)^\top \in \Psi(y^0)$. Note that $x_2 \equiv 0$ on the feasible set of the upper level. Hence, the objective function of the upper level problem is equal to

$$\mathcal{F}(x(y), y) = y_1^2 + y_2^2$$

on the feasible set and the point (x^0, y^0) is the unique global optimal solution. Here, straightforward calculations give

$$
\begin{aligned}
\Omega(r) \quad &:= \quad \{d : d \text{ equals the optimal solution of problem} \\
&\qquad QP(\lambda, \mu, r) \text{ for some vertex } (\lambda, \mu) \in \Lambda(x^0, y^0)\} \\[2mm]
&= \quad \begin{cases} \{(0, 0)^\top\}, & \text{if } r_2 < 0, \\ \{(-r_2, -r_1)^\top\}, & \text{if } r_2 > 0, \\ \{(0, 0)^\top, (0, -r_1)^\top\}, & \text{if } r_2 = 0. \end{cases}
\end{aligned}
$$

This set denotes some kind of a generalized directional derivative for point-to-set mappings, the so-called contingent derivative [10] or upper Dini derivative [79]. Knowing only the sets $\Omega(\cdot)$ it is not possible to detect the true directional derivative. This implies that (5.11) does not give a true necessary optimality condition for the bilevel programming problem under assumption (A) even in the regular case.

For the directional derivative of the lower level optimal solution we get $x'(y^0; r) = (0,0)^\top$ for $r = (r_1, 0)^\top$ for each r_1. This directional derivative can be used in this case for a valid necessary optimality condition. In this example, setting

$$H(y) = F(x(y), y),$$

we have

$$H'(y^0; r) = \begin{cases} 0, & r_2 \leq 0, \\ -r_1, & r_2 > 0. \end{cases}$$

Note, that the linearized upper level constraint at the point y^0 (i.e. the constraint in problem (5.11)) is $2r_2 \leq 0$.

On the other hand, for the *lower and the upper directional Dini derivatives* [282] of $H(\cdot)$ at $y = y^0$ into direction $r = (1, 0)^\top$ we get

$$
\begin{aligned}
H_+(y^0; r) &:= \liminf_{t \to 0+, r' \to r} t^{-1}[H(y^0 + tr') - H(y^0)] \\
&= \min\{\nabla_x F(x^0, y^0)d \mid d \in \Omega(r)\} + \nabla_y F(x^0, y^0)r = -1
\end{aligned}
$$

and

$$
\begin{aligned}
H^+(y^0; r) &:= \limsup_{t \to 0+, r' \to r} t^{-1}[H(y^0 + tr') - H(y^0)] \\
&= \max\{\nabla_x F(x^0, y^0)d \mid d \in \Omega(r)\} + \nabla_y F(x^0, y^0)r = 0.
\end{aligned}
$$

The necessity for the inclusion of the convergence $r' \to r$ into the definition of the upper and the lower Dini directional derivatives results from the missing Lipschitz continuity of the function H (cf. formulae (4.25) and (4.26)). A closer look at problems (5.10) and (5.11) shows that both problems use the lower directional Dini derivative. Hence, using (5.11) under assumption (A) is not possible since it does not detect the global minimum (even in the regular case). The use of the upper directional Dini derivative is of course possible but in general too weak for this problem. In [67] it is shown that the use of problem (5.10) leads to a valid necessary optimality condition also under assumption (A) □

The following more restrictive assumption gives a sufficient optimality condition of first order:

THEOREM 5.5 ([67]) *Let (x^0, y^0) be a feasible solution of the bilevel programming problem (5.1), (5.2) with a convex lower level problem. Assume that the lower level problem (5.1) satisfies the conditions (MFCQ), (SSOC), and (CRCQ) at (x^0, y^0). If the optimal function value v_1 of the problem*

$$
\begin{aligned}
\nabla_x F(x^0, y^0)x'(y^0; r) + \nabla_y F(x^0, y^0)r &\to \min \\
\nabla G_i(y^0)r &\leq 0, \forall\, i : G_i(y^0) = 0, \quad (5.15) \\
\|r\| &= 1
\end{aligned}
$$

is strongly greater than zero $v_1 > 0$, then (x^0, y^0) is a strict local optimal solution of the bilevel problem, i.e. for arbitrary $c \in (0, v_1)$ there is $\varepsilon > 0$ such that

$$F(x(y), y) \geq F(x^0, y^0) + c\|y - y^0\|$$

for all y satisfying $\|y - y^0\| \leq \varepsilon$, $G(y) \leq 0$.

Also the proof of this theorem is standard and will therefore be omitted (see e.g. [270]). Both Theorems 5.4 and 5.5 give conditions for the so-called Bouligand stationarity of feasible points (x^0, y^0) involving the directional derivative (or Bouligand derivative) of the optimal solution of the lower level problem.

DEFINITION 5.4 *Let $u : \mathbb{R}^p \to \mathbb{R}$ be a function and $\bar{z} \in \mathbb{R}^p$. If u is directionally differentiable, then $\bar{z}$ is called a Bouligand stationary point if $u'(\bar{z}; r) \geq 0$ for all directions $r \in \mathbb{R}^p$. If u is locally Lipschitz continuous, $\bar{z}$ is a Clarke stationary point, provided that $0 \in \partial u(\bar{z})$. If u is pseudodifferentiable, the point $\bar{z}$ is called pseudostationary if $0 \in \Gamma_u(z)$.*

All three notions of stationary points can similarly also be applied to constrained optimization problems. A Bouligand stationary point of a locally Lipschitz continuous and directionally differentiable function is also Clarke stationary, the opposite implication is in general not true. Also, a Clarke stationary point of a pseudodifferentiable function is a pseudostationary point and this is again in general not valid in the opposite direction.

Conditions using Clarke's generalized gradients of the objective function of the auxiliary problem are weaker ones than that of Theorems 5.4, 5.5 but can also be given by applying the results in Chapter 4:

THEOREM 5.6 *Let (x^0, y^0) be a local optimal solution to problem (5.1), (5.2) and assume that the assumptions (MFCQ), (SSOC), and (CRCQ) are satisfied for (5.1) at (x^0, y^0) as well as (ULR). Let the lower level problem be convex. Then, there exist a vector $\kappa \geq 0$ such that*

$$0 \in \nabla_x F(x^0, y^0)\partial x(y^0) + \nabla_y F(x^0, y^0) + \kappa^{\top} \nabla G(y^0),$$

$$\kappa^{\top} G(y^0) = 0.$$

This result is a straightforward implication of the Clarke necessary optimality conditions [61] and Theorems 4.6 and 4.10. It should be noted that the differentiability assumptions for the functions F, G_i can be weakened to Lipschitz continuity [61]. Moreover, Theorems 4.12 and 4.13 can be used to give more explicit results. Also, Corollary 4.1 can be used to derive a necessary optimality condition of first order involving the pseudodifferential of the lower level optimal solution function.

THEOREM 5.7 *Under the same assumptions as in Theorem 5.6 we get the existence of a vector $\kappa \geq 0$ such that*

$$0 \in \nabla_x F(x^0, y^0) \Gamma_x(y^0) + \nabla_y F(x^0, y^0) + \kappa^\top \nabla G(y^0),$$

$$\kappa^\top G(y^0) = 0.$$

Optimality conditions of second order based on the implicit function approach presented in this section can be found in [94]. Optimality conditions of second order for bilevel problems without lower level constraints can be found in [297].

5.3 THE CONTINGENT CONE TO THE FEASIBLE SET

As shown in [188, 229] the contingent cone to the feasible set

$$\mathcal{M} := \{(x, y) : x \in \Psi(y), G(y) \leq 0\}$$

is a polyhedral cone provided that certain regularity conditions are satisfied. Here, a *polyhedral cone* is defined as the union of a finite number of convex polyhedral cones.

The following results originate from [229] where they have been obtained for mathematical programs with equilibrium constraints. For introducing them we will start here with the restrictive assumption of a strongly stable lower level solution. This assumption will dropped later on. If the assumptions (MFCQ), (CRCQ), and (SSOC) are satisfied for the lower level problem (5.1) and (ULR) for the upper level problem (5.2) then this contingent cone is given as

$$\widehat{K}_\mathcal{M}(x^0, y^0) = \{(d, r) : d = x'(y^0; r), \nabla G_j(y^0) r \leq 0, j \in \{i : G_i(y^0) = 0\}\}.$$

Using the formulae for the computation of the directional derivative of the lower level local solution function $x(\cdot)$ given in Theorem 4.11 we derive that $\widehat{K}_\mathcal{M}(x^0, y^0)$ is equal to the projection of the following cone on $\mathbb{R}^n \times \mathbb{R}^m$:

$$K_\mathcal{M}(x^0, y^0) = \bigcup_{(\lambda, \mu) \in E\Lambda(x^0, y^0)} \bigcup_{I \in \mathcal{I}(\lambda)} K_{I, \lambda, \mu}(x^0, y^0), \qquad (5.16)$$

where $K_{I, \lambda, \mu}(x^0, y^0)$ is the solution set of the following system of equations and inequalities (setting $z^0 = (x^0, y^0)$):

$$\nabla G_i(y^0) r \leq 0, \forall\, i : G_i(y^0) = 0$$
$$\nabla^2_{xx} L(z^0, \lambda, \mu) d + \nabla^2_{yx} L(z^0, \lambda, \mu) r + \nabla^\top_x g(z^0) \gamma + \nabla^\top_x h(z^0) \eta = 0$$

$$\nabla_x g_i(z^0)d + \nabla_y g_i(z^0)r = 0, \ i \in I, \qquad (5.17)$$
$$\nabla_x g_i(z^0)d + \nabla_y g_i(z^0)r \leq 0, \ i \in I(z^0) \setminus I,$$
$$\nabla_x h_j(z^0)d + \nabla_y h_j(z^0)r = 0, \ j = 1, \ldots, q,$$
$$\gamma_i \geq 0, \ i \in I \setminus J(\lambda), \ \gamma_i = 0, \ i \notin I.$$

Now, drop the restrictive assumption, that the lower level optimal solution is strongly stable and consider the following problem where the lower level problem has been replaced by its Karush-Kuhn-Tucker conditions:

$$F(x, y) \to \min_{x,y,\lambda,\mu}$$
$$G(y) \leq 0,$$
$$\nabla_x L(x, y, \lambda, \mu) = 0, \qquad (5.18)$$
$$g(x, y) \leq 0, \ \lambda \geq 0,$$
$$h(x, y) = 0,$$
$$\lambda^\top g(x, y) = 0.$$

Note that this problem is equivalent to the bilevel programming problem (5.1), (5.2) provided that the optimistic position (5.3) is used and the lower level problem is a convex parametric one for which a regularity assumption is satisfied and the optimal solution is uniquely determined and strongly stable. We will come back to this in Theorem 5.15 and the subsequent example. Note also that this equivalence relation is strongly connected with the search for global optimal solutions in both levels. If a local optimistic optimal solution of problem (5.1), (5.2) is searched for, this is in general no longer valid (cf. [140]). For now, the definition of a local optimal solution of an optimization problem is given:

DEFINITION 5.5 *Consider an optimization problem*

$$\min_{x,y}\{f(x, y) : g(x, y) \leq 0, h(x, y) = 0\} \qquad (5.19)$$

and let (x^0, y^0) *satisfying* $g(x^0, y^0) \leq 0, h(x^0, y^0) = 0$ *be a feasible solution of this problem. Then,* (x^0, y^0) *is called a local optimal solution of problem (5.19) if there exists an open neighborhood* $W_\varepsilon(x^0, y^0), \varepsilon > 0$ *of* (x^0, y^0) *such that*

$$f(x, y) \geq f(x^0, y^0) \ \forall \ (x, y) \in W_\varepsilon(x^0, y^0) \ with \ g(x, y) \leq 0, h(x, y) = 0.$$

Using an active-set strategy, problem (5.18) decomposes into a family of problems

$$F(x, y) \to \min_{x,y,\lambda,\mu}$$

$$G(y) \leq 0,$$
$$\nabla_x L(x, y, \lambda, \mu) = 0,$$
$$g_i(x, y) \leq 0, \ \lambda_i = 0, \ i \notin I, \qquad (5.20)$$
$$g_i(x, y) = 0, \ \lambda_i \geq 0, \ i \in I,$$
$$h(x, y) = 0,$$

where, around some given point (x^0, y^0), $x^0 \in \Psi(y^0)$, the family of all interesting sets I can be restricted to the union of all the sets $\mathcal{I}(\lambda)$ for the vertices $(\lambda, \mu) \in \Lambda(x^0, y^0)$ if the (CRCQ) is satisfied (cf. the proof of Theorem 4.10). If the (CRCQ) is dropped, limit Lagrange multipliers to optimal solutions to perturbed lower level problems are in general no longer vertices of the set $\Lambda(x^0, y^0)$ (cf. the Example on page 76). This implies that the sets I will in general not satisfy the condition (C2) on page 78 but the restriction to the vertices of $\Lambda(x^0, y^0)$ is again allowed. But, nevertheless, the number of different possible sets I is finite. Now, if we consider the contingent cone to the feasible set of one of the problems (5.20) we obtain the solution set of the system (5.17) which shows that formula (5.16) remains valid (with the possible exception of condition (C2) in case of violation of (CRCQ)).

REMARK 5.1 *The solution sets of (5.17) are convex polyhedral cones and it is easy to see that the projections of these cones onto $\mathbb{R}^n \times \mathbb{R}^m$ are convex, polyhedral cones again. Hence, the sets $\widehat{K}_{\mathcal{M}}(x^0, y^0)$ are polyhedral cones.*

Using these polyhedral cones, the necessary condition in Theorem 5.4 reads as

$$\nabla_x F(x^0, y^0) d + \nabla_y F(x^0, y^0) r \geq 0 \ \forall \ (d, r, \gamma, \eta)^\top \in K_{\mathcal{M}}(x^0, y^0).$$

And the sufficient optimality condition in Theorem 5.5 is equivalent to

$$\nabla_x F(x^0, y^0) d + \nabla_y F(x^0, y^0) r > 0 \ \forall \ (d, r, \gamma, \eta)^\top \in K_{\mathcal{M}}(x^0, y^0),$$

where in both cases the assumptions of the respective theorems are assumed to be satisfied.

DEFINITION 5.6 *Let $C \subseteq \mathbb{R}^n$ be a cone. The dual cone C^* to C is defined as*

$$C^* = \{v \in \mathbb{R}^n : \langle v, d \rangle \geq 0 \ \forall \ d \in C\}.$$

For a convex polyhedral cone $C = \{d : Ad = 0, \ Bd \leq 0\}$ the dual cone is

$$C^* = \{v = -A^\top \lambda - B^\top \mu : \mu \geq 0\}.$$

Now, the necessary condition in Theorem 5.4 is equivalent to each of the two conditions

$$(\nabla_x F(x^0, y^0), \nabla_y F(x^0, y^0), 0, 0)^\top \in K_{\mathcal{M}}(x^0, y^0)^*$$

and

$$(\nabla_x F(x^0, y^0), \nabla_y F(x^0, y^0))^\top \in \widehat{K}_{\mathcal{M}}(x^0, y^0)^*$$

or: There exist a Lagrange multiplier $\nu^0 = (\lambda^0, \mu^0) \in E\Lambda(x^0, y^0)$ and a set $I \in \mathcal{I}(\lambda^0)$ (or a set I satisfying condition (C1) if (CRCQ) is violated) such that the following system has a solution:

$$
\begin{aligned}
\nabla F(z^0) + \kappa^\top(0, \nabla_y G(y^0)) + \nabla(\nabla_x L(z^0, \nu^0)\omega) & \\
+ \zeta^\top \nabla g(z^0) + \tau^\top \nabla h(z^0) &= 0 \\
(\nabla_x g(z^0), \nabla_x h(z^0))^\top \omega - (\xi, 0)^\top &= 0 \\
g_i(z^0)\zeta_i &= 0, \; \forall i \\
\zeta_i &\geq 0, \; \forall i \in I(z^0) \setminus I \\
\xi_i &\geq 0, \; \forall i \in I \setminus J(\lambda^0) \\
\xi_i &= 0, \; \forall i \in J(\lambda^0) \\
\kappa^\top G(y^0) &= 0, \\
\kappa &\geq 0, \quad\quad (5.21)
\end{aligned}
$$

where ∇F denotes the (partial) Jacobi matrix of the vector valued function F with respect to $z = (x, y)$ only. Note that the condition $\nabla_x g_i(z^0)\omega = \xi_i$ implies that $\nabla_x g_i(z^0)\omega$ is essentially unconstrained for $i \notin I$ which is equivalent to non-existence of that condition.

REMARK 5.2 *The sufficient optimality condition in Theorem 5.5 reads as*

$$(\nabla_x F(x^0, y^0), \nabla_y F(x^0, y^0), 0, 0)^\top \in \text{int } K_{\mathcal{M}}(x^0, y^0)^*.$$

The main question is it now to describe a way to compute the dual cone $K_{\mathcal{M}}(x^0, y^0)^*$.

For $I^0 = \bigcap_{(\lambda,\mu) \in E\Lambda(x^0, y^0)} J(\lambda)$ consider the relaxed problem to (5.18):

$$
\begin{aligned}
F(x, y) &\to \min_{x, y, \lambda, \mu} \\
\nabla_x L(x, y, \lambda, \mu) &= 0, \\
g_i(x, y) &= 0, \; \forall \, i \in I^0 \\
g_i(x, y) &\leq 0, \; \forall \, i \notin I^0 \\
\lambda_i &= 0, \; \forall \, i \notin I(x^0, y^0) \\
\lambda_i &\geq 0, \; \forall i \in I(x^0, y^0) \\
h(x, y) &= 0, \; G(y) \leq 0.
\end{aligned}
\quad\quad (5.22)
$$

Then, if the (MFCQ) is valid for the lower level problem (5.1), each point $(x^0, y^0, \lambda^0, \mu^0)$ with $G(y^0) \leq 0, x^0 \in \Psi(y^0), (\lambda^0, \mu^0) \in E\Lambda(x^0, y^0)$ is feasible for (5.22) and, by $I^0 \subseteq J(\lambda) \subseteq I$ for all $(\lambda, \mu) \in E\Lambda(x^0, y^0), I \in \mathcal{I}(\lambda)$ we have

$$K_{\mathcal{M}}(x^0, y^0) \subseteq K^0(x^0, y^0), \tag{5.23}$$

where $K^0(x^0, y^0)$ denotes the contingent cone to the feasible set of problem (5.22) at (x^0, y^0) which is given by the solution set of the following system:

$$\nabla G_i(y^0)r \leq 0, \forall\, i : G_i(y^0) = 0$$
$$\nabla^2_{xx}L(z^0, \lambda, \mu)d + \nabla^2_{yx}L(z^0, \lambda, \mu)r + \nabla^\top_x g(z^0)\gamma + \nabla^\top_x h(z^0)\eta = 0$$
$$\nabla_x g_i(z^0)d + \nabla_y g_i(z^0)r = 0, \; i \in I^0,$$
$$\nabla_x g_i(z^0)d + \nabla_y g_i(z^0)r \leq 0, \; i \in I(z^0) \setminus I^0,$$
$$\nabla_x h_j(z^0)d + \nabla_y h_j(z^0)r = 0, \; j = 1, \ldots, q,$$
$$\gamma_i \geq 0, \; i \in I(z^0) \setminus J(\lambda^0)$$
$$\gamma_i = 0, \; i \notin I(z^0).$$

Then, using the definition of the dual cone, it is easy to see that

$$K_{\mathcal{M}}(x^0, y^0)^* \supseteq K^0(x^0, y^0)^*. \tag{5.24}$$

The questions whether formulae (5.23) or (5.24) are satisfied as equations and, if not, if there are cases where the cone $K_{\mathcal{M}}(x^0, y^0)^*$ can be represented "constructively" are investigated in [229].

We borrow the following two examples from [229] to illustrate these questions in the case of a mathematical program with equilibrium constraints (MPEC):

Example: Consider the complementarity problem

$$x \geq 0, \; \lambda \geq 0, \; x\lambda = 0$$

at the point $(x, \lambda) = (0, 0)^\top$. Then

$$\mathcal{M} = \{(x, \lambda) \geq 0 : x = 0\} \cup \{(x, \lambda) \geq 0 : \lambda = 0\}, K_{\mathcal{M}}(0, 0) = \mathcal{M}$$

but the relaxed problem has the feasible set $\{(x, \lambda) : x \geq 0, \lambda \geq 0\}$ with the contingent cone

$$K^0(0, 0) = \mathbb{R}^2_+.$$

Here, both cones are different but their duals coincide:

$$K_{\mathcal{M}}(0, 0)^* = K^0(0, 0)^* = \mathbb{R}^2_+.$$

□

Example: Consider the complementarity problem

$$z_1 \geq 0, \ z_1 + z_2 \geq 0, \ z_1(z_1 + z_2) = 0, \ z_2 \leq 0$$

again at the point $z = (0,0)^\top$. Then, the feasible set of this problem reduces to

$$\mathcal{M} = \{z : z_2 \leq 0, z_1 + z_2 = 0\},$$

and we get for the contingent cone

$$K_\mathcal{M}(0,0) = \{d : d_2 \leq 0, \ d_1 + d_2 = 0\}.$$

The relaxed feasible set and its contingent cone are

$$M = \{z : z_1 \geq 0, z_2 \leq 0, z_1 + z_2 \geq 0\},$$

$$K_M(0,0) = \{d : d_1 + d_2 \geq 0, d_2 \leq 0\}.$$

Considering the dual cones, we derive

$$K_\mathcal{M}(0,0)^* = \{v : v_1 - v_2 \geq 0\}, K_M(0,0)^* = \{v : v_1 \geq 0, \ v_1 - v_2 \geq 0\}.$$

In this example both the primal and the dual cones are different.   □

REMARK 5.3 *The distinction in the construction of the two cones* $K_\mathcal{M}(x^0, y^0)$ *and* $K^0(x^0, y^0)$ *with the corresponding cones in [229] should be noticed which results from a different role played by the Lagrange multipliers in both approaches: While the lower level optimal solution* x *and the corresponding multiplier* $(\lambda, \mu) \in \Lambda(x, y)$ *are considered uniformly in [229], they play different parts in bilevel programming. This implies that we have to define different neighborhoods of feasible solutions in both problems. Let* $x^0 \in \Psi(y^0)$, $(\lambda^0, \mu^0) \in \Lambda(x^0, y^0)$, *i.e. let* (x^0, λ^0, μ^0) *satisfy the Karush-Kuhn-Tucker conditions for the lower level problem (5.1) for* $y = y^0$. *Then, considered as an (MPEC), in each sufficiently small open neighborhood of* $(x^0, y^0, \lambda^0, \mu^0)$ *positive components of* λ^0 *remain positive. But, considered as bilevel programming problem, only* x *is the lower level variable, and in arbitrarily small open neighborhoods of* (x^0, y^0) *the corresponding values of the corresponding Lagrange multipliers can be extremely different.*

This can be seen in the following example.

Example: Consider the following parametric optimization problem which replaces the lower level problem of a bilevel programming problem:

$$\min\{x_1^2 + (x_2 - 1)^2 : x_1^2 + (x_2 + 1)^2 \leq y, \ x_1^2 + (x_2 + 1)^2 \leq 2 - y\}.$$

The Karush-Kuhn-Tucker conditions of this problem are:

$$2x_1 + 2\lambda_1 x_1 + 2\lambda_2 x_1 = 0$$
$$2(x_2 - 1) + 2\lambda_1(x_2 + 1) + 2\lambda_2(x_2 + 1) = 0$$
$$g_1(x, y) = x_1^2 + (x_2 + 1)^2 - y \leq 0$$
$$g_2(x, y) = x_1^2 + (x_2 + 1)^2 - 2 + y \leq 0$$
$$\lambda_1 \geq 0, \ \lambda_2 \geq 0$$
$$\lambda_i g_i(x, y) = 0, \ i = 1, 2.$$

Then, for $y^0 = 1$, the point $x^0 = (0,0)^\top$ is the only optimal solution of the optimization problem with

$$\Lambda(x^0, y^0) = \mathrm{conv} \ \{(1,0)^\top, \ (0,1)^\top\}.$$

Let $\lambda^0 = (1,0)^\top$. Considered as an (MPEC) in each sufficiently small open neighborhood of the point (x^0, y^0, λ^0) we have $\lambda_1 > 0$. But considered as a bilevel programming problem we have to bear in mind that for $x \in \Psi(y)$ and y sufficiently close to y^0 the corresponding Lagrange multiplier λ can approach $(0,1)^\top$ with $\lambda_1 = 0$. Here this is the case for $y \in (1,2]$. Hence, the following two systems are to be used to describe the cone $K_{\mathcal{M}}(x^0, y^0)$:

$$2x_1 + 2\lambda_1 x_1 + 2\lambda_2 x_1 = 0$$
$$2(x_2 - 1) + 2\lambda_1(x_2 + 1) + 2\lambda_2(x_2 + 1) = 0$$
$$g_1(x, y) = x_1^2 + (x_2 + 1)^2 - y = 0$$
$$g_2(x, y) = x_1^2 + (x_2 + 1)^2 - 2 + y \leq 0$$
$$\lambda_1 \geq 0, \ \lambda_2 = 0$$

and

$$2x_1 + 2\lambda_1 x_1 + 2\lambda_2 x_1 = 0$$
$$2(x_2 - 1) + 2\lambda_1(x_2 + 1) + 2\lambda_2(x_2 + 1) = 0$$
$$g_1(x, y) = x_1^2 + (x_2 + 1)^2 - y \leq 0$$
$$g_2(x, y) = x_1^2 + (x_2 + 1)^2 - 2 + y = 0$$
$$\lambda_1 = 0, \ \lambda_2 \geq 0.$$

In distinction, the subsequent two problems can be used in [229]:

$$2x_1 + 2\lambda_1 x_1 + 2\lambda_2 x_1 = 0$$
$$2(x_2 - 1) + 2\lambda_1(x_2 + 1) + 2\lambda_2(x_2 + 1) = 0$$
$$g_1(x, y) = x_1^2 + (x_2 + 1)^2 - y = 0$$
$$g_2(x, y) = x_1^2 + (x_2 + 1)^2 - 2 + y \leq 0$$
$$\lambda_1 \geq 0, \ \lambda_2 = 0$$

and

$$2x_1 + 2\lambda_1 x_1 + 2\lambda_2 x_1 = 0$$
$$2(x_2 - 1) + 2\lambda_1(x_2 + 1) + 2\lambda_2(x_2 + 1) = 0$$
$$g_1(x, y) = x_1^2 + (x_2 + 1)^2 - y = 0$$
$$g_2(x, y) = x_1^2 + (x_2 + 1)^2 - 2 + y = 0$$
$$\lambda_1 \geq 0, \ \lambda_2 \geq 0.$$

□

The following result shows a simple convexity property of the dual cone $K_{\mathcal{M}}(x^0, y^0)^*$:

THEOREM 5.8 ([229]) *Consider the bilevel programming problem (5.1), (5.5) and let the assumptions (MFCQ) at the point (x^0, y^0) as well as (ULR) be satisfied. Then, $K_{\mathcal{M}}(x^0, y^0)^*$ is a convex polyhedral cone.*

This theorem gives us the principal possibility to describe the cone $K_{\mathcal{M}}(x^0, y^0)^*$ via a system of linear (in)equalities. The above examples indicate that this cannot be done without a closer look at the constraints defining the primal cone $K_{\mathcal{M}}(x^0, y^0)$. Especially the second example above seem to indicate that inequalities which can not be satisfied strictly play a special role in these investigations. This leads to the following notion of a nondegenerate inequality in the system defining the cone $K^0(x^0, y^0)$:

DEFINITION 5.7 *Let the system $Ax = a, Bx \leq b$ of linear (in)equalities be given. An inequality $\langle B^i, x \rangle \leq b_i$ is nondegenerate if there is a solution $\widehat{x}$ of this system satisfying $\langle B^i, \widehat{x} \rangle < b_i$.*

Let a feasible point $(x^0, y^0, \lambda^0, \mu^0)$ of the bilevel programming problem be given and consider the convex polyhedral cone $K^0(x^0, y^0)$. Put

$$\alpha_g = \{i \in I(z^0) \setminus I^0 : \text{inequality } \nabla g_i(z^0)(d, r)^\top \leq 0 \text{ is nondegenerate}\}$$

and

$$\alpha_\lambda = \{i \in I(x^0, y^0) \setminus J(\lambda^0) : \text{inequality } \gamma_i \geq 0 \text{ is nondegenerate}\}.$$

Note, that by convexity of the cone $K^0(x^0, y^0)$, inclusion of an index $i_0 \in I(x^0, y^0) \setminus I^0$ into one of the sets α_g and α_λ can be checked by solving two systems of linear (in)equalities:

$$(\nabla(\nabla_x L(x^0, y^0, \lambda^0, \mu^0)), \nabla_x^\top g(x^0, y^0), \nabla_x^\top h(x^0, y^0))(d, r, \gamma, \eta)^\top = 0,$$
$$\nabla g_i(x^0, y^0)(d, r)^\top = 0, \ \forall \ i \in I^0$$

$$\nabla g_i(x^0, y^0)(d, r)^\top \le 0, \ \forall \, i \in I(x^0, y^0) \setminus (I^0 \cup \{i_0\})$$
$$\nabla g_{i_0}(x^0, y^0)(d, r)^\top \le -1$$
$$\gamma_i = 0, \ \forall \, i \notin I(x^0, y^0)$$
$$\gamma_i \ge 0, \ \forall i \in I(x^0, y^0) \setminus J(\lambda^0)$$
$$\nabla h(x^0, y^0)(d, r)^\top = 0,$$
$$\nabla G_i(y)r \le 0, \ \forall i : G_i(y^0) = 0.$$

and

$$(\nabla(\nabla_x L(x^0, y^0, \lambda^0, \mu^0)), \nabla_x^\top g(x^0, y^0), \nabla_x^\top h(x^0, y^0))(d, r, \gamma, \eta)^\top = 0,$$
$$\nabla g_i(x^0, y^0)(d, r)^\top = 0, \ \forall \, i \in I^0$$
$$\nabla g_i(x^0, y^0)(d, r)^\top \le 0, \ \forall \, i \in I(x^0, y^0) \setminus I^0$$
$$\gamma_i = 0, \ \forall \, i \notin I(x^0, y^0)$$
$$\gamma_i \ge 0, \ \forall i \in I(x^0, y^0) \setminus (J(\lambda^0) \cup \{i_0\})$$
$$\gamma_{i_0} \ge 1$$
$$\nabla h(x^0, y^0)(d, r)^\top = 0,$$
$$\nabla G_i(y)r \le 0, \ \forall i : G_i(y^0) = 0.$$

Hence, computing both sets α_g and α_λ requires the solution of no more than $2(|I(x^0, y^0)| - |I^0|)$ linear programs which can be done in polynomial time [148]. The following Theorem has its roots in [229].

THEOREM 5.9 *Consider the optimistic bilevel programming problem (5.1), (5.5) and let assumptions (MFCQ) and (ULR) be satisfied at the point (x^0, y^0). Then, equality is satisfied in (5.23) if and only if $\alpha_g \cap \alpha_\lambda = \emptyset$.*

Roughly speaking, condition $\alpha_g \cap \alpha_\lambda = \emptyset$ implies that in the lower level problem (5.1) the (LICQ) together with the strict complementarity slackness condition are satisfied. Having a look at the opposite implication we see that (LICQ) together with the strict complementarity slackness condition imply that $I^0 = I(x^0, y^0)$ and, thus, that the set $\alpha_g \cap \alpha_\lambda = \emptyset$. Assume for the interpretation that the condition (SSOC) is also valid. Then problem (5.1), (5.5) is equivalent to the differentiable program

$$\min_y \{F(x(y), y) : G(y) \le 0\}$$

[97] and, by (ULR), there exists a vector $\kappa \ge 0$, $\kappa^\top G(y^0) = 0$ with

$$\nabla_x F(x^0, y^0)\nabla x(y^0) + \nabla_y F(x^0, y^0) + \kappa^\top \nabla_y G(y^0) = 0.$$

Now we come to the general case where $\alpha_g \cap \alpha_\lambda$ can be non-empty. Theorem 5.8 gives us the principal possibility to describe the dual cone

$K_{\mathcal{M}}(x^0, y^0)$ via a finite system of linear equations and inequalities. On the other hand, Theorem 5.9 says that the inequalities with index in the set $\alpha_g \cap \alpha_\lambda$ can make the formulation of this system difficult. In the following we use two more assumptions. To formulate them we apply the abbreviation

$$\alpha_0 = \alpha_g \cap \alpha_\lambda, \ \alpha_1 = I(x^0, y^0) \setminus (I^0 \cup \alpha_0).$$

The set α_0 is the index set of all inequality constraints in problem (5.22) which are nondegenerate at the same time with respect to both inequalities. In the following we use the formula $(\alpha_2, \alpha_3) \in \mathcal{P}(\alpha_0)$ to denote a pair of sets satisfying

$$\alpha_2 \cup \alpha_3 = \alpha_0, \alpha_2 \cap \alpha_3 = \emptyset.$$

(PFC) Condition (PFC) is satisfied if there exists a pair $(\alpha_2, \alpha_3) \in \mathcal{P}(\alpha_0)$ such that the following conditions are both satisfied:

- the gradients in

  $$\mathcal{A}_1 := \{(\nabla g_i(x^0, y^0), 0, 0) : i \in \alpha_2\} \cup \{(0, 0, e_i, 0) : i \in \alpha_3\}$$

  are linearly independent,

- Let

  $$\begin{aligned}
  \mathcal{A}_2 \ := \ & \{(\nabla g_i(x^0, y^0), 0, 0) : i \in \alpha_1 \cup \alpha_3 \cup I^0\} \\
  \cup \ & \{(0, 0, e_i^\top, 0) : i \in \alpha_1 \cup \alpha_2 \cup J(-g(x^0, y^0))\} \\
  \cup \ & \{(0, \nabla_y G_i(y^0), 0, 0), \forall i : G_i(y^0) = 0\} \cup \{(\nabla h(x^0, y^0), 0, 0)\} \\
  \cup \ & \{(\nabla(\nabla_x L(x^0, y^0, \lambda^0, \mu^0)), \nabla_x^\top g(x^0, y^0), \nabla_x^\top h(x^0, y^0))\}.
  \end{aligned}$$

  Then, span $\mathcal{A}_1 \cap$ span $\mathcal{A}_2 = \{0\}$.

Here, for a finite set of vectors A, span A is the set of all vectors obtained as linear combination of the vectors in A. The following lemma is used to illustrate this regularity condition.

LEMMA 5.1 *The following three conditions are equivalent:*

1. *Condition (PFC) is valid for a set $(\alpha_2, \alpha_3) \in \mathcal{P}(\alpha_0)$.*

2. *The following implication holds:*

$$\begin{aligned}
& \kappa^\top (0, \nabla_y G(y^0)) + \nabla(\nabla_x L(z^0, \nu^0)\omega) + \zeta^\top \nabla g(z^0) + \tau^\top \nabla h(z^0) = 0 \\
& (\nabla_x g(z^0), \nabla_x h(z^0))^\top \omega - (\xi, 0)^\top = 0 \\
& g_i(z^0)\zeta_i = 0, \ \forall i \\
& \xi_i = 0, \ \forall i \in I^0 \\
& \kappa^\top G(y^0) = 0,
\end{aligned} \tag{5.25}$$

implies $\zeta_i = 0$, $i \in \alpha_2$ *and* $\xi_i = 0$, $i \in \alpha_3$.

3. *The following conditions hold simultaneously:*

   ▪ *For all* $i_0 \in \alpha_2$ *there is a solution of the following system:*

   $$(\nabla(\nabla_x L(x^0, y^0, \lambda^0, \mu^0)), \nabla_x^\top g(x^0, y^0), \nabla_x^\top h(x^0, y^0))(d, r, \gamma, \eta)^\top = 0,$$
   $$\nabla g_i(x^0, y^0)(d, r)^\top = 0, \ \forall \ i \in I(x^0, y^0) \setminus \{i_0\}$$
   $$\nabla g_{i_0}(x^0, y^0)(d, r)^\top < 0,$$
   $$\gamma_i = 0, \ \forall i \notin J(\lambda^0)$$
   $$\nabla h(x^0, y^0)(d, r)^\top = 0,$$
   $$\nabla G_i(y) r = 0, \ \forall i : G_i(y^0) = 0$$

   and

   ▪ *for all* $i_0 \in \alpha_3$ *there is a solution of the following system:*

   $$(\nabla(\nabla_x L(x^0, y^0, \lambda^0, \mu^0)), \nabla_x^\top g(x^0, y^0), \nabla_x^\top h(x^0, y^0))(d, r, \gamma, \eta)^\top = 0,$$
   $$\nabla g_i(x^0, y^0)(d, r)^\top = 0, \ \forall \ i \in I(x^0, y^0)$$
   $$\gamma_i = 0, \ \forall i \notin J(\lambda^0), i \neq i_0$$
   $$\gamma_{i_0} > 0$$
   $$\nabla h(x^0, y^0)(d, r)^\top = 0,$$
   $$\nabla G_i(y) r = 0, \ \forall i : G_i(y^0) = 0.$$

The following theorem having its roots in [229] gives a condition for coincidence of the dual cones to the feasible sets of the original and the relaxed optimistic bilevel problems.

THEOREM 5.10 *Consider the bilevel problem (5.1), (5.5) under assumptions (MFCQ) and (ULR) at the point* (x^0, y^0). *Let* $(\lambda^0, \mu^0) \in E\Lambda(x^0, y^0)$. *If also (PFC) is satisfied with respect to a pair of sets* $(\alpha_2, \alpha_3) \in \mathcal{P}(\alpha_0)$ *satisfying*

$$\alpha_2 \cup \left(I(x^0, y^0) \setminus \alpha_g \right) \supseteq J(\lambda^0),$$

then $K^0(x^0, y^0)^* = K_\mathcal{M}(x^0, y^0)^*$.

5.4 REGULARITY

In the Introduction we have seen several possible reformulations of the bilevel programming problem into ordinary one-level problems. One uses the Karush-Kuhn-Tucker conditions of the lower level problem to replace them:

$$
\begin{aligned}
F(x, y) &\to \min_{x, y, \lambda, \mu} \\
G(y) &\leq 0 \\
\nabla_x L(x, y, \lambda, \mu) &= 0 \\
g(x, y) \leq 0, \lambda &\geq 0 \\
\lambda^\top g(x, y) &= 0 \\
h(x, y) &= 0.
\end{aligned}
\tag{5.26}
$$

Note that (5.1), (5.2) and (5.26) are equivalent provided that the lower level programming problem (5.1) is a convex parametric optimization problem satisfying (MFCQ) at all feasible points $x \in M(y)$, $G(y) \leq 0$ and we settle upon the computation of global optimistic optimal solutions. Without convexity assumption, problem (5.26) has a larger feasible set including not only global optimal solutions of the lower level problem but also all local optimal solutions and also all stationary points. Hence, the optimal function value of (5.26) is never larger than that of the bilevel programming problem (5.1), (5.2). Moreover, problem (5.26) is a smooth optimization problem which could be used as an indication for an easier treatment. But this is not completely correct since at least the regularity assumptions which are needed for successfully handling smooth optimization problems are never satisfied. To find a better approach, a nonsmooth equivalent of the bilevel programming problem (5.1), (5.2) is formulated below. Then, respective regularity assumptions for nonsmooth optimization problems can be satisfied.

The main difficulty concerning the reformulation (5.26) is the violation of most of the usual constraint qualifications.

THEOREM 5.11 ([249]) *If the Karush-Kuhn-Tucker conditions of problem (5.1) are part of the constraints of an optimization problem, then the Mangasarian-Fromowitz constraint qualification is violated at every feasible point.*

The same unpleasant result has been shown in [59] for the Arrow-Hurwicz-Uzawa constraint qualification [8]. Moreover, even the bilevel programming problem with a convex quadratic lower level problem is most likely not Kuhn-Tucker regular [162].

To circumvent the resulting difficulties for the construction of Karush-Kuhn-Tucker type necessary optimality conditions for the bilevel programming problem, in [249] a nonsmooth version of the KKT reformulation of the bilevel programming problem in the optimistic case is constructed:

$$F(x, y) \rightarrow \min_{x,y,\lambda,\mu}$$
$$G(y) \leq 0$$
$$\nabla_x L(x, y, \lambda, \mu) = 0 \qquad (5.27)$$
$$\min\{-g(x, y), \lambda\} = 0$$
$$h(x, y) = 0.$$

Here, for $a, b \in \mathbb{R}^n$, the formula $\min\{a, b\} = 0$ is understood component wise.

For problem (5.27) the following generalized variant of the linear independence constraint qualification [253] can be defined:

(PLICQ) The *piecewise linear independence constraint qualification* is satisfied for problem (5.27) at a point $(x^0, y^0, \lambda^0, \mu^0)$ if the gradients of all the vanishing components of the constraint functions $G(y), \nabla_x L(x, y, \lambda, \mu), g(x, y), \lambda, h(x, y)$ are linearly independent.

Problem (5.27) can be investigated by considering the following patchwork of nonlinear programs for fixed set I:

$$F(x, y) \rightarrow \min_{x,y,\lambda,\mu}$$
$$G(y) \leq 0$$
$$\nabla_x L(x, y, \lambda, \mu) = 0$$
$$g_i(x, y) = 0, \text{ for } i \in I$$
$$g_i(x, y) \leq 0, \text{ for } i \notin I \qquad (5.28)$$
$$\lambda_i = 0, \text{ for } i \notin I$$
$$\lambda_i \geq 0, \text{ for } i \in I$$
$$h(x, y) = 0.$$

Then, the piecewise linear independence constraint qualification is valid for problem (5.27) at some point $(x^0, y^0, \lambda^0, \mu^0)$ if and only if it is satisfied for each of the problems (5.28) for all sets $J(\lambda^0) \subseteq I \subseteq I(x^0, y^0)$.

REMARK 5.4 *Note that in (PLICQ) the gradients are all taken with respect to all variables x, y, λ, μ. Hence, this condition does not imply strong stability of the solution of the lower level problem for which a similar assumption with gradients taken with respect to x, λ, μ only would be sufficient [157]. Note the strong relation of (PLICQ) with (FRR) if the upper level constraints $G(y) \leq 0$ are absent.*

Note that this is a more general consideration than that in Theorem 5.11 since the complementarity condition has been replaced here. Our aim is it now to show that the assumption (PLICQ) can generically be assumed to hold for problem (5.27). The usual way to do that is to allow for arbitrarily small perturbations of the constraint functions in problem

(5.27) and to show that the assumption (PLICQ) can be forced to hold for perturbed problems. Hence we have to do two things: First define a distance of functions to determine the notion of small perturbations. This is done using the so-called Whitney topology. Second, define a subset of all problems (5.27) for which the (PLICQ) is satisfied at all feasible points. To reach the desired property it is then shown that this set is open and dense in the set of all problems (5.27) with respect to this Whitney topology. For this, Sard's Lemma is used guaranteeing that sufficiently smooth systems of equations are locally invertible at all of its solutions for almost all right-hand sides. To illustrate the bounded value of this theorem for systems with a large number of equations we cite a result that such systems have no solution for almost all right-hand side vectors. Coming back to the initial bilevel programming problem (5.1), (5.2) we see that (PLICQ) can be assumed to hold for almost all problems in which the number of active inequalities in the lower level problem is small.

We start with the definition of a neighborhood of a problem. The distance of two problems is defined via the distance of the functions defining the constraints and the objective functions of both problems. Hence, the distance of two problems of the type (5.27) will be measured in the Whitney C^k topology. A neighbourhood of a data map (i.e. a set of functions) $D \in C^l(\mathbb{R}^r, \mathbb{R}^t), k \leq l$, consists of all maps which are close to D together with all derivatives up to order k: a zero neighborhood in this topology is indexed by a positive continuous function $\varepsilon : \mathbb{R}^r \to \mathbb{R}$ and contains all functions $f \in C^k(\mathbb{R}^r, \mathbb{R}^t)$ with: each component function f_i together with all its partial derivatives up to order k is bounded by the function ε. Details can be found in [125].

Define the set

$$\mathcal{H}_B^l = \{ (F, G, f, g, h) \in C^l(\mathbb{R}^{m+n}, \mathbb{R}^{1+s+1+p+q}) : \text{(PLICQ) is satisfied} \\ \text{at each feasible point of (5.27) with } \max\{\|\lambda\|_\infty, \|\mu\|_\infty\} \leq B\}$$

for an arbitrary constant $0 < B < \infty$, $\|\lambda\|_\infty = \max\{|\lambda_i| : 1 \leq i \leq p\}$ is the L_∞-norm of a vector $\lambda \in \mathbb{R}^p$ and $l \geq 2$. The following theorem says that, if the dimensions n and m of the variables x and y are large enough in comparison with the number of constraints, then for almost all bilevel programming problems the (PLICQ) is satisfied at all feasible solutions.

THEOREM 5.12 ([253]) *The set $\mathcal{H}_B^l$ is open in the C^k-topology for each $2 \leq k \leq l$. Moreover, for $l > \max\{1, m - q\}$, the set $\mathcal{H}_B^l$ is also dense in the C^k-topology for all $2 \leq k \leq l$.*

In the proof of this theorem, which is given in Section 5.8, Sard's Theorem is used:

DEFINITION 5.8 *A point* $y \in \mathbb{R}^t$ *is a regular value of a smooth function* $H : \mathbb{R}^r \to \mathbb{R}^t$ *if* $\nabla H(x)$ *has full row rank for every* x *with* $H(x) = y$.

THEOREM 5.13 (SARD'S THEOREM,[125, 268]) *If* $H : \mathbb{R}^r \to \mathbb{R}^t$ *is* C^k *with* $k > \max\{0, r - t\}$, *then almost all* $y \in \mathbb{R}^t$ *are regular values of* H.

Sard's Theorem says that for almost all right-hand sides y at all solutions of the system $H(x) = y$ of nonlinear equations the assumptions of an implicit function theorem are satisfied. Consequently, the solution of the system $H(x) = y$ is a smooth mapping of the right-hand side in a sufficiently small open neighborhood of y.

If there are more equations than variables $(t > r)$ then the following result says that almost all $y \in \mathbb{R}^t$ are regular values since the system $H(x) = y$ is not solvable. To formulate the result, define the image of $\mathbb{R}^r$ via the mapping H by $H(\mathbb{R}^r) = \{w : \exists v \in \mathbb{R}^r \text{ with } w = H(v)\}$.

THEOREM 5.14 ([125]) *Let* $s > r$ *and consider a continuously differentiable mapping* $H : \mathbb{R}^r \to \mathbb{R}^s$. *Then,* $\mathbb{R}^s \setminus H(\mathbb{R}^r)$ *is dense.*

Applying this result to the ideas in the proof of Theorem 5.12 we derive that, if the number of active inequalities of solvable bilevel programming problems is too large then the projection of the feasible solutions $(x, y) \in \mathbb{R}^n \times \mathbb{R}^m$ with $x \in \Psi(y), G(y) \leq 0$ onto $\mathbb{R}^m$ is of Lebesgue measure zero.

5.5 OPTIMALITY CONDITIONS USING THE KKT REFORMULATION

Now we come back to the investigation of optimality conditions for bilevel programming problems.

THEOREM 5.15 *Consider the optimistic bilevel programming problem (5.1), (5.3), (5.4) and assume that the lower level problem is a convex parametric optimization problem for which (MFCQ) is satisfied at all feasible points* (x^0, y^0) *with* $G(y^0) \leq 0, x^0 \in \Psi(y^0)$. *Then, each optimistic local optimal solution of (5.1), (5.3), (5.4) corresponds to a local optimal solution for (5.27).*

The following example shows that the opposite direction in this theorem is not true in general:

Example: Consider the simple linear bilevel programming problem

$$\min\{y : x \in \Psi_L(y), -1 \leq y \leq 1\},$$

where

$$\Psi_L(y) = \underset{x}{\mathrm{Argmin}} \ \{xy : 0 \leq x \leq 1\}$$

at the point $(x^0, y^0) = (0, 0)$. Then, this point is a local optimal solution according to problem (5.27), i.e. there exists an open neighborhood $W_\varepsilon(0,0) = (-\varepsilon, \varepsilon) \times (-\varepsilon, \varepsilon)$ with $0 < \varepsilon < 1$ such that $y \geq 0$ for all $(x, y) \in W_\varepsilon(0,0)$ with $x \in \Psi_L(y)$ and $-1 \leq y \leq 1$. The simple reason for this is that there is no $-\varepsilon < x < \varepsilon$ with $x \in \Psi_L(y)$ for $y < 0$ since $\Psi_L(y) = \{1\}$ for $y < 0$. But if we consider the definition of a local optimistic optimal solution in Definition 5.2 then the point $(0, 0)$ is not a local optimistic solution since $y^0 = 0$ is not a local minimum of the function $\varphi_o(y) = y$. □

Clearly, if the point (x^0, y^0) is a global optimal solution of problem (5.27) then it is also an optimistic optimal solution of the bilevel programming problem (5.1), (5.5). It should also be mentioned that, if the lower level problem is a regular convex one and its optimal solution is uniquely determined and strongly stable for all parameter values then the bilevel programming problem (5.1), (5.2) is equivalent to its reformulations (5.18) and (5.27). Without convexity even an optimistic global optimal solution of (5.1), (5.5) need not to be global optimal for (5.27). The following example shows even that a local optimum of the bilevel programming problem (5.1), (5.2) need not be a stationary point of the problem (5.27):

Example: Consider the problem

$$\min\{-x : x \in \Psi(y), y \geq 0\},$$

where

$$\Psi(y) = \underset{x}{\mathrm{Argmin}} \ \{x^4 - yx^3 + 0.28y^2x^2\}.$$

The objective function of the lower level problem is plotted in Figure 5.4. Then, for all y the global lower level optimal solution is situated at the point $x^0 = 0$ but the points $x = (3/8 \pm 1/1600)y$ are also stationary. Hence, the bilevel programming problem has the global optimal solution $y^0 = 0$ with lower level optimal solution x^0 but this point is not stationary for problem (5.27) since there exist feasible solutions $(x, y) = ((3/8 \pm 1/1600)y, y)$ with strictly negative upper level objective function value. □

The last theorem can now be used as initial point for the formulation of necessary optimality conditions for the optimistic bilevel programming problem. Applying the necessary optimality conditions of [249] for an

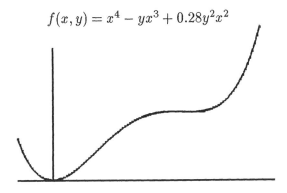

$$f(x,y) = x^4 - yx^3 + 0.28y^2x^2$$

Figure 5.4. Lower level objective function in the Example on page 150

(MPEC) to the bilevel program, a necessary optimality condition of Fritz John type can be derived without additional regularity assumptions:

THEOREM 5.16 *If (z^0, ν^0) with $z^0 = (x^0, y^0), \nu^0 = (\lambda^0, \mu^0)$ is a local optimal solution of problem (5.27) then there exists a non-vanishing vector $(\kappa_0, \kappa, \omega, \zeta, \tau, \xi)$ satisfying*

$$\kappa_0 \nabla F(z^0) + \kappa^\top (0, \nabla_y G(y^0)) + \nabla(\nabla_x L(z^0, \nu^0)\omega)$$
$$+\zeta^\top \nabla g(z^0) + \tau^\top \nabla h(z^0) = 0$$
$$(\nabla_x g(z^0), \nabla_x h(z^0))^\top \omega - (\xi, 0)^\top = 0$$
$$\kappa_0, \kappa \geq 0$$
$$g_i(z^0)\zeta_i = 0, \forall i$$
$$\lambda_i^0 \xi_i = 0, \forall i$$
$$\zeta_i \xi_i \geq 0, \forall i \in K$$
$$\kappa^\top G(y^0) = 0,$$

$$(5.29)$$

where the set $K = \{i : g_i(x^0, y^0) = \lambda_i^0 = 0\}$ and ∇ denotes the row vector of the partial derivatives with respect to x, y only.

To interpret the F. John conditions in Theorem 5.16 consider the F. John necessary optimality conditions of problem (5.28) for $I = I(x^0, y^0)$ which clearly are also valid. Then, the conditions (5.29) are weaker ones since the non-negativity of the multipliers to active inequality constraints is replaced by the demand that these multipliers have the same sign for all indices i for which strict complementarity slackness is violated. In distinction to Theorem 5.4 the result in Theorem 5.29 is not of a combinatorial nature. Also note that, due to Theorem 5.11, no constraint qualification can be verified for problem (5.26). Here the situation is substantially better with respect to the nonsmooth problem (5.27) as shown in Theorem 5.12. A weaker regularity condition than the (PLICQ) is a (strict) Mangasarian-Fromowitz constraint qualification as formulated below.

Let $(x^0, y^0, \lambda^0, \mu^0)$ be a feasible point for problem (5.27). Then, the following problem gives an approximation of (5.27) locally around $(x^0, y^0, \lambda^0, \mu^0)$:

$$
\begin{aligned}
F(x, y) \quad &\to \quad \min_{x, y, \lambda, \mu} \\
\nabla_x L(x, y, \lambda, \mu) \quad &= \quad 0 \\
g_i(x, y) \quad &= \quad 0 \text{ for } g_i(x^0, y^0) = 0 \\
\lambda_i \quad &= \quad 0 \text{ for } \lambda_i^0 = 0 \\
g_i(x, y) \quad &\leq \quad 0 \text{ for } g_i(x^0, y^0) < 0 \\
\lambda_i \quad &\geq \quad 0 \text{ for } \lambda_i^0 > 0 \\
h(x, y) \quad &= \quad 0 \\
G(y) \quad &\leq \quad 0.
\end{aligned}
\tag{5.30}
$$

Problem (5.30) models a certain section of the feasible set of problem (5.27). It is easy to see that the point $(x^0, y^0, \lambda^0, \mu^0)$ is feasible for problem (5.30) and that each feasible point for (5.30) is also feasible for (5.27). Hence, if $(x^0, y^0, \lambda^0, \mu^0)$ is a local optimal solution of problem (5.27) then it is also a local optimal solution of (5.30). In the following theorem we need a constraint qualification lying between the (LICQ) and the (MFCQ):

(MFCQ) The *Mangasarian-Fromowitz constraint qualification* is fulfilled for the problem

$$
\min\{f(x) : g(x) \leq 0, \ h(x) = 0\}
\tag{5.31}
$$

at the point x^0 if there exists a direction d satisfying

$$
\begin{aligned}
\nabla g_i(x^0)d < 0, \quad &\text{for each } i \in I(x^0) := \{j : g_j(x^0) = 0\}, \\
\nabla h_j(x^0)d = 0, \quad &\text{for each } j = 1, \dots, q
\end{aligned}
\tag{5.32}
$$

and the gradients $\{\nabla g_i(x^0) : i \in I(x^0)\} \cup \{\nabla_x h_j(x^0) : j = 1, \dots, q\}$ are linearly independent.

(SMFCQ) We say that the *strict Mangasarian-Fromowitz constraint qualification* (SMFCQ) is satisfied at x^0 for problem (5.31) if there exists a Lagrange multiplier (λ, μ),

$$
\lambda \geq 0, \lambda^\top g(x^0) = 0, \nabla f(x^0) + \lambda^\top \nabla g(x^0) + \mu^\top \nabla h(x^0) = 0,
$$

as well as a direction d satisfying

$$\begin{aligned}
\nabla g_i(x^0)d < 0, & \quad \text{for each } i \in I(x^0) \setminus J(\lambda), \\
\nabla g_i(x^0)d = 0, & \quad \text{for each } i \in J(\lambda), \\
\nabla h_j(x^0)d = 0, & \quad \text{for each } j = 1, \ldots, q,
\end{aligned} \tag{5.33}$$

and the gradients $\{\nabla g_i(x^0) : i \in J(\lambda)\} \cup \{\nabla_x h_j(x^0) : j = 1, \ldots, q\}$ are linearly independent.

(LICQ) The *linear independence constraint qualification* is fulfilled for the problem (5.31) at the point x^0 if the gradients $\{\nabla g_i(x^0) : i \in I(x^0)\} \cup \{\nabla_x h_j(x^0) : j = 1, \ldots, q\}$ are linearly independent.

If x^0 is a local optimal solution, then the Mangasarian-Fromowitz constraint qualification implies that the set of Lagrange multipliers is not empty and compact and the strict Mangasarian-Fromowitz constraint qualification is equivalent to the existence of a unique Lagrange multiplier [165].

In the following theorem, which is an application of the results in [249] to bilevel programming, we obtain that (MFCQ) for (5.30) implies that $\kappa_0 = 1$ can be taken in (5.29). If the stronger (SMFCQ) is valid for this problem than we can finally show necessary optimality conditions in a familiar form.

THEOREM 5.17 *Let* $(x^0, y^0, \lambda^0, \mu^0)$ *be a local minimizer of problem (5.27).*

- *If the (MFCQ) is valid for problem (5.30) at* $(x^0, y^0, \lambda^0, \mu^0)$, *then there exist multipliers* $(\kappa, \omega, \zeta, \tau, \xi)$ *satisfying*

$$\begin{aligned}
\nabla F(z^0) + \kappa^\top (0, \nabla_y G(y^0)) + \nabla(\nabla_x L(z^0, \nu^0)\omega) & \\
+ \zeta^\top \nabla g(z^0) + \tau^\top \nabla h(z^0) &= 0 \\
(\nabla_x g(z^0), \nabla_x h(z^0))^\top \omega - (\xi, 0)^\top &= 0 \\
g_i(z^0)\zeta_i &= 0, \ \forall i \\
\lambda_i^0 \xi_i &= 0, \ \forall i \\
\zeta_i \xi_i &\geq 0, \ i \in K \\
\kappa^\top G(y^0) &= 0, \\
\kappa &\geq 0,
\end{aligned} \tag{5.34}$$

where again $K = \{i : g_i(x^0, y^0) = \lambda_i^0 = 0\}$.

- *If the (SMFCQ) is fulfilled for the problem (5.30), then there exists unique multipliers $(\kappa, \omega, \zeta, \tau, \xi)$ solving*

$$
\begin{aligned}
\nabla F(z^0) + \kappa^\top (0, \nabla_y G(y^0)) + \nabla(\nabla_x L(z^0, \nu^0)\omega) \\
+ \zeta^\top \nabla g(z^0) + \tau^\top \nabla h(z^0) &= 0 \\
(\nabla_x g(z^0), \nabla_x h(z^0))^\top \omega - (\xi, 0)^\top &= 0 \\
g_i(z^0)\zeta_i &= 0, \ \forall i \\
\lambda_i^0 \xi_i &= 0, \ \forall i \\
\zeta_i \geq 0, \xi_i &\geq 0, \ i \in K \\
\kappa^\top G(y^0) &= 0, \\
\kappa &\geq 0,
\end{aligned}
$$

$$(5.35)$$

The following example from [229] for the more general mathematical program with equilibrium constraints (MPEC) shows that the last result is not valid if the (SMFCQ) is replaced with the (MFCQ):

Example: Consider the problem

$$
\begin{aligned}
-z_2 + z_3 &\to \min \\
-z_1 + z_2 - z_3 \leq 0, \ -z_3 &\leq 0, \\
-z_1 \leq 0, \ -z_1 - z_2 \leq 0 \ z_1(z_2 + z_3) &= 0.
\end{aligned}
$$

Optimal solutions of this problem are $z_1^* = 0$, $z_2^* = z_3^* \geq 0$. Consider the point $z^* = (0, 0, 0)^\top$. Then, the (MFCQ) is satisfied. For this problem, the system (5.35) reads as

$$
\begin{aligned}
-\zeta - \kappa_1 - \xi &= 0 \\
-1 - \xi + \kappa_1 &= 0 \\
1 - \kappa_1 - \kappa_2 &= 0 \\
\kappa_1(-z_1^* + z_2^* - z_3^*) &= \kappa_2(-z_3^*) = 0 \\
\zeta(-z_1^*) &= \xi(-z_1^* - z_2^*) = 0 \\
\zeta \geq 0, \xi \geq 0, \kappa &\geq 0
\end{aligned}
$$

This system has no solution since the first equation together with the nonnegativity conditions imply $\zeta = \xi = \kappa_1 = 0$ contradicting the second equation.   □

We will mention that also necessary and sufficient optimality conditions of second order can be derived using this nonsmooth reformulation of the bilevel programming problem. This has been done in [249].

5.6 THE APPROACH VIA THE LOWER LEVEL OPTIMAL VALUE FUNCTION

In the optimistic case, the bilevel programming problem (5.1), (5.5) can be transformed into an ordinary one-level optimization problem using the optimal value function $\varphi(y)$ of the lower level problem (5.1). This leads to the following equivalent problem:

$$
\begin{aligned}
F(x,y) &\to \min_{x,y} \\
f(x,y) - \varphi(y) &\le 0 \\
g(x,y) &\le 0 \\
h(x,y) &= 0 \\
G(y) &\le 0.
\end{aligned}
\tag{5.36}
$$

With respect to the relations between the optimistic problem (5.1), (5.5) and problem (5.36) we can repeat Theorem 5.15 but now we do not need the convexity assumption: A local optimistic optimal solution of the bilevel programming problem (5.1), (5.5) is a local optimal solution of problem (5.36). Again the opposite implication is in general not true.

Under suitable assumptions, the optimal value function is locally Lipschitz continuous (cf. e.g. Theorem 4.14). Then, if all functions in problem (5.36) are continuously differentiable (or at least locally Lipschitz continuous), necessary optimality conditions can be formulated using Clarke's generalized derivative. The difficult point here again is the regularity condition. The feasible set of this problem cannot have an inner point since the first inequality is satisfied as equation for all feasible points. Having a short look at the formulae for the directional derivative or the generalized gradient of the function $\varphi(\cdot)$ (cf. Theorems 4.16 and 4.17) it is easy to see that also generalized variants of the (MFCQ) or of the (LICQ) cannot be satisfied. Consider the optimization problem

$$
\min\{f(z) : g(z) \le 0, \ h(z) = 0\}
\tag{5.37}
$$

with locally Lipschitz continuous problem functions $g : \mathbb{R}^n \to \mathbb{R}^p$, $h : \mathbb{R}^n \to \mathbb{R}^q$.

(NLICQ) The *(nonsmooth) linear independence constraint qualification* holds at a feasible point z of the optimization problem (5.37) if each vectors $\eta^k \in \partial g_k(z)$, $k \in \{k : g_k(z) = 0\}$, $\xi^j \in \partial h_j(z)$, $j = 1, \ldots, q$, are linearly independent.

(NMFCQ) The *(nonsmooth) Mangasarian-Fromowitz constraint qualification* is said to be satisfied for the problem (5.37) if, for each vectors $\eta^k \in \partial g_k(z)$, $k \in \{k : g_k(z) = 0\}$, $\xi^j \in \partial h_j(z)$, $j = 1, \ldots, q$, the vectors in the set $\{\xi^j : j = 1, \ldots, q\}$ are linearly independent and there exists a vector $d \in \mathbb{R}^n$ satisfying

$$\eta^k d < 0, \ \forall k \in \{k : g_k(z) = 0\}$$
$$\xi^j d = 0, \ \forall j = 1, \ldots, q.$$

Both definitions can be found in [122]. Also, analogously to the smooth case, the nonsmooth linear independence constraint qualification implies the nonsmooth Mangasarian-Fromowitz constraint qualification. Let $L(z, \lambda, \mu) = f(z) + \lambda^\top g(z) + \mu^\top h(z)$ denote the Lagrangian of problem (5.37). The set

$$\Lambda^1(z) = \{(\lambda, \mu) \in \mathbb{R}^p_+ \times \mathbb{R}^q : 0 \in \partial_z L(z, \lambda, \mu), \lambda^\top g(z) = 0\} \qquad (5.38)$$

is the set of regular multipliers whereas the set

$$\Lambda^0(z) = \{(\lambda, \mu) \in \mathbb{R}^p_+ \times \mathbb{R}^q : 0 \in \lambda^\top \partial g(z) + \mu^\top \partial h(z), \lambda^\top g(z) = 0\} \qquad (5.39)$$

denotes the set of abnormal multipliers. As in the classical case, the nonsmooth Mangasarian-Fromowitz constraint qualification is equivalent to $\Lambda^0(z) = \{0\}$:

THEOREM 5.18 ([297]) *Let z be a feasible point for problem (5.37). Then, the nonsmooth Mangasarian-Fromowitz constraint qualification holds at z if and only if $\Lambda^0(z) = \{0\}$.*

Now we come back to the bilevel programming reformulation with the aid of the optimal value function of the lower level problem (5.36). The following theorem shows that a nonsmooth variant of a regularity condition can also not be satisfied for the reformulation (5.36) of the bilevel programming problem.

THEOREM 5.19 ([297]) *Consider problem (5.36) at a point (x^0, y^0) with $x^0 \in \Psi(y^0)$ and let the assumptions (C) and (MFCQ) be satisfied at each (global) optimal solution $x^0 \in \Psi(y^0)$ of the lower level problem and assume that equality holds in (4.27) in Chapter 4. Then, there exists a nontrivial abnormal multiplier for problem (5.36).*

This implies that no nonsmooth variant of neither the (MFCQ) nor the (LICQ) can be satisfied for the problem (5.36). Note that the condition on equality in equation (4.27) in Chapter 4 is satisfied if (LICQ) is valid for the lower level problem or if the lower level problem is either jointly convex or it is convex and has a unique global optimal solution (cf. Corollaries 4.5 – 4.7). Hence, this assumption is generically satisfied provided the problem functions are at least three times continuously differentiable and the Whitney topology is used [113].

Consider the following perturbed variant of problem (5.36):

$$\begin{aligned} F(x,y) &\to \min_{x,y} \\ f(x,y) - \varphi(y) + u &= 0 \\ g(x,y) &\leq 0 \\ h(x,y) &= 0 \\ G(y) &\leq 0. \end{aligned} \qquad (5.40)$$

Here, u is a parameter and both problems (5.36) and (5.40) coincide for $u = 0$.

DEFINITION 5.9 *Let (x^0, y^0) be an optimal solution of problem (5.36). Then, this problem is called partially calm at (x^0, y^0) if there exist $\kappa > 0$ and an open neighborhood $W_\omega(x^0, y^0, 0) \subset \mathbb{R}^n \times \mathbb{R}^m \times \mathbb{R}$, $\omega > 0$, such that for each $(x', y', u') \in W_\omega(x^0, y^0, 0)$ with (x', y') being feasible for problem (5.40) for $u = u'$ the inequality*

$$F(x', y') - F(x^0, y^0) + \kappa |u'| \geq 0$$

holds.

Partial calmness is a property closely related to calmness of a Lipschitz optimization problem [61]. Sufficient conditions for partial calmness of problem (5.36) at local optimal solutions are the following [297]:

- The lower level problem is a parametric linear programming problem.

- The function $F(\cdot, \cdot)$ is locally Lipschitz continuous with respect to x uniformly in y and the lower level problem has a uniformly weak sharp minimum.

Here, problem (5.1) has a uniformly weak sharp minimum if there exists $\kappa > 0$ such that

$$f(z, y) - \varphi(y) \geq \kappa \varrho(z, \Psi(y)) \ \forall z \in M(y) \ \forall y : G(y) \leq 0.$$

Now we are prepared to formulate first a necessary optimality of Fritz John type and then one of Karush-Kuhn-Tucker type.

THEOREM 5.20 ([297]) *Let (x^0, y^0) be a local optimal solution of problem (5.36). Let (MFCQ) and (C) be satisfied for the lower level problem at all points $x \in \Psi(y^0)$. Then, there exist $\kappa_0 \in \{0, 1\}$, $\mu_0 \geq 0$, $\xi \in \mathbb{R}_+^s$, $\gamma \in \mathbb{R}_+^p, \eta \in \mathbb{R}^q$ as well as $x^k \in \Psi(y^0)$, $(\lambda^k, \mu^k) \in \Lambda(x^k, y^0)$, $\zeta_k \geq 0$*

$k = 1, \ldots, m + 1, \ \sum\limits_{k=1}^{m+1} \zeta_k = 1 \ such \ that$

$$
\begin{aligned}
0 \ = \ & \kappa_0 \nabla_x F(x^0, y^0) + \mu_0 \nabla_x f(x^0, y^0) + \gamma^\top \nabla_x g(x^0, y^0) \\
& + \ \eta^\top \nabla_x h(x^0, y^0),
\end{aligned}
$$

$$
\begin{aligned}
0 \ = \ & \kappa_0 \nabla_y F(x^0, y^0) + \mu_0 \left(\nabla_y f(x^0, y^0) - \sum\limits_{k=1}^{m+1} \zeta_k \nabla_y L(x^k, y^0, \lambda^k, \mu^k) \right) \\
& + \ \gamma^\top \nabla_y g(x^0, y^0) + \eta^\top \nabla_y h(x^0, y^0) + \xi^\top \nabla_y G(y^0), \\
& \xi^\top G(y^0) = 0, \ \gamma^\top g(y^0, y^0) = 0.
\end{aligned}
$$

Using the partial calmness assumption we can guarantee that the leading multiplier $\kappa_0 \neq 0$.

THEOREM 5.21 ([297]) *Let the assumptions of the preceeding theorem be satisfied and let problem (5.36) be partially calm at a local optimistic optimal solution (x^0, y^0) of the bilevel programming problem (5.1), (5.2). Assume that equality holds in (4.27) and that (ULR) is valid for the upper level problem. Then, the conclusion of Theorem 5.20 holds with $\kappa_0 = 1$ and $\mu_0 > 0$.*

Different reformulations of the bilevel programming problem (5.1), (5.2) can be used as starting points for the construction of exact penalty function approaches. Some examples for these approaches are given in Section 6.3. The resulting problems are (nonsmooth) one-level programming problems which clearly can be used to derive necessary and sufficient optimality conditions for the bilevel programming problem. Since this is more or less standard in optimization we left it for the interested reader.

5.7   GENERALIZED PC¹ FUNCTIONS

In this section an approach for deriving optimality conditions is developed which can be used for both the optimistic and the pessimistic local optimal solutions. It is closely related to the implicit function approach presented in Section 5.2. We used the main ideas of this approach already in the optimality conditions in Subsection 3.5.2. Recall that, if all the lower level optimal solutions are strongly stable, then inserting them into the upper level objective function, a family of continuous functions arises. Depending on the chosen optimistic respectively pessimistic position for treating the ambiguity resulting from non-unique lower level solutions, either the pointwise maximum or the pointwise minimum function of this family is to be minimized in solving the bilevel problem. If this function would be continuous a PC^1-function arises but, unfortunately, this is in general not the case. This leads to the following considerations.

5.7.1 DEFINITION

First we will give a simple example to motivate the following definition.

Example: Consider the following bilevel problem with a linear parametric optimization problem as the lower level. Let $x, y \in \mathbb{R}^2$ and

$$\Psi(y) = \underset{x}{\text{Argmin}} \ \{\langle x, y \rangle : x \geq 0; x_1 + x_2 \leq y_1^2 + y_2^2\}.$$

As the objective function in the upper level we choose

$$F(x, y) = x_2 - x_1^2 y_1^2.$$

Obviously the solution set of the lower level is in general not single-valued; so we take (5.5) as a regularization. By simple calculations we obtain

$$\varphi_o(y) = \begin{cases} 0 & =: \varphi_1(y) & \text{for } y \geq 0, \\ -y_1^2(y_1^2 + y_2^2) & =: \varphi_2(y) & \text{for } y_1 < 0, y_1 \leq y_2, \\ y_1^2 + y_2^2 & =: \varphi_3(y) & \text{for } y_2 < 0, y_1 > y_2. \end{cases}$$

□

Here, φ_o is a selection of continuously differentiable functions φ_i, $1 \leq i \leq 3$, however, it is not continuous itself as PC^1-functions are. This motivates the following definition.

DEFINITION 5.10 ([77]) *A function $\alpha : \mathbb{R}^m \to \mathbb{R}^n$ is called generalized PC^1-function (or shortly GPC^1-function) at $x^0 \in \mathbb{R}^m$ if the following conditions are fulfilled:*

- *There exist an open neighborhood $U_\varepsilon(x^0)$ of x^0 and a finite number of continuously differentiable functions $\alpha^i : U_\varepsilon(x^0) \to \mathbb{R}^n, i = 1, \ldots, k$, with*

$$\alpha(z) \in \{\alpha^i(z)\}_{i \in \{1, \ldots, k\}} \ \forall z \in U_\varepsilon(x^0).$$

- *The interior set int $\text{Supp}(\alpha, \alpha^i)$ of the support set is connected and $\text{Supp}(\alpha, \alpha^i) \subseteq \text{cl int Supp}(\alpha, \alpha^i)$ is satisfied for $i = 1, \ldots, k$.*

- *For the contingent cones $K_{\text{Supp}(\alpha, \alpha^i)}(x^0)$ to the sets $\text{Supp}(\alpha, \alpha^i)$ the condition $K_{\text{Supp}(\alpha, \alpha^i)}(x^0) \subseteq \text{cl int } K_{\text{Supp}(\alpha, \alpha^i)}(x^0)$ holds, $i = 1, \ldots, k$.*

- *There exists $\delta > 0$ such that the following is true for all vectors $r \in K_{\mathrm{Supp}(\alpha,\alpha^i)}(x^0) \cap K_{\mathrm{Supp}(\alpha,\alpha^j)}(x^0)$, $\|r\| = 1$ with $i \neq j$:*
  $\exists t_0 = t(r) \geq \delta$ *with*

$$x^0 + tr \in \mathrm{Supp}(\alpha, \alpha^i) \ \text{or} \ x^0 + tr \in \mathrm{Supp}(\alpha, \alpha^j) \ \forall t \in (0, t_0).$$

- *For $i \neq j$ we have*

$$\mathrm{int} \ K_{\mathrm{Supp}(\alpha,\alpha^i)}(x^0) \cap \mathrm{int} \ K_{\mathrm{Supp}(\alpha,\alpha^j)}(x^0) = \emptyset.$$

α is called a GPC^1-function on $U \subseteq \mathbb{R}^m$ if it is GPC^1 at each point $x^0 \in U$.

Of course, for the investigation of GPC^1-functions and the possible implementation of an algorithm minimizing such a function, we need a special kind of a derivative. The classical gradient or directional derivative is not suited because GPC^1-functions do not need to be continuous. It turns out that the radial-directional derivative introduced by Recht [236] fits very well with the above definition of a generalized PC^1-function. Therefore before we explain the individual demands in Definition 5.10 we will repeat the definition of the radial-directional derivative in the sense of Recht which makes the following explanations a little bit clearer.

DEFINITION 5.11 *Let $U \subseteq \mathbb{R}^m$ be an open set, $x^0 \in U$ and $\alpha : U \to \mathbb{R}$. We say that α is radial-continuous at x^0 in direction $r \in \mathbb{R}^m, \|r\| = 1$, if there exists a real number $\alpha_r(x^0)$ such that*

$$\lim_{t \downarrow 0} \alpha(x^0 + tr) = \alpha_r(x^0).$$

If the radial limit $\alpha_r(x^0)$ exists for all $r \in \mathbb{R}^m, \|r\| = 1$, α is called radial-continuous at x^0.
α is radial-directionally differentiable at x^0, if there exists a positively homogeneous function $d\alpha(x^0; \cdot) : \mathbb{R}^m \to \mathbb{R}$ such that

$$\alpha(x^0 + tr) - \alpha_r(x^0) = t d\alpha(x^0; r) + o(x^0, tr)$$

with $\lim_{t \downarrow 0} \frac{o(x^0, tr)}{t} = 0$ holds for all $r \in \mathbb{R}^m, \|r\| = 1$, and all $t > 0$.

Obviously, the vector $d\alpha(x^0; \cdot)$ is uniquely defined and is called the radial-directional derivative of α at x^0.

This definition is originally given just for real-valued functions. We will generalize it for vector-valued functions in the following way:

DEFINITION 5.12 *A function* $\alpha = (\alpha^1, \ldots, \alpha^n) : U \subseteq \mathbb{R}^m \to \mathbb{R}^n$ *is called radial-continuous if every component function* α_i *has this property. It is radial-directionally differentiable at* $x^0 \in U$, *if all functions* $\alpha_i, i = 1, \ldots, n$, *are radial-directionally differentiable at* x^0. *The vector* $(d\alpha_1(x^0; r), \ldots, d\alpha_n(x^0; r))^\top$ *is said to be the radial-directional derivative* $d\alpha(x^0; r)$ *of* α *in direction* r.

Now we will make the something strange demands in definition 5.10 clearer. It should be noted that the number k of involved functions α^i is chosen to be the smallest of all possible such that all the other demands are satisfied. Condition $K_{\text{Supp}(\alpha,\alpha^i)}(x^0) \subseteq \text{cl int } K_{\text{Supp}(\alpha,\alpha^i)}(x^0)$ is essential for radial-directional differentiability while $\text{Supp}(\alpha, \alpha^i) \subseteq \text{cl int Supp}(\alpha, \alpha^i)$ is needed to make several conclusions from this differentiability possible.

Note that the set of all GPC^1-functions does not create an algebraic structure since the sum of two such functions is not necessarily GPC^1.

Example: [77] Consider the following two GPC^1-functions:

$$\alpha^1(x) = \begin{cases} x_1^2 + x_2^2, & x \geq 0, \\ 0, & \text{else,} \end{cases} \qquad \alpha^2(x) = \begin{cases} x_1^2 + x_2^2, & x_1 \geq 0, \ x_2 \leq 0, \\ 0, & \text{else.} \end{cases}$$

Then,

$$\alpha^1(x) + \alpha^2(x) = \begin{cases} 2(x_1^2 + x_2^2), & x_1 \geq 0, \ x_2 = 0, \\ x_1^2 + x_2^2, & x_1 \geq 0, \ x_2 \neq 0, \\ 0, & \text{else.} \end{cases}$$

□

In the following we derive some simple properties of GPC^1-functions.

THEOREM 5.22 ([77]) *Directly from the definition one derives the following properties:*

- *The sets* $K_{\text{Supp}(\alpha,\alpha^i)}(x^0)$ *are closed cones.*

- $\displaystyle\bigcup_{1 \leq i \leq k} K_{\text{Supp}(\alpha,\alpha^i)}(x^0) = \mathbb{R}^m$

- *If* $r \in K_{\text{Supp}(\alpha,\alpha^i)}(x^0)$ *and* $r \notin K_{\text{Supp}(\alpha,\alpha^j)}(x^0) \ \forall j \neq i$ *then*

$$\exists \varepsilon > 0 : x^0 + tr \in \text{Supp}(\alpha, \alpha^i) \ \forall 0 < t < \varepsilon. \qquad (5.41)$$

THEOREM 5.23 ([77]) *Generalized PC^1-functions α are both radial-continuous and radial-directionally differentiable.*

The following small examples should underline that the properties from Definition 5.10 are essential for the investigations using the radial-directional derivative.

Example: [77] Let $\alpha : \mathbb{R}^2 \to \mathbb{R}$ be defined as

$$\alpha(x_1, x_2) = \begin{cases} 0, & (x_1 - 1)^2 + x_2^2 = 1, x_1 \neq 0, \\ 1, & x_1 = x_2 = 0, \\ 1 + |x_1| + |x_2|, & \text{otherwise.} \end{cases}$$

Here, the function values on $\mathrm{Supp}(\alpha, \alpha^1) = \{x : (x_1 - 1)^2 + x_2^2 = 1, x_1 \neq 0\}$ do not have any influence on the radial-directional derivative because the corresponding tangent cone has empty interior. Hence, this function does not satisfy the third condition of Definition 5.10 at $x^0 = 0$. This means that the function values of this function can be changed arbitrarily on the set $\mathrm{Supp}(\alpha, \alpha^1)$. Such functions are not suitable for investigations relying on the derivative in the sense of Recht. Especially the following necessary optimality condition is satisfied for this function independent of its function values on $\mathrm{Supp}(\alpha, \alpha^1)$. $\square$

Example: [77] Consider the function

$$\alpha(x_1, x_2) = \begin{cases} x_2, & \text{if } x_2 \geq x_1 \sin(1/x_1), \ x_1 > 0, \\ x_1, & \text{if } x_2 \leq x_1 \sin(1/x_1), \ x_1 > 0, \\ 0, & \text{if } x_1 \leq 0. \end{cases}$$

This function is radial-continuous at zero but has no radial-directional derivative in the direction $r = (1,0)^\top$ at $x = 0$. By a slight modification of the function values we can derive a function which is not radial-continuous. In this example the 4th property of the Definition 5.10 is not satisfied. $\square$

5.7.2   CRITERIA BASED ON THE RADIAL-DIRECTIONAL DERIVATIVE

In the sequel we will deal with criteria for optimality. We start with the following necessary one. Here Ψ_{loc} denotes the set of local minima of the function $\alpha(\cdot)$.

THEOREM 5.24 ([77]) *Let $\alpha : \mathbb{R}^m \to \mathbb{R}$ be a GPC^1-function and $x^0 \in \mathbb{R}^m$ a fixed point. If there exists $r \in \mathbb{R}^m$ such that one of the following two conditions is satisfied then $x^0 \notin \Psi_{loc}$.*

- $d\alpha(x^0; r) < 0$ *and* $\alpha_r(x^0) \leq \alpha(x^0)$

- $\alpha_r(x^0) < \alpha(x^0)$.

It should be noticed that Theorem 5.24 is valid for general radial-continuous and radial-differentiable functions, the restriction to GPC^1-functions is not necessary. Especially the demands on the contingent cones to the support sets of the selection functions are not needed in this theorem which is essential for the application to bilevel programming problems.

In the following we will call points x^0 satisfying the foregoing necessary optimality condition of first order stationary. We go on with a sufficient criterion.

THEOREM 5.25 ([77]) *Let* $\alpha : \mathbb{R}^m \to \mathbb{R}$ *be a generalized PC^1-function and x^0 a fixed point which satisfies one of the following two conditions.*

- $\alpha(x^0) < \alpha_r(x^0) \ \forall r \in \mathbb{R}^m$

- $\alpha(x^0) \leq \alpha_r(x^0) \ \forall r$ *and* $d\alpha(x^0; r) > 0 \ \forall r : \alpha(x^0) = \alpha_r(x^0), \|r\| = 1$.

Then, α achieves a local minimum at x^0.

The following example shows that the assumptions of this theorem are necessary.

Example: Let $\alpha : \mathbb{R}^2 \to \mathbb{R}$ be defined as

$$\alpha(x_1, x_2) = \begin{cases} 0, & (x_1 - 1)^2 + x_2^2 = 1, x_1 \neq 0, \\ 1, & x_1 = x_2 = 0, \\ 2 + |x_1| + |x_2|, & \text{otherwise,} \end{cases}$$

and consider the point $x^0 = (0,0)^\top$. Then, the first condition of Theorem 5.25 is satisfied for each direction r but x^0 is not a local minimum. Now consider the modification of this function given in Example on page 162. Then, for each direction the condition 2. of Theorem 5.25 is valid but x^0 is again no local optimum. □

5.7.3 RADIAL SUBDIFFERENTIAL

We now want to introduce the new notion of a radial subdifferential which will be used for further investigations in connection with GPC^1-functions.

DEFINITION 5.13 *Let* $U \in \mathbb{R}^m$, $x^0 \in U$, *and* $\alpha : U \to \mathbb{R}$ *be radial-directional differentiable at x^0. We say $d^\top \in \mathbb{R}^m$ is a radial subgradient of α at x^0 if*

$$\alpha(x^0) + dr \leq \alpha_r(x^0) + d\alpha(x^0; r) \qquad (5.42)$$

is satisfied for all r with $\alpha(x^0) \geq \alpha_r(x^0)$.

The set of all radial subgradients is called radial subdifferential and denoted by $\partial_{rad}\alpha(x^0)$.

The following necessary criterion for the existence of a radial subgradient is valid:

THEOREM 5.26 ([77]) *If there exists $r \in \mathbb{R}^m$ such that $\alpha_r(x^0) < \alpha(x^0)$ then $\partial_{rad}\alpha(x^0) = \emptyset$.*

Arguing by contradiction, the proof is easy done by using positive homogeneity of the scalar product and the radial-directional derivative.

With this theorem we get the following equivalent definition of a radial subgradient:

$$\partial_{rad}\alpha(x^0) = \{d^\top \in \mathbb{R}^m : dr \leq d\alpha(x^0; r) \; \forall r \text{ satisfying } \alpha(x^0) = \alpha_r(x^0)\},$$

if there is no direction such that the radial limit in this direction is less than the function value.

We now want to present some ideas for finding radial subgradients. Consider any index i. For each $r \in \text{int } K_{\text{Supp}(\alpha,\alpha^i)}(x^0)$ we have $d\alpha(x^0; r) = \nabla\alpha^i(x^0)r$ which, together with the necessary condition according to Theorem 5.26, leads to

$$\alpha(x^0) + dr \leq \alpha_r(x^0) + \nabla\alpha^i(x^0)r \; \forall r \in \text{int } K_{\text{Supp}(\alpha,\alpha^i)}(x^0).$$

Clearly, a vector d satisfying this inequality can only belong to the radial subdifferential if it satisfies the respective inequality also for all other tangent cones $K_{\text{Supp}(\alpha,\alpha^j)}(x^0)$ for which the condition $\alpha^j(x^0) = \alpha(x^0)$ holds for. Let $I_\alpha(x^0) := \{j : \alpha(x^0) = \alpha^j(x^0)\}$. With this definition we have:

$$d \in \partial_{rad}\alpha(x^0) \iff (d - \nabla\alpha^j(x^0))r \leq 0 \; \forall r \in \bigcup_{j \in I_\alpha(x^0)} \text{int } K_{\text{Supp}(\alpha,\alpha^j)}(x^0).$$

In the following special cases it is easy to compute elements of the radial subdifferential:

THEOREM 5.27 ([77]) *Consider a GPC^1-function $\alpha : \mathbb{R}^m \to \mathbb{R}$ at a point x^0.*

- *If $\alpha(x^0) < \alpha_r(x^0)$ for all r, then $\partial_{rad}(x^0) = \mathbb{R}^m$.*

- *If $\alpha(x^0) = \alpha_r(x^0)$ if and only if $r \in K_{\text{Supp}(\alpha,\alpha^i)}(x^0)$ and $\alpha(x^0) < \alpha_r(x^0)$ for all $r \notin K_{\text{Supp}(\alpha,\alpha^i)}(x^0)$ for some fixed i, then $\nabla\alpha^i(x^0) \in \partial_{rad}(x^0)$.*

- *If α is continuous on a certain open neighborhood of x^0, it is a special PC^1-function and, hence, locally Lipschitz [114]. In this case, the radial subdifferential coincides with Clarke's generalized gradient.*

THEOREM 5.28 ([77]) *The radial subdifferential of GPC^1-functions has the following properties:*

- $\partial_{rad}\alpha(x^0)$ *is a closed set.*

- $\partial_{rad}(\lambda\alpha)(x^0) = \lambda\partial_{rad}\alpha(x^0) \; \forall\lambda > 0$

- $\partial_{rad}(\alpha^1 + \alpha^2)(x^0) \supseteq \partial_{rad}\alpha^1(x^0) + \partial_{rad}\alpha^2(x^0)$

- $\sum_i \lambda_i\partial_{rad}\alpha^i(x^0) \subseteq \partial_{rad}\left(\sum_i \lambda_i\alpha^i(x^0)\right) \; \forall\lambda_i \geq 0$

- $\partial_{rad}\alpha(x^0)$ *is convex.*

- *If α is continuous at x^0 then $\partial_{rad}\alpha(x^0)$ is bounded.*

- *Let $r^0 \neq 0$ be such that $\alpha(x^0) = \alpha_{r^0}(x^0)$. Then,*

$$d\alpha(x^0; r^0) \geq \sup\{dr^0 : d \in \partial_{rad}\alpha(x^0)\}.$$

The proof of this theorem is straightforward. It should be noted that the statements 5.28, 5.28, 5.28 may in general not be strengthened. Now we derive optimality criteria in connection with the radial subdifferential.

THEOREM 5.29 ([77]) *Let $\alpha : \mathbb{R}^m \to \mathbb{R}$ be a GPC^1-function. If $x^0 \in \Psi_{loc}$ then $0 \in \partial_{rad}\alpha(x^0)$.*

THEOREM 5.30 ([77]) *Let $\alpha : \mathbb{R}^m \to \mathbb{R}$ be a GPC^1-function. If $0 \in$ int $\partial_{rad}\alpha(x^0)$ then α achieves at x^0 a local minimum.*

It should be noted that this last theorem is valid only for GPC^1-functions. This can easily be seen considering the function $\alpha(\cdot)$ in Example 5.7.1. For this function $\partial_{rad}\alpha(x^0) = \{d : -1 \leq d_i \leq 1, \; i = 1, 2\}$. Hence, $0 \in$ int $\partial_{rad}\alpha(x^0)$, but x^0 is not a local minimum.

5.7.4 APPLICATION TO BILEVEL PROGRAMMING

In this subsection we will turn back to bilevel programming problems with the above mentioned regularization according to (5.5) or (5.6).

First consider (mixed-discrete) linear lower level problems with parameters in the right-hand side and the objective function. For linear problems the sets of all parameter values for which one optimal solution respective the corresponding basic matrix remains constant (the so-called region of stability, cf. Definition 3.3) are polyhedral [220]. This opens the way for a direct application of the above results e.g. to linear bilevel programming problems. The resulting properties are given in Theorems 3.8 to 3.10.

Now consider a mixed-discrete linear optimization problem

$$\min\{c^{\top}x : Ax \leq b, x \text{ integer}\}$$

with parameters b, c. Then, the region of stability $\mathcal{R}(x^0)$ has the structure

$$\mathcal{R}(x^0) = U(c) \times \left(D(b) \setminus \bigcup_{x \in \Delta_1(b)} D^x\right),$$

where $U(c) \subseteq \mathbb{R}^n, D(b) \subseteq \mathbb{R}^p, D^x \subseteq \mathbb{R}^p$ are polyhedral cones and $\Delta_1(b)$ is a finite set [15, 16]. These are the main points for validating the conditions for the (contingent cones to the) support sets in the definition of generalized PC^1-functions.

For a fixed parameter value, the (convex hull of the) set of optimal solutions of the lower level problem is also a polyhedral set. Since linear functions have minima at vertices of polyhedra, the optimal solutions of the problems (5.4) and (5.7) form piecewise affine-linear functions and, hence, the functions $\varphi_o(\cdot)$ and $\varphi_p(\cdot)$ are itself generalized PC^1-functions in these cases. This makes the place free for a direct application of the necessary and sufficient optimality conditions for minimizing generalized PC^1-functions. We have seen the corresponding results for linear lower and upper level problems in Subsection 3.5.2. Linear bilevel problems with discrete lower level problems will be considered in Section 8.4.

Second we consider bilevel programming problems in which all stationary solutions in the lower level problem are strongly stable in the sense of Kojima [157]. If the assumptions in the Theorem 4.4 are satisfied for all stationary points of (5.1) for all feasible parameter values, then there exists a finite number of continuous functions $x^j(\cdot)$, $j = 1, \ldots, k$ such that

$$\Psi(y) \subseteq \{x^1(y), \ldots, x^k(y)\} \; \forall y \text{ with } G(y) \leq 0.$$

This implies that the functions $\varphi_o(\cdot)$ and $\varphi_p(\cdot)$ are selections of the continuous (and directionally differentiable by the remarks after Theorem 4.11) functions

$$\varphi_o(y) \in \{F(x^1(y), y), \ldots, F(x^k(y), y)\} \; \forall y \text{ with } G(y) \leq 0 \qquad (5.43)$$

and

$$\varphi_p(y) \in \{F(x^1(y), y), \ldots, F(x^k(y), y)\} \; \forall y \text{ with } G(y) \leq 0. \qquad (5.44)$$

It has been shown in [141] that continuity of the functions $\varphi_o(\cdot)$ and $\varphi_p(\cdot)$ can only be guaranteed if the optimal solution of the lower level problem (5.1) is uniquely determined. Since this is in general not the case if $k > 1$, both functions are generally not continuous.

Now, add the assumption (CRCQ) to (MFCQ) and (SSOC). Then the functions (5.43) or (5.44) are finite selections of locally Lipschitz functions. But due to the more difficult structure of the regions of stability than in the linear case, we will in general not be able to prove that the functions (5.43) or (5.44) are generalized PC^1-functions. Nevertheless, we can take advantage of the ideas in this Section to derive necessary and sufficient optimality conditions for the bilevel programming problem in this case.

The main drawback resulting from the vacancy of the properties of generalized PC^1-functions is the lost of directional differentiability of the functions $\varphi_o(\cdot)$ and $\varphi_p(\cdot)$.

We start with necessary and sufficient conditions for optimistic optimality.

THEOREM 5.31 *Consider the bilevel programming problem (5.1), (5.2). Let the assumptions (C), (CRCQ), (MFCQ), (SSOC) together with (ULR) be satisfied at all feasible points (x, y^0) with $G(y^0) \leq 0$, $x \in \Psi(y^0)$ and denote the lower level globally optimal solutions by $\Psi(y^0) = \{x^1(y^0), \ldots, x^k(y^0)\}$. Let $T = \{i \in \{1, \ldots, k\} : F(x^i(y^0), y^0) = \varphi_o(y^0)\}$ denote the subset of the global lower level optimal solutions which are feasible for the optimistic bilevel problem.*

- *If (x^0, y^0), $x^0 \in \Psi(y^0)$ is a local optimistic optimal solution then, for all $i \in T$ the following system of inequalities has no solution r:*

$$\nabla_y G_j(y^0)r \leq 0, \text{ for all } j : G_j(y^0) = 0 \tag{5.45}$$

$$\nabla_x F(x^i(y^0), y^0)x^{i'}(y^0; r) + \nabla_y F(x^i(y^0), y^0)r < 0 \tag{5.46}$$

$$\nabla_x f(x^i(y^0), y^0)x^{i'}(y^0; r) + \nabla_y f(x^i(y^0), y^0)r$$
$$< \nabla_x f(x^j(y^0), y^0)x^{j'}(y^0; r) + \nabla_y f(x^j(y^0), y^0)r \ \forall \ j \in T \setminus \{i\}. \tag{5.47}$$

- *Let (x^0, y^0), $x^0 \in \Psi(y^0)$ be feasible for the optimistic problem, i.e. let $F(x^0, y^0) = \varphi_o(y^0)$. If for all $i \in T$ the following system of inequalities has no nontrivial solution r:*

$$\nabla_y G_j(y^0)r \leq 0, \text{ for all } j : G_j(y^0) = 0 \tag{5.48}$$

$$\nabla_x F(x^i(y^0), y^0)x^{i'}(y^0; r) + \nabla_y F(x^i(y^0), y^0)r \leq 0 \tag{5.49}$$

$$\nabla_x f(x^i(y^0), y^0)x^{i'}(y^0; r) + \nabla_y f(x^i(y^0), y^0)r$$
$$\leq \nabla_x f(x^j(y^0), y^0)x^{j'}(y^0; r) + \nabla_y f(x^j(y^0), y^0)r \ \forall \ j \in T \setminus \{i\}. \tag{5.50}$$

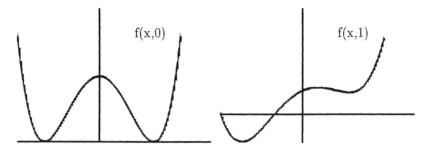

Figure 5.5.   The lower level objective function in the Example on page 168

then (x^0, y^0) is locally optimistic optimal for the bilevel programming problem.

The following examples should illustrate the optimality conditions.

Example: We consider a lower level problem without constraints:

$$\Psi(y) = \text{Argmin } \{f(x, y) = x^4 - 2x^2 + xy + 1\}$$

and the bilevel programming problem

$$\min\{x^2 : x \in \Psi(y)\}$$

The objective function in the lower level problem is plotted in Figure 5.5. Then, since we are searching for a global minimum in the lower level problem, $f(x(y), y) \leq 0$ and it can be shown that $x^2(y) \geq 1$ for all $y, x(y) \in \Psi(y)$. Hence, the optimal solutions of the bilevel programming problem are $(-1, 0)$ and $(1, 0)$. Both local solution functions $x^i(y)$ are differentiable at the point $y = 0$ with the common derivative $x^{i'}(0) = -0.125$, the directional derivatives of the auxiliary objective functions $f(x^i(y), y)$ at $y^0 = 0$ are

$$f_x(x^i(y^0), y^0)x^{i'}(y^0; r) + f_y(x^i, y^0)r$$
$$= (4x^3 - 4x + y)x^{i'}(0)r + x^i(0)r = \begin{cases} -r & \text{if } x^i(0) = -1 \\ r & \text{if } x^i(0) = 1 \end{cases}$$

and

$$F_x(x^i(y^0), y^0)x^{i'}(y^0; r) + F_y(x^i(y^0), y^0)r = 2xx^{i'}(0)r$$
$$= \begin{cases} 0.25r & \text{if } x^i(0) = -1 \\ -0.25r & \text{if } x^i(0) = 1 \end{cases}$$

Hence, the necessary optimality condition in Theorem 5.31 is the following: There does not exist directions r satisfying at least one of the following systems:

$$0.25r < 0 \atop -r < r \Biggr\} \qquad -0.25r < 0 \atop r < -r \Biggr\}$$

Clearly, this is true. Note that the points $(-1,0)$ and $(1,0)$ are not local optimal solutions for the problem (5.26) since local and global lower level solutions are not distinguished in this problem. □

Example: Now, change the lower level problem to

$$\Psi(y) = \text{Argmin } \{\hat{f}(x,y) = x^4 - 2x^2 + xy^3 + 1\}.$$

Then, since $\hat{f}''(x,0) > 0$ for $x^0 = \pm 1$, the local minima $x^0 = \pm 1$ are strongly stable local minima for y near zero. Let $x^1(y)$ and $x^2(y)$ denote the (differentiable) local optimal solution functions of the lower level problem with $x^1(0) = -1, x^2(0) = 1$. It is easy to see that it is $x^{i'}(0) = 0$ for $i = 1, 2$. The directional derivative of the lower level objective function is now

$$f_x(x^i(y^0), y^0)x^{i'}(y^0; r) + f_y(x^i, y^0)r = 0$$

for both $i = 1$ and $i = 2$ for all directions. The optimal solutions of the bilevel programming problem remain unchanged, the necessary optimality condition in Theorem 5.31 is valid since there is no direction r such that

$$f_x(-1,0)x^{1'}(0; r) + f_y(-1,0)r < f_x(1,0)x^{2'}(0; r) + f_y(1,0)r.$$

The stronger sufficient condition is not satisfied. □

Example: Consider again the lower level problem

$$\Psi(y) = \text{Argmin } \{\hat{f}(x,y) = x^4 - 2x^2 + xy^3 + 1\}$$

and search for a minimum of the function

$$F(x,y) = -xy$$

subject to $x \in \Psi(y)$. A closer look at both local minima shows that the global minimum of $\hat{f}(x,y)$ is near the point $x^0 = -1$ for $y > 0$ and near the point $x^0 = 1$ for $y < 0$ and y sufficiently close to zero. This implies that the product $-xy \geq 0$ for $x \in \Psi(y)$ and y near zero. Hence, the points $(x^0, y^0) = (-1, 0)$ and $(x^0, y^0) = (1, 0)$ are local optima and

$y^0 = 0$ minimizes the function $\varphi_0(y)$ locally. Considering the upper level objective function we obtain

$$F_x(x^i(y^0), y^0)x^{i'}(y^0; r) + F_y(x^i(y^0), y^0)r = -x^i(y^0)r.$$

The necessary optimality condition in Theorem 5.31 is satisfied at $(x^0, y^0) = (-1, 0)$ since there is no direction r such that

$$f_x(-1, 0)x^{1'}(0; r) + f_y(-1, 0)r < f_x(1, 0)x^{2'}(0; r) + f_y(1, 0)r$$

and

$$F_x(-1, 0)x^{1'}(0; r) + F_y(-1, 0)r < 0.$$

But the system

$$F_x(-1, 0)x^{1'}(0; r) + F_y(-1, 0)r < 0$$
$$f_x(-1, 0)x^{1'}(0; r) + f_y(-1, 0)r \leq f_x(1, 0)x^{2'}(0; r) + f_y(1, 0)r$$

has nonvanishing solutions. This shows that it is not possible to replace the respective strong inequalities in the necessary conditions by inequalities. □

Now we turn over to the pessimistic optimal solution. The main difference to the previous theorem is that we have to guarantee that the upper level variable point y^0 cannot be reached via a sequence $\{y^k\}_{k=1}^{\infty}$ for which $\lim_{k \to \infty} \varphi_p(y^k) < \varphi_p(y^0)$.

THEOREM 5.32 *Consider the bilevel programming problem (5.1), (5.2). Let the assumptions (C), (CRCQ), (MFCQ), (SSOC) together with (ULR) be satisfied at all feasible points (x, y^0) with $G(y^0) \leq 0$, $x \in \Psi(y^0)$ and denote the lower level globally optimal solutions by $\Psi(y^0) = \{x^1(y^0), \ldots, x^k(y^0)\}$. Let $T = \{i \in \{1, \ldots, k\} : F(x^i(y^0), y^0) = \varphi_p(y^0)\}$ denote the subset of the global lower level optimal solutions which are feasible for the pessimistic bilevel problem. The point (x^0, y^0) is not a local pessimistic optimal solution if one of the following conditions is satisfied:*

■ $K_{\{y:G(y) \leq 0\}}(y^0) \not\subseteq \bigcup_{i \in T} K_{\text{Supp}(x, x^j)}(y^0).$

■ *There exists $i \in T$ such that the following system of inequalities has a solution r:*

$$\nabla_y G_j(y^0)r \leq 0, \text{ for all } j : G_j(y^0) = 0 \tag{5.51}$$
$$\nabla_x F(x^i(y^0), y^0)x^{i'}(y^0; r) + \nabla_y F(x^i(y^0), y^0)r < 0 \tag{5.52}$$

$$\nabla_x f(x^i(y^0), y^0) x^{i'}(y^0; r) + \nabla_y f(x^i(y^0), y^0) r$$
$$> \nabla_x f(x^j(y^0), y^0) x^{j'}(y^0; r) + \nabla_y f(x^j(y^0), y^0) r \ \forall \ j \in T \setminus \{i\}.$$
$$(5.53)$$

REMARK 5.5 *The condition* $K_{\{y:G(y)\leq 0\}}(y^0) \subseteq \bigcup_{i \in T} K_{\mathrm{Supp}(x,x^i)}(y^0)$ *is clearly satisfied in any one of the following situations:*

- $\varphi_p(y^0) = \varphi_o(y^0)$

- $|\Psi(y^0)| = 1$

- $\{i : F(x^i(y^0), y^0) = \varphi_p(y^0)\} = \{x^i(y^0)\}$ *and there exists an open neighborhood $V_\delta(y^0)$ of y^0 with $x^i(y) \in \Psi(y)$ for all $y \in V_\delta(y^0)$*

As an example illustrating the first condition consider again the Example on page 168 but now with another upper level objective function.

Example: We consider a lower level problem without constraints:

$$\Psi(y) = \mathrm{Argmin} \ \{f(x, y) = x^4 - 2x^2 + xy + 1\}$$

and the bilevel programming problem

$$\min\{\varphi_p(y) : -1 \leq y \leq 1\}$$

with $\varphi_p(y) = \max\{x^2 + x : x \in \Psi(y)\}$. Then again, there are two differentiable functions $x^i(y)$ describing local optimal solutions of the lower level problem in an open neighborhood of $y = y^0 = 0$. At the point $y = 0$ we have $\Psi(0) = \{1, -1\}$ with $\varphi_p(0) = 1$ attained by the lower level optimal solution $x^2(0) = 1$. For $y < 0$ the lower level optimal solution is larger than one, for $y > 0$ it is smaller that -1. Hence the infimum of the function $\varphi_p(y)$ is zero and we have $0 = F(x^1(0), 0)$ but $\varphi_p(0) = 2$. Hence, the bilevel problem has no solution. In this example we have

$$K_{\{y:G(y)\leq 0\}}(0) = \mathbb{R} \not\subseteq K_{\mathrm{Supp}(x,x^2)}(0) = \{y : y \leq 0\}.$$

$\square$

The following theorem gives a sufficient optimality condition. A proof of this theorem is not necessary since it is very similar to the second part of the proof of Theorem 5.31.

THEOREM 5.33 *Consider the bilevel programming problem (5.1), (5.2) and let the assumptions of Theorem 5.32 be satisfied. If the following* *two conditions are satisfied then the point (x^0, y^0) is a local optimal pessimistic solution:*

- $K_{\{y:G(y)\leq 0\}}(y^0) \subseteq \bigcup_{i \in T} K_{\text{Supp}(x,x^j)}(y^0).$

- *For all $i \in T$, the following system of inequalities has no nontrivial solution r:*

$$\nabla_y G_j(y^0)r \leq 0, \text{ for all } j : G_j(y^0) = 0 \qquad (5.54)$$

$$\nabla_x F(x^i(y^0), y^0)x^{i'}(y^0; r) + \nabla_y F(x^i(y^0), y^0)r \leq 0 \qquad (5.55)$$

$$\nabla_x f(x^i(y^0), y^0)x^{i'}(y^0; r) + \nabla_y f(x^i(y^0), y^0)r$$
$$\geq \nabla_x f(x^j(y^0), y^0)x^{j'}(y^0; r) + \nabla_y f(x^j(y^0), y^0)r \ \forall \ j \in T \setminus \{i\}. \qquad (5.56)$$

5.8   PROOFS

PROOF OF THEOREM 5.1: By uniqueness of the optimal solution of the problem (5.1), problem (5.1), (5.2) is equivalent to

$$\min\{F(x(y), y) : y \in Y\}.$$

Because of assumptions (C), (MFCQ) in connection with Theorem 4.3, the objective function of this problem is continuous. Hence, the theorem follows from the famous Weierstraß' Theorem.   □

PROOF OF THEOREM 5.3: Due to lower semicontinuity of the point-to-set mapping $\Psi(\cdot)$ the optimal value function $\varphi_p(\cdot)$ is lower semicontinuous [17]. Hence, this function attains its minimum on the compact set Y provided this set is non-empty.   □

PROOF OF THEOREM 5.8: The cone $K_{\mathcal{M}}(x^0, y^0)$ is a polyhedral cone as finite union of the convex polyhedral cones $K_{I,\lambda,\mu}(x^0, y^0)$ and formula (5.16). Then, the result follows from the following facts:

- The dual of the convex hull of a polyhedral cone is equal to the dual of the polyhedral cone itself.

- The convex hull of a polyhedral cone is a convex polyhedral cone.

- The dual of a convex polyhedral cone is a convex polyhedral cone again.

□

PROOF OF THEOREM 5.9: Assume first that $\alpha_g \cap \alpha_\lambda = \emptyset$. Consider the cone $K^0(x^0, y^0)$ being the solution set of the following system of (in)equalities:

$$\nabla G_i(y^0)r \leq 0, \forall i : G_i(y^0) = 0$$
$$\nabla_{xx}^2 L(z^0, \lambda, \mu)d + \nabla_{yx}^2 L(z^0, \lambda, \mu)r + \nabla_x^\top g(z^0)\gamma + \nabla_x^\top h(z^0)\eta = 0$$
$$\nabla_x g_i(z^0)d + \nabla_y g_i(z^0)r = 0, \ i \in I^0,$$
$$\nabla_x g_i(z^0)d + \nabla_y g_i(z^0)r \leq 0, \ i \in I(z^0) \setminus I^0,$$
$$\nabla_x h_j(z^0)d + \nabla_y h_j(z^0)r = 0, \ j = 1, \ldots, q,$$
$$\gamma_i \geq 0, \ i \in I(z^0) \setminus J(\lambda^0), \ \gamma_i = 0, \ i \notin I(z^0).$$

Now, since $\alpha_g \cap \alpha_\lambda = \emptyset$ the set I^0 can be replaced by $I^0 \cup \alpha_\lambda$ and $J(\lambda^0)$ by $J(\lambda^0) \cup \alpha_g$. Hence, the complementarity slackness conditions are satisfied in the inequality system describing the cone $K^0(x^0, y^0)$, which in fact shows that this cone coincides with the one of the cones forming $K_{\mathcal{M}}(x^0, y^0)$. Hence

$$K^0(x^0, y^0) = K_{I^0 \cup \alpha_\lambda, \lambda, \mu}(x^0, y^0) \subseteq K_{\mathcal{M}}(x^0, y^0) \subseteq K^0(x^0, y^0)$$

by (5.23).

Assume that the set $\alpha_g \cap \alpha_\lambda \neq \emptyset$ and that $i \in \alpha_g \cap \alpha_\lambda$. Then there exists a solution $(d^0, r^0, \gamma^0, \eta^0)$ of the system defining $K^0(x^0, y^0)$ satisfying both inequalities

$$\nabla_x g_i(x^0, y^0)d + \nabla_y(x^0, y^0)r \leq 0, \ \gamma_i \geq 0$$

as strong inequalities. This vector cannot belong to any of the cones $K_{I,\lambda,\mu}(x^0, y^0)$ comprising the cone $K_{\mathcal{M}}(x^0, y^0)$. $\qquad \square$

PROOF OF LEMMA 5.1:

- Suppose Condition 1. holds. Let the equations (5.25) be satisfied. Then,

$$\sum_{i \notin \alpha_2} \zeta_i \nabla g_i(z^0) + \kappa^\top (0, \nabla_y G(y^0)) + \nabla(\nabla_x L(z^0, \nu^0)\omega) + \tau^\top \nabla h(z^0) =$$
$$- \sum_{i \in \alpha_2} \zeta_i \nabla g_i(z^0)$$

$$- \sum_{i \notin \alpha_3} \xi_i(e_i, 0) + (\nabla_x g(z^0), \nabla_x h(z^0))^\top \omega = \sum_{i \in \alpha_3} \xi_i(e_i, 0)$$
$$g_i(z^0)\zeta_i = 0, \ \forall i$$
$$\xi_i = 0, \ \forall i \in I^0$$
$$\kappa^\top G(y^0) = 0,$$

The vector in the left–hand side belongs to the span of $\mathcal{A}_2$ while the vector in the right–hand side is an element of the span of $\mathcal{A}_1$. Hence, by (PFC), both vectors must be equal to zero. Applying (PFC) again, we obtain Condition 2.

- Equivalence of Conditions 2. and 3. is due to the famous Farkas theorem of linear algebra.

- Suppose now that the Condition 3. is true. If the first condition in (PFC) is violated, the vectors $\{\nabla g_i(x^0, y^0) : i \in \alpha_2\}$ are linearly dependent and the first system in Condition 3. would be inconsistent. Now, if the second condition in (PFC) is not valid, similarly to the first part of the proof, the existing nonzero vector in span $\mathcal{A}_1 \cap$ span $\mathcal{A}_2$ can easily be used to show that either one of the systems in Condition 3. cannot have a solution.

$\square$

PROOF OF THEOREM 5.10: Since, by the definitions of the cones and their duals $K^0(x^0, y^0)^* \subseteq K_{\mathcal{M}}(x^0, y^0)^*$, we have only to show the opposite inclusion. Take $v \in K_{\mathcal{M}}(x^0, y^0)^*$. Since

$$K_{\mathcal{M}}(x^0, y^0) = \bigcup_{(\lambda,\mu)\in E\Lambda(x^0,y^0)} \bigcup_{I\in\mathcal{I}(\lambda)} K_{I,\lambda,\mu}(x^0, y^0),$$

by (5.16) and

$$\left(\bigcup_{(\lambda,\mu)\in E\Lambda(x^0,y^0)} \bigcup_{I\in\mathcal{I}(\lambda)} K_{I,\lambda,\mu}(x^0, y^0) \right)^* = \bigcap_{(\lambda,\mu)\in E\Lambda(x^0,y^0)} \bigcap_{I\in\mathcal{I}(\lambda)} K_{I,\lambda,\mu}(x^0, y^0)^*$$

by the definition of the dual cone, we have

$$v = (v_1, v_2, v_3) \in K_{I,\lambda,\mu}(x^0, y^0)^*$$

for all $(\lambda, \mu) \in E\Lambda(x^0, y^0)$ and all $I \in \mathcal{I}(\lambda)$. Note that the family of sets $\mathcal{I}(\lambda)$ in this formula can be replaced by any larger family and that the single essential property is that only sets $I \supseteq J(\lambda)$ are used. Fix

$$I = \alpha_2 \cup \left(I(x^0, y^0) \setminus \alpha_g \right)$$

which is possible by the assumptions. Then, by the definition of the dual cone $K_{I,\lambda,\mu}(x^0, y^0)^*$ there exist $(\kappa, \omega, \zeta, \tau, \xi)$ such that the following system is satisfied:

$$v_1 + \kappa^\top (0, \nabla_y G(y^0)) + \nabla(\nabla_x L(z^0, \nu^0)\omega)$$

$$+\zeta^{\mathsf{T}}\nabla g(z^0) + \tau^{\mathsf{T}}\nabla h(z^0) = 0$$

$$(v_2, v_3) + (\nabla_x g(z^0), \nabla_x h(z^0))^{\mathsf{T}}\omega - (\xi, 0)^{\mathsf{T}} = 0$$

$$g_i(z^0)\zeta_i = 0, \; \forall i$$

$$\zeta_i \geq 0, \; \forall i \in I(z^0) \setminus I$$

$$\xi_i \geq 0, \; \forall i \in I \setminus J(\lambda^0)$$

$$\xi_i = 0, \; \forall i \in J(\lambda^0)$$

$$\kappa^{\mathsf{T}}G(y^0) = 0,$$

$$\kappa \geq 0. \tag{5.57}$$

Considering the dual cone to $K^0(x^0, y^0)$ and taking the definition of the sets α_g, α_λ into account, we see that we intend to show that the following system has a solution $(\kappa, \omega, \zeta, \tau, \xi)$:

$$v_1 + \kappa^{\mathsf{T}}(0, \nabla_y G(y^0)) + \nabla(\nabla_x L(z^0, \nu^0)\omega)$$

$$+\zeta^{\mathsf{T}}\nabla g(z^0) + \tau^{\mathsf{T}}\nabla h(z^0) = 0$$

$$(v_2, v_3) + (\nabla_x g(z^0), \nabla_x h(z^0))^{\mathsf{T}}\omega - (\xi, 0)^{\mathsf{T}} = 0$$

$$g_i(z^0)\zeta_i = 0, \quad \forall i$$

$$\zeta_i \geq 0, \quad \forall i \in (I(z^0) \setminus I) \cap \alpha_g$$

$$\xi_i \geq 0, \quad \forall i \in (I \setminus J(\lambda^0)) \cap \alpha_\lambda$$

$$\xi_i = 0, \quad \forall i \in J(\lambda^0) \text{ or } i \notin \alpha_\lambda$$

$$\kappa^{\mathsf{T}}G(y^0) = 0,$$

$$\kappa \geq 0. \tag{5.58}$$

Since each of the indices $i \in I(x^0, y^0) \setminus I^0$ belongs to either of the sets

$$i \in \alpha_g \setminus \alpha_\lambda, \text{ or } i \in \alpha_\lambda \setminus \alpha_g, \text{ or } i \in \alpha_0,$$

the inequalities

$$\zeta_i \geq 0, \; \forall i \in \alpha_2, \; \xi_i \geq 0, \; \forall i \in \alpha_3$$

are to be shown. We show this for the inequality $\zeta_i \geq 0$ for some $i \in \alpha_2$. The other condition can be proved similarly.

Let this inequality not be satisfied, i.e. let $\zeta_{i_0} < 0$ for some $i_0 \in \alpha_2$. Let $(d^0, r^0, \gamma^0, \eta^0)$ be the vector satisfying the corresponding system in part 3(a) in Lemma 5.1. Then, we can derive the following sequence of equations:

$$\langle v_1, (d^0, r^0)^{\mathsf{T}}\rangle$$

$$= -\kappa^{\mathsf{T}}(0, \nabla_y G(y^0))(d^0, r^0)^{\mathsf{T}} - \nabla(\nabla_x L(z^0, \nu^0)\omega)(d^0, r^0)^{\mathsf{T}}$$

$$-\zeta^{\mathsf{T}}\nabla g(z^0)(d^0, r^0)^{\mathsf{T}} - \tau^{\mathsf{T}}\nabla h(z^0)(d^0, r^0)^{\mathsf{T}}$$

$$= -\zeta_{i_0}\nabla g_{i_0}(z^0)(d^0, r^0)^{\mathsf{T}} + \gamma^{0\mathsf{T}}\nabla_x g(z^0)\omega + \eta^{0\mathsf{T}}\nabla_x h(z^0)\omega$$

$$= -\langle(v_2, v_3)^{\mathsf{T}}, (\gamma^0, \eta^0)^{\mathsf{T}}\rangle + \xi^{\mathsf{T}}\gamma^0 - \zeta_{i_0}\nabla g_{i_0}(z^0)(d^0, r^0)^{\mathsf{T}}.$$

Hence, by $\xi^\top \gamma^0 = 0$, we derive

$$\langle v_1, (d^0, r^0)^\top \rangle + \langle (v_2, v_3)^\top, (\gamma^0, \eta^0)^\top \rangle < 0.$$

Obviously, the vector $(d^0, r^0, \gamma^0, \eta^0) \in K_{I,\lambda,\mu}(x^0, y^0)$ for all multipliers $(\lambda, \mu) \in E\Lambda(x^0, y^0)$ and all sets $I \in \mathcal{I}(\lambda)$ with $i_0 \notin I$. Since $v \in K_{I,\lambda,\mu}(x^0, y^0)^*$ we must have

$$\langle v_1, (d^0, r^0)^\top \rangle + \langle (v_2, v_3)^\top, (\gamma^0, \eta^0)^\top \rangle \geq 0$$

which is the desired contradiction.   □

PROOF OF THEOREM 5.11: Set $w(x, y) := -g(x, y)$ and let the Karush-Kuhn-Tucker conditions of (5.1)

$$\nabla_x L(x, y, \lambda, \mu) = 0$$
$$w(x, y) := -g(x, y) \geq 0, \lambda \geq 0, \lambda^\top w(x, y) = 0$$
$$h(x, y) = 0$$

be part of the constraints of of an optimization problem.   If the Mangasarian-Fromowitz constraint qualification holds at a feasible point $(x^0, y^0, \lambda^0, \mu^0)$ then the second line in these conditions give:
    The gradient of the equation $\lambda^\top g(x, y) = 0$ is not vanishing:

$$(\lambda^{0\top} \nabla_x w(x^0, y^0), \lambda^{0\top} \nabla_y w(x^0, y^0)^\top, w(x^0, y^0), 0)^\top \neq 0 \qquad (5.59)$$

and there is a vector (d, r, γ) satisfying

$$\lambda^{0\top} \nabla_x w(x^0, y^0)d + \lambda^{0\top} \nabla_y w(x^0, y^0)r + \gamma^\top w(x^0, y^0) = 0, \qquad (5.60)$$
$$\nabla_x w_i(x^0, y^0)d + \nabla_y w_i(x^0, y^0)r > 0, \; \forall \, i : w_i(x^0, y^0) = 0, \qquad (5.61)$$
$$\gamma_i > 0, \; \forall \, i : \lambda_i = 0. \qquad (5.62)$$

Now, if $\nabla_x w_i(x^0, y^0)d + \nabla_y w_i(x^0, y^0)r \leq 0$ then $w_i(x^0, y^0) > 0$ and by complementarity slackness $\lambda_i^0 = 0$. Analogously, $\gamma_i \leq 0$ implies $w_i(x^0, y^0) = 0$. Thus

$$\lambda^{0\top} \nabla_x w(x^0, y^0)d + \lambda^{0\top} \nabla_y w(x^0, y^0)r + \gamma w(x^0, y^0) \geq 0. \qquad (5.63)$$

On the other hand, if $w_i(x^0, y^0) > 0$ then $\lambda_i^0 = 0$ or by (5.62) $\gamma_i > 0$. Hence, inequality (5.63) holds as a strict inequality which contradicts (5.60).   Consequently, $w(x^0, y^0) = 0$.   Analogously, $\lambda_i^0 > 0$ implies $\nabla_x w_i(x^0, y^0)d + \nabla_y w_i(x^0, y^0)r > 0$ which again contradicts (5.60). Hence, $\lambda^0 = 0$ which together with $w(x^0, y^0) = 0$ is a contradiction to (5.59). Thus, the Mangasarian-Fromowitz constraint qualification cannot hold.   □

PROOF OF THEOREM 5.12: We start with showing openness. Since the C^k-topology is strictly finer than the $C^{k'}$-topology for $k > k'$ we have to show the theorem only for $k = 2$. Fix any index sets $I_G \subseteq \{1, \ldots, s\}$, $I_g \subseteq \{1, \ldots, p\}$, $I_\lambda \subseteq \{1, \ldots, p\}$ and a point $(\bar{z}, \bar{\nu}) = (\bar{x}, \bar{y}, \bar{\lambda}, \bar{\mu})$. Define the matrix $M(I_G, I_g, I_\lambda, \bar{x}, \bar{y}, \bar{\lambda}, \bar{\mu})$

$$
\begin{pmatrix}
0 & \nabla_y G_{I_G}(\bar{y}) & 0 & 0 \\
\nabla^2_{xx} L_{I_g}(\bar{z}, \bar{\nu}) & \nabla^2_{yx} L_{I_g}(\bar{z}, \bar{\nu}) & \nabla^\top_x g_{I_\lambda}(\bar{z}) & \nabla^\top_x h(\bar{z}) \\
\nabla_x g_{I_g}(\bar{z}) & \nabla_y g_{I_g}(\bar{z}) & 0 & 0 \\
\nabla_x h(\bar{z}) & \nabla_y h(\bar{z}) & 0 & 0
\end{pmatrix}
$$

with $L_I(x, y, \lambda, \mu) = f(x, y) + \sum_{i \in I} \lambda_i g_i(x, y) + \sum_{j=1}^q \mu_j h_j(x, y)$ and $G_{I_G} = (G_i)_{i \in I_G}$ is the subvector of G consisting of the functions $G_i, i \in I_G$ only. It is easy to see that the (PLICQ) is satisfied for problem (5.27) at the point $(\bar{x}, \bar{y}, \bar{\lambda}, \bar{\mu})$ if and only if the matrix $M = M(I_G, I_g, I_\lambda, \bar{x}, \bar{y}, \bar{\lambda}, \bar{\mu})$ is has full row rank $|I_G| + n + |I_g| + q$ for $I_G = \{i : G_i(\bar{y}) = 0\}$, $I_g = I(\bar{x}, \bar{y})$, $I_\lambda = J(\bar{\lambda})$. Now introduce the continuous function

$$
\chi(x, y) =
$$
$$
\min_{I_G, I_g, I_\lambda} \ \min_{\substack{(\lambda, \mu): \\ \lambda_{I_\lambda^c} = 0}} \ \Big\{ \sum \{|\det A| : A \text{ is maximal square submatrix of } M\}
$$
$$
+ \max\{ \ \|G_{I_G}(x, y)\|, G_j(x, y), \|\nabla_x L_{I_g}(x, y, \lambda, \mu)\|,
$$
$$
\| \min\{-g(x, y), \lambda\}\|, \|g_{I_g}(x, y)\|, g_i(x, y),
$$
$$
\|h(x, y)\|, \|(\lambda, \mu)\|_\infty - B\}\}
$$

where $I_\lambda^c = \{1, \ldots, p\} \setminus I_\lambda$. Notice that $\chi(x, y) > 0$ if and only if for every (λ, μ) with $\|(\lambda, \mu)\|_\infty \le B$ either (x, y, λ, μ) is infeasible or (PLICQ) is satisfied. Moreover, $\chi(x, y) \ge 0$. Now the openness proof follows since $\chi(x, y)$ involves only derivatives of the data up to second order.

To verify the denseness part of the theorem, we need Sard's theorem together with a corollary.

DEFINITION 5.14 *A system of equations and inequalities*

$$
g(x) \le 0
$$
$$
h(x) = 0 \tag{5.64}
$$
$$
\min\{G(x), H(x)\} = 0
$$

is called regular if zero is a regular value of each of the following systems:

$$
g_i(x) = 0 \ \forall \ i : g_i(\bar{x}) = 0
$$
$$
h(x) = 0
$$
$$
G_j(x) = 0 \ \forall \ j : G_j(\bar{x}) = 0
$$
$$
H_j(x) = 0 \ \forall \ j : H_j(\bar{x}) = 0,
$$

where $\bar{x}$ is a solution of the system (5.64).

Sard's Theorem can be applied to a system of equations and inequalities as follows:

COROLLARY 5.1 ([253]) *Let $(g, h, G, H) : \mathbb{R}^n \to \mathbb{R}^p \times \mathbb{R}^q \times \mathbb{R}^m \times \mathbb{R}^m$ be k times continuously differentiable functions with $k > \max\{0, n-q-m\}$. Then, for almost all $(a, b, c, d) \in \mathbb{R}^{p+q+m+m}$, the system*

$$g(x) \leq a$$
$$h(x) = b \tag{5.65}$$
$$\min\{G(x) - c, H(x) - d\} = 0$$

is regular.

PROOF OF COROLLARY 5.1: Fix index sets $I \subseteq \{1, \ldots, q\}$, $J, K \subseteq \{1, \ldots, m\}$ with $J \cup K \supseteq \{1, \ldots, m\}$. Then, by Sard's Theorem, for almost all (a, b, c, d) the restriction (a_I, b, c_J, d_K) is a regular value of the following system:

$$g_I(x) \leq a_I$$
$$h(x) = b \tag{5.66}$$
$$G_J(x) = c_J$$
$$H_K(x) = d_K.$$

This implies the result since each feasible solution of (5.65) is feasible for (5.66) for suitably chosen sets I, J, K and the number of different selections of I, J, K is finite.   □

Now we proceed with the proof of the denseness part in Theorem 5.12. Having a closer look at the Jacobian of a system

$$G_I(y) \leq a_I$$
$$\nabla_x L(x, y, \lambda, \mu) = b$$
$$g_J(x, y) = c_J$$
$$\lambda_K = 0$$
$$h(x, y) = d_K$$

it is easy to see that the constraint systems of (5.27) are also regular for almost all (a, b, c, d) but now the right-hand side of the equations $\lambda_K = 0$ has not to be perturbed.

Now we outline the proof of the density property of the set $\mathcal{H}_B^l$. For details omitted by the sake of shortness the reader is referred to [125] and [139].

Again we can assume that $k \geq 2$. Let the mapping

$$(G, f, g, h) \in C^l(\mathbb{R}^{n+m}, \mathbb{R}^{s+1+p+q})$$

be arbitrarily chosen and take a neighborhood V of (G, f, g, h) in the C^k-topology. Choose two countable open coverings $\{U^j\}_{j=1}^\infty$ and $\{\text{int supp}\,\Theta^j\}_{j=1}^\infty$ of $\mathbb{R}^{m+n}$ with compact closures cl U^j and partitions $\{\theta^j\}_{j=1}^\infty$ respectively $\{\chi^j\}_{j=1}^\infty$ of unity subordinate to these coverings. Here, $\text{supp}\,\Theta = \{x : \Theta(x) > 0\}$ denotes the support set of a function Θ. Then, there exist smooth functions $\{\xi^j\}_{j=1}^\infty$ satisfying

$$0 \leq \xi^j(x, y) \leq 1 \; \forall \; (x, y) \in \mathbb{R}^{m+n}$$
$$\xi^j(x, y) = 1 \text{ on a neighbourhood of } \text{supp}\,\chi^j$$
$$\text{supp}\,\xi^j \subset \text{int supp}\,\theta^j,$$

cf. [138]. For shortness call the mapping (G, f, g, h) regular on a set S if PLICQ is satisfied at every feasible point $(x, y, \lambda, \mu) \in S$ with $\max\{\|\lambda\|_\infty, \|\mu\|_\infty\} \leq B$. By the above considerations in Corollary 5.1 we can restore regularity of (G, f, g, h) on cl $\text{supp}\,\chi^1$ without leaving V by choosing a suitably small (a_1, b_1, c_1, d_1) and defining the function

$$(G_1, f_1, g_1, h_1)(x, y) = (G, f, g, h)(x, y) - \xi^1(x, y)(a_1, b_1, c_1, d_1).$$

Now we define the function

$$(G_2, f_2, g_2, h_2)(x, y) = (G_1, f_1, g_1, h_1)(x, y) - \xi^2(x, y)(a_2, b_2, c_2, d_2)$$

again by use of the above considerations, where (a_2, b_2, c_2, d_2) is chosen such that $(G_2, f_2, g_2, h_2) \in V$ is regular, but now on cl $\text{supp}\,\chi^1 \cup$ cl $\text{supp}\,\chi^2$. To guarantee regularity of $(G_2, f_2, g_2, h_2) \in V$ on cl $\text{supp}\,\chi^1$ use that regularity on a set S is equivalent to the condition that the function χ used in the openness part of this proof supplemented by the characteristic function of S is positive. Then, the compactness of cl $\text{supp}\,\chi^j$ guarantees that the functions (G_2, f_2, g_2, h_2) remain regular for sufficiently small (a_2, b_2, c_2, d_2) on cl $\text{supp}\,\chi^1$. We proceed in the same way to get the sequence of functions

$$\{(G_j, f_j, g_j, h_j)\}_{j=1}^\infty.$$

Since a partition of unity is locally finite, the pointwise limit

$$(G^*, f^*, g^*, h^*) = \lim_{j \to \infty} (G_j, f_j, g_j, h_j)$$

exists and is in V. Moreover, since the sets $\{U^j\}_{j=1}^\infty$ form an open covering of $\mathbb{R}^{n+m}$ we conclude that $(G^*, f^*, g^*, h^*) \in \mathcal{H}_B^l$. Note that regularity of the mapping (G^*, f^*, g^*, h^*) is maintained since, in each step in the above proof, the mapping (G_j, f_j, g_j, h_j) remains unchanged at all the points where it has been regularized in previous steps, i.e. in each step it is regularized on a new set of points.   □

PROOF OF THEOREM 5.15: Let y^0 be a local optimistic optimal solution of problem (5.1), (5.2), i.e. there is an open neighborhood $U_\delta(y^0), \delta > 0$, such that $\varphi_o(y) \geq \varphi_o(y^0)$ for all $y \in U_\delta(y^0)$ with $G(y) \leq 0$. Since the Karush-Kuhn-Tucker conditions are sufficient and necessary for optimality in the lower level problem, we have that

$$\varphi_o(y^0) \leq F(x, y)$$

for all (x, y) with $y \in U_\delta(y^0)$ and

$$\nabla_x L(x, y, \lambda, \mu) = 0$$
$$g(x, y) \leq 0, \ \lambda \geq 0, \ \lambda^\top g(x, y) = 0$$
$$h(x, y) = 0.$$

This implies the existence of an open neighborhood $W_\epsilon(x^0, y^0)$ such that

$$\varphi_o(y^0) \leq F(x, y) \ \forall \ (x, y) \in W_\epsilon(x^0, y^0)$$

satisfying

$$\nabla_x L(x, y, \lambda, \mu) = 0$$
$$g(x, y) \leq 0, \ \lambda \geq 0, \ \lambda^\top g(x, y) = 0$$
$$h(x, y) = 0$$

and the theorem follows by the definition of $\varphi_0(y^0)$ together with a repeated application of the Karush-Kuhn-Tucker Theorem.   □

PROOF OF THEOREM 5.16: By [60] there exists a non-vanishing vector $(\kappa_0, \kappa, \omega, \tau, u)^\top$ satisfying

$$\kappa_0 \nabla F(z^0) + \kappa^\top (0, \nabla_y G(y^0)) + \nabla(\nabla_x L(z^0, \nu^0)\omega)$$
$$-u^\top d + \tau^\top \nabla h(z^0) = 0$$
$$(\nabla_x g(z^0), \nabla_x h(z^0))^\top \omega - u^\top (r, 0) = 0 \qquad (5.67)$$
$$\kappa_0, \kappa \geq 0$$
$$\kappa^\top G(y^0) = 0,$$

where $d \in \partial_z \min\{-g(z^0), \lambda^0\}$ denotes the Clarke generalized gradient of the min-function with respect to (x, y) and $r \in \partial_\lambda \min\{-g(z^0), \lambda^0\}$. By the formulae for computing the Clarke generalized gradient

$$(d_i, r_i) \in \text{conv} \ \{(-\nabla g_i(z^0), 0), (0, 0, 1)\}$$

if $g_i(x^0, y^0) = \lambda_i^0 = 0$, i.e.

$$d_i = -\alpha_i \nabla g_i(z^0), \quad r_i = (1 - \alpha_i)$$

for some $\alpha_i \in (0, 1)$. Note that $(u_i\alpha_i)(u_i(1 - \alpha_i)) \geq 0$ and that at least one of $u_i\alpha_i$ and $u_i(1 - \alpha_i)$ does not vanish if $u_i \neq 0$. Now, if we set

$$\zeta_i = \begin{cases} u_i & \text{if } \lambda_i^0 > 0, \\ 0 & \text{if } g_i(z^0) < 0 \\ \alpha_i u_i & \text{if } g_i(z^0) = \lambda_i^0 = 0 \end{cases} \qquad \xi_i = \begin{cases} u_i & \text{if } g_i(z^0) < 0, \\ 0 & \text{if } \lambda_i^0 > 0 \\ (1 - \alpha_i)u_i & \text{if } g_i(z^0) = \lambda_i^0 = 0 \end{cases}$$

the assertion follows. $\qquad\qquad\qquad\qquad\qquad\qquad\qquad\qquad\qquad\qquad\quad \square$

PROOF OF THEOREM 5.17:

- By Theorem 5.16 every local minimizer of problem (5.27) satisfies conditions (5.29) for some non-vanishing vector $(\kappa_0, \kappa, \omega, \zeta, \tau, \xi)$. Let $\kappa_0 = 0$. Then,

$$\kappa^\top (0, \nabla_y G(y^0)) = -\nabla(\nabla_x L(z^0, \nu^0)\omega) - \zeta^\top \nabla g(z^0) - \tau^\top \nabla h(z^0), \tag{5.68}$$

$$0 = (\nabla_x g(z^0), \nabla_x h(z^0))^\top \omega - (\xi, 0)^\top. \tag{5.69}$$

Let $L = \{1, \ldots, p\} \setminus J(\lambda^0)$. Recall that (MFCQ) holds for (5.30) if the gradients

$$(\nabla^2_{xx} L(z^0, \nu^0), \nabla^2_{yx} L(z^0, \nu^0), \nabla_x g(z^0), \nabla_x h(z^0)),$$

and

$$(\nabla_x g_{I(z^0)}(z^0), \nabla_y g_{I(z^0)}(z^0), 0, 0), (0, 0, e_L, 0), (\nabla_x h(z^0), \nabla_y h(z^0), 0, 0)$$

are linearly independent and there exists a vector (d, r, γ, η) satisfying

$$\nabla(\nabla_x L(z^0, \nu^0))(d, r)^\top + \gamma^\top \nabla_x g(z^0) + \eta^\top \nabla_x h(z^0) = 0$$
$$\nabla_x g_{I(z^0)}(z^0)d + \nabla_y g_{I(z^0)}(z^0)r = 0 \tag{5.70}$$
$$\nabla_x h(z^0)d + \nabla_y h(z^0)r = 0 \tag{5.71}$$
$$\gamma_L = 0$$
$$\nabla G_i(y^0)r < 0, \text{ for } G_i(y^0) = 0 \tag{5.72}$$

Note that the other inequalities $g_i(x, y) \leq 0$ are inactive for $i \notin I(x^0, y^0)$ and we have $\gamma_i = 0$ for these indices. Multiplication of (5.69) by $(\gamma^\top \mu^\top)$ from left implies $\gamma^\top \nabla_x g(z^0)\omega + \mu^\top \nabla_x h(z^0)\omega = 0$ since $\xi_i = 0$ for $\lambda_i^0 > 0$ and $\gamma_i = 0$ for $\lambda_i^0 = 0$. Then, the (MFCQ) gives that $\nabla(\nabla_x L(z^0, \nu^0)\omega)(d\,r)^\top = 0$. Now, it is easy to see that by (5.70), (5.71)

$$-\nabla(\nabla_x L(z^0, \nu^0)\omega)(d, r)^\top - \zeta^\top \nabla g(z^0)(d, r)^\top - \tau^\top \nabla h(z^0)(d, r)^\top$$
$$= \kappa^\top \nabla G(y^0) r = 0.$$

Since $\kappa^\top G(y^0) = 0$ and $\nabla G_i(y^0) r < 0$ for $G_i(y^0) = 0$ this implies $\kappa = 0$. Recall that

$$\nabla(\nabla_x L(z, \nu)) = (\nabla^2_{xx} L(z, \nu), \nabla^2_{yx} L(z, \nu)).$$

Then, by linear independence and (5.68), (5.69) this implies that all multipliers vanish which contradicts Theorem 5.16. Hence, $\kappa_0 \neq 0$ and can be assumed to be equal to one.

■ Let $\kappa, \omega, \zeta, \tau, \xi$ be Lagrange multipliers corresponding to a stationary point of (5.30) and suppose that $\zeta_i < 0$ for some $i \in K \subseteq I(x^0, y^0)$. Let without loss of generality $1 \in K$ and $\zeta_1 < 0$. By (SMFCQ), the gradients

$$\left(\nabla(\nabla_x L(x^0, y^0, \lambda^0, \mu^0)), \nabla_x g(x^0, y^0), \nabla_x h(x^0, y^0)\right)$$
$$\left(\nabla g_i(x^0, y^0), 0, 0\right) \text{ for } g_i(x^0, y^0) = 0$$
$$(0, 0, e_i, 0) \text{ for } \lambda^0_i = 0$$
$$\left(\nabla h(x^0, y^0), 0, 0\right)$$
$$\left(0, \nabla_y G_i(y^0), 0, 0\right) \text{ for } \kappa_i > 0$$

are linearly independent and there exists a vector (d, r, γ, η) satisfying (where we use the abbreviation $z^0 = (x^0, y^0)$)

$$\nabla(\nabla_x L(z^0, \lambda^0, \mu^0))(d, r)^\top + \gamma^\top \nabla_x g(z^0) + \eta^\top \nabla_x h(z^0) = 0$$
$$\nabla g_i(z^0)(d, r)^\top = 0 \text{ for } g_i(x^0, y^0) = 0$$
$$\gamma_i = 0 \text{ for } \overline{\lambda}^0_i = 0$$
$$\nabla h(z^0)(d, r)^\top = 0$$
$$\nabla_y G_i(y^0) r = 0 \text{ for } \kappa_i > 0$$
$$\nabla_y G_i(y^0) r < 0 \text{ for } \kappa_i = G_i(y^0) = 0$$

By the linear independence there exists a vector $(d^0, r^0, \gamma^0, \eta^0)$ satisfying

$$\nabla(\nabla_x L(z^0, \lambda^0, \mu^0))(d^0, r^0)^\top + \gamma^{0\top} \nabla_x g(z^0) + \eta^{0\top} \nabla_x h(z^0) = 0$$
$$(5.73)$$

$$\nabla g_1(z^0)(d^0, r^0)^\top = -1 \tag{5.74}$$

$$\nabla g_i(z^0)(d^0, r^0)^\top = 0 \text{ for } g_i(x^0, y^0) = 0, i \neq 1 \tag{5.75}$$

$$\gamma_i^0 = 0 \text{ for } \lambda_i^0 = 0 \tag{5.76}$$

$$\nabla h(z^0)(d^0, r^0)^\top = 0 \tag{5.77}$$

$$\nabla_y G_i(y^0)r^0 = 0 \text{ for } \kappa_i > 0 \tag{5.78}$$

where the abbreviation $z^0 = (x^0, y^0)$ has been used. Passing to $(d^0, r^0, \gamma^0, \eta^0)^\top + \alpha(d, r, \gamma, \eta)^\top$ with $\alpha > 0$ sufficiently large if necessary the last set of equations is also satisfied and we can thus assume that $\nabla_y G_i(y^0)r^0 < 0$ for $\kappa_i = G_i(y^0) = 0$.

By (5.69) we get $\gamma^{0\top}\nabla_x g(x^0, y^0)\omega + \eta^{0\top}\nabla_x h(x^0, y^0)\omega = 0$, since again $\gamma^{0\top}\xi = 0$. Hence, by (5.73) we derive $\nabla(\nabla_x L(z^0, \lambda^0, \mu^0)\omega)(d^0, r^0)^\top = 0$. By the first part of this theorem and $\zeta_1 < 0$ this implies $\nabla F(x^0, y^0)(d^0 \ r^0)^\top = \zeta_1 < 0$. Hence, $(x^0, y^0, \lambda^0, \mu^0)$ is not a Bouligand stationary point of the problem

$$
\begin{aligned}
F(x,y) \quad &\to \quad \min_{x,y,\lambda,\mu} \\
\nabla_x L(x, y, \lambda, \mu) &= \quad 0 \\
g_i(x, y) = 0 \quad &\text{for} \quad g_i(x^0, y^0) = 0, i \neq 1 \\
\lambda_i = 0 \quad &\text{for} \quad \lambda_i^0 = 0 \\
g_i(x, y) \leq 0 \quad &\text{for} \quad g_i(x^0, y^0) < 0 \\
g_1(x, y) \leq 0 \quad & \\
\lambda_i \geq 0 \quad &\text{for} \quad \lambda_i^0 > 0 \\
h(x, y) &= \quad 0 \\
G(y) &\leq \quad 0.
\end{aligned}
$$

Since the (SMFCQ) for problem (5.30) implies that the (MFCQ) is satisfied for this problem, the point $(x^0, y^0, \lambda^0, \mu^0)$ is not a local minimizer of this problem. This, however contradicts our assumption. Hence, $\zeta_1 \geq 0$. Analogously, we can also show that $\xi_i \geq 0$ for all $i \in K$.

$\square$

PROOF OF THEOREM 5.18: To prove necessity, let the nonsmooth Mangasarian-Fromowitz constraint qualification be satisfied. Assume that there exists an abnormal multiplier vector (λ, μ). First, let $\lambda_i > 0$ for some index i with $g_i(z) = 0$. Let the vector d satisfy the conditions in the nonsmooth Mangasarian-Fromowitz constraint qualification. Then, by the conditions defining $\Lambda^0(z)$, for some vectors $\eta^k \in \partial g_k(z)$, $k \in \{k :$

$g_k(z) = 0\}$, $\xi^j \in \partial h_j(z)$, $j = 1, \ldots, q$, we have

$$0 = 0^\top d = \left(\sum_{k=1}^{p} \lambda_k \eta^k + \sum_{j=1}^{q} \mu_j \xi^j \right) d \leq \lambda_i \eta^i d < 0.$$

This contradiction proves that $\lambda = 0$. But then, $\mu = 0$ is a consequence of the linear independence of the vectors $\{\xi^j : j = 1, \ldots, q\}$. This shows the first part of the proof.

Suppose now that $\Lambda^0(z) = \{0\}$. Let $\eta^k \in \partial g_k(z)$, $k \in \{k : g_k(z) = 0\}$, $\xi^j \in \partial h_j(z)$, $j = 1, \ldots, q$, be arbitrary vectors in the generalized gradients of the constraint functions. If $\{\xi^j : j = 1, \ldots, q\}$ would not be linearly independent there exists a non-vanishing vector $(0, \mu)$ in $\Lambda^0(z)$ with $\sum_{j=1}^{q} \mu_j \xi^j = 0$. Also, if the system

$$\eta^k d < 0, \ \forall k : g_k(z) = 0, \xi^j d = 0, \ \forall j = 1, \ldots, q$$

would have no solution then by the famous Motzkin Theorem of the alternative there exists a vector (λ, μ) with

$$0 = \sum_{k:g_k(z)=0} \lambda_k \eta^k + \sum_{j=1}^{q} \mu_j \xi^j, \ \lambda \geq 0$$

with $\lambda \neq 0$. Since both cases contradict the assumption, the nonsmooth Mangasarian-Fromowitz constraint qualification (NMFCQ) holds. □

PROOF OF THEOREM 5.19: Since (MFCQ) is satisfied for the lower level problem, there is a Lagrange multiplier vector $(\lambda, \mu) \in \Lambda(x^0, y^0)$ satisfying

$$\nabla_x f(x^0, y^0) + \lambda^\top \nabla_x g(x^0, y^0) + \mu^\top \nabla_x h(x^0, y^0) = 0,$$

$$\lambda \geq 0, \ \lambda^\top g(x^0, y^0) = 0.$$

Since

$$\nabla_y f(x^0, y^0) + \lambda^\top \nabla_y g(x^0, y^0) + \mu^\top \nabla_y h(x^0, y^0) \in \partial \varphi(y^0)$$

by our assumptions and Theorem 4.17 we have

$$0 \in \nabla f(x^0, y^0) + \lambda^\top \nabla g(x^0, y^0) + \mu^\top \nabla h(x^0, y^0) - \{0\} \times \partial \varphi(y^0),$$

$$\lambda \geq 0, \ \lambda^\top g(x^0, y^0) = 0.$$

But this means that $(1, \lambda, \mu)$ is a nontrivial abnormal multiplier for problem (5.36). □

PROOF OF THEOREM 5.20: Applying the Lagrange multiplier rule for locally Lipschitz optimization problems [61] we get the existence of $\kappa_0 \in \{0,1\}$, $\mu_0 \geq 0$, $\xi \in \mathbb{R}^s_+$, $\gamma \in \mathbb{R}^p_+$, $\eta \in \mathbb{R}^q$ such that

$$0 \in \kappa_0 \nabla F(x^0, y^0) + \mu_0 \partial \left(f(x^0, y^0) - \varphi(y^0) \right) + \gamma^\top \nabla g(x^0, y^0)$$
$$+ \eta^\top \nabla h(x^0, y^0) + \xi^\top \left(0 \ \nabla G(y^0) \right)$$
$$\xi^\top G(y^0) = 0, \ \gamma^\top g(x^0, y^0) = 0.$$

Now the proof follows from Theorem 4.17.   □

In the following proof Proposition 6.4.4 in [61] is used which says that the Karush-Kuhn-Tucker conditions are necessary optimality conditions for a Lipschitz optimization problem provided that the problem is calm at the local optimal solution under investigation. Consider problem

$$\min\{f(z) : g(z) + u \leq 0, \ h(z) + v = 0\}. \tag{5.79}$$

Then, problem (5.37) is calm at some feasible point $\bar{z}$ provided that there exist $\kappa > 0$, $\omega > 0$ and an open neighborhood $W_\omega(\bar{z})$ such that for each feasible point $z' \in W_\omega(\bar{z})$ for problem (5.79) with $\|(u, v)\| \leq \omega$ the inequality

$$f(z') - f(\bar{z}) + \kappa\|(u, v)\| \geq 0$$

holds.   It is well known that a problem with differentiable constraints and a Lipschitz continuous objective function which satisfies the Mangasarian-Fromowitz constraint qualification at a local optimal solution is calm there [61].

PROOF OF THEOREM 5.21: Using Theorem 6.7 there is $\mu_0 > 0$ such that (x^0, y^0) is an optimal solution of the problem

$$F(x, y) + \mu_0(f(x, y) - \varphi(y)) \to \min_{x,y}$$
$$g(x, y) \leq 0$$
$$h(x, y) = 0$$
$$G(y) \leq 0.$$

We show that this problem satisfies the Mangasarian-Fromowitz constraint qualification: First the gradients $\{\nabla h_j(x^0, y^0) : j = 1, \ldots, q\}$ are linearly independent due to (MFCQ) for the lower level problem at (x^0, y^0). Let d be a vector satisfying

$$\nabla_x g_i(x^0, y^0)d < 0, i \in I(x^0, y^0) \text{ and } \nabla_x h_j(x^0, y^0)d = 0, \ \forall j \tag{5.80}$$

and let r satisfy

$$\nabla G_k(y^0)r < 0, \text{ for all } k : G_k(y^0) = 0. \qquad (5.81)$$

Let $\overline{d}$ be a vector solving the system of linear equations

$$\nabla h(x^0, y^0)(\overline{d}, r) = 0$$

which exists due to linear independence. Then, there exists $\alpha > 0$ such that the vector $(d + \alpha\overline{d}, \alpha r)$ satisfies

$$\nabla g_i(x^0, y^0)(d + \alpha\overline{d}, \alpha r)^\top < 0, \ \forall i \in I(x^0, y^0),$$

$$\nabla h(x^0, y^0)(d + \alpha\overline{d}, \alpha r)^\top = 0,$$

$$\nabla G_k(y^0)r < 0, \text{ for all } k : G_k(y^0) = 0.$$

Hence, the Karush-Kuhn-Tucker necessary optimality condition [61] for locally Lipschitz optimization problems applies to this penalty function problem. Together with Theorem 4.17 this implies the conclusion.   □

PROOF OF THEOREM 5.22: The deduction of the first two properties is straightforward and therefore omitted here. We will outline the proof of the last one.

Let $r \in K_{\text{Supp}(\alpha,\alpha^i)}(x^0)$ and $r \notin K_{\text{Supp}(\alpha,\alpha^j)}(x^0) \ \forall j \neq i$. Because of the finite number of the tangent cones and properties 1. and 2. it follows that r belongs to int $K_{\text{Supp}(\alpha,\alpha^i)}(x^0)$. This together with $r \notin K_{\text{Supp}(\alpha,\alpha^j)}(x^0)$ has the consequence that for any sequences $\{x^s\}_{s=1}^\infty$, $\lim_{s\to\infty} x^s = x^0$ and $\{t_s\}_{s=1}^\infty, t_s > 0 \ \forall s, \ \lim_{s\to\infty} t_s = 0$ with the property $r = \lim_{s\to\infty} t_s^{-1}(x^s - x^0)$ we have $x^s \in \text{Supp}(\alpha, \alpha^i)$ for sufficiently large k. Since $\{x^0 + t_s r\}_{s=1}^\infty$ is such a sequence if $t_s \downarrow 0$ it follows that $x^0 + t_s r \in \text{Supp}(\alpha, \alpha^i) \ \forall k \geq k_0$ for any sequence $t_s \downarrow 0$. This implies the proof.   □

PROOF OF THEOREM 5.23: Let $\alpha : \mathbb{R}^m \to \mathbb{R}$ be a GPC^1-function and $x^0 \in \mathbb{R}^m$ a fixed point. According to Definition 5.10 there exist $U_\varepsilon(x^0)$ and k continuously differentiable functions α^i defined on $U_\varepsilon(x^0)$ such that

$$\alpha(x) \in \{\alpha^i(x)\}_{i \in \{1,\dots,k\}} \ \forall x \in U_\varepsilon(x^0).$$

Let $r \in \mathbb{R}^m$ be any direction. If $r \in K_{\text{Supp}(\alpha,\alpha^i)}(x^0) \cap K_{\text{Supp}(\alpha,\alpha^j)}(x^0)$ for some $i \neq j$ we conclude directly from Definition 5.10 that

$$\alpha(x^0 + tr) = \alpha^i(x^0 + tr) \ \forall t \in (0, t_0)$$

with some positive $t_0 \geq \delta$. Continuity of α^i implies the existence of the radial limit

$$\alpha_r(x^0) := \lim_{t\downarrow 0} \alpha(x^0 + tr) = \lim_{t\downarrow 0} \alpha^i(x^0 + tr) = \alpha^i(x^0).$$

Otherwise, if $r \in K_{\mathrm{Supp}(\alpha,\alpha^i)}(x^0)$ and $r \notin K_{\mathrm{Supp}(\alpha,\alpha^j)}(x^0)$ $\forall j \neq i$ it follows from (5.41) that the same equality $\alpha_r(x^0) = \alpha^i(x^0)$ holds. This establishes radial-continuity.

With exactly the same argumentation – remember that $\alpha^i \in C^1$ – we get

$$
\begin{aligned}
\mathrm{d}\alpha(x^0; r) &= \lim_{t \downarrow 0} \frac{\alpha(x^0 + tr) - \alpha_r(x^0)}{t} \\
&= \lim_{t \downarrow 0} \frac{\alpha^i(x^0 + tr) - \alpha^i(x^0)}{t} = \nabla \alpha^i(x^0) r
\end{aligned}
$$

This completes the proof.   □

PROOF OF THEOREM 5.24: Let the vector r^0 with $\|r^0\| = 1$ satisfy the first condition. That means

$$
\mathrm{d}\alpha(x^0; r^0) = \lim_{t \downarrow 0} t^{-1}(\alpha(x^0 + tr^0) - \alpha_{r^0}(x^0)) < 0,
$$

which has as a direct consequence the existence of a positive real number t_0 such that

$$
\alpha(x^0 + tr^0) < \alpha_{r^0}(x^0) \quad \forall t \in (0, t_0).
$$

Because of $\alpha_{r^0}(x^0) \leq \alpha(x^0)$ we have $\alpha(x^0 + tr^0) < \alpha(x^0)$ for all these t, too. This means that x^0 is no local minimum.

Now let w.l.o.g. the second condition for r^0 with $\|r^0\| = 1$ be satisfied. Hence,

$$
\delta := \alpha(x^0) - \alpha_{r^0}(x^0) = \alpha(x^0) - \lim_{t \downarrow 0} \alpha(x^0 + tr^0) > 0.
$$

Assuming the existence of a positive sequence $\{t_s\}_{s=1}^\infty$ with $t_s \downarrow 0$ and $\alpha(x^0 + t_s r^0) \geq \alpha(x^0)$ we get an immediate contradiction; thus, there exists $t_0 > 0$ with

$$
\alpha(x^0 + tr^0) < \alpha(x^0) \quad \forall t \in (0, t_0).
$$

Continuing as in the first part we get the desired result.   □

PROOF OF THEOREM 5.25: Arguing by contradiction we assume that there is a sequence $\{x^l\}_{l=1}^\infty$ with $x^l \to x^0$ and $\alpha(x^l) < \alpha(x^0)$ $\forall l \geq 1$. Since α is a selection of only a finite number of C^1-functions α^i there is an index i such that the set $\mathrm{Supp}(\alpha, \alpha^i)$ contains infinitely many of the terms x^l. In the following we consider the sequence $\{x^l\}_{l=1}^\infty \cap \mathrm{Supp}(\alpha, \alpha^i)$ and denote it by $\{x^l\}_{l=1}^\infty$ again. Due to the continuity of α^i we derive

$$
\alpha^i(x^0) = \lim_{l \to \infty} \alpha^i(x^l) = \lim_{l \to \infty} \alpha(x^l) \leq \alpha(x^0).
$$

We now define $r^0 := \lim_{l \to \infty} \frac{x^l - x^0}{\|x^l - x^0\|}$. Hence, due to $x^l \in \text{Supp}(\alpha, \alpha^i)$ we have $r^0 \in K_{\text{Supp}(\alpha, \alpha^i)}(x^0)$. Because of

$$K_{\text{Supp}(\alpha, \alpha^i)}(x^0) \subseteq \text{cl int } K_{\text{Supp}(\alpha, \alpha^i)}(x^0)$$

there is $\hat{r} \in \text{int } K_{\text{Supp}(\alpha, \alpha^i)}(x^0)$ arbitrarily close to r^0 with $\|\hat{r}\| = 1$. Now, let $\{t_l\}_{l=1}^{\infty}$ be any sequence converging to zero from above and consider $\{x^0 + t_l \hat{r}\}_{l=1}^{\infty}$. Since $\hat{r} \in \text{int } K_{\text{Supp}(\alpha, \alpha^i)}(x^0)$ we have both

$$x^0 + t_l \hat{r} \in \text{Supp}(\alpha, \alpha^i) \text{ for sufficiently large } l$$

(by Theorem 5.22) and

$$\|(x^0 + t_l \hat{r}) - x^l\| \to 0 \text{ for } l \to \infty.$$

Continuity of α^i leads to

$$\alpha_{\hat{r}}(x^0) = \lim_{l \to \infty} \alpha(x^0 + t_l \hat{r}) = \lim_{l \to \infty} \alpha^i(x^l) = \alpha^i(x^0) \leq \alpha(x^0).$$

Hence, the first condition in the Theorem cannot be valid. Thus, $\alpha(x^0) = \alpha_{\hat{r}}(x^0)$.

Due to

$$\text{Supp}(\alpha, \alpha^i) \subseteq \text{cl int } \text{Supp}(\alpha, \alpha^i),$$

for any l, there is an open neighborhood $U^1(x^l)$ such that

$$U^1(x^l) \cap \text{int } \text{Supp}(\alpha, \alpha^i) \neq \emptyset.$$

On the other hand, owing to the initial assumption and the continuity of α^i there is another neighborhood $U^2(x^l)$ with

$$\alpha^i(x) < \alpha(x^0) \ \forall x \in U^2(x^l) \cap \text{Supp}(\alpha, \alpha^i).$$

Set $U_\varepsilon(x^l) := U^1(x^l) \cap \text{int } \text{Supp}(\alpha, \alpha^i) \cap U^2(x^l)$. Then, $U_\varepsilon(x^l) \neq \emptyset$ as well as

$$\alpha(x) = \alpha^i(x) < \alpha(x^0) \ \forall x \in U_\varepsilon(x^l).$$

Now, take a sequence $\{\hat{x}^l\}_{l=1}^{\infty}$ converging to x^0 with $\hat{x}^l \in U_\varepsilon(x^l) \ \forall l$. Let $\hat{r}$ be an accumulation point of the sequence $\{\hat{r}^l\}_{l=1}$ defined by

$$\hat{r}^l = \frac{\hat{x}^l - x^0}{\|\hat{x}^l - x^0\|}.$$

Then, as in the first part we get

$$\lim_{l \to \infty} \alpha_{\hat{r}^l}(x^0) = \alpha_{\hat{r}}(x^0) = \alpha^i(x^0) \leq \alpha(x^0) \leq \alpha_{\hat{r}}(x^0) \tag{5.82}$$

since the first condition of the Theorem holds. Now, $\hat{x}^l = x^0 + \|\hat{x}^l - x^0\|\hat{r}^l$ for all l. This leads to

$$\alpha(\hat{x}^l) = \alpha^i(\hat{x}^l) = \alpha^i(x^0) + \|\hat{x}^l - x^0\|\nabla\alpha^i(x^0)\hat{r}^l + o(\|\hat{x}^l - x^0\|) < \alpha(x^0)$$

by the initial assumption for sufficiently small $\|x^l - \hat{x}^l\|$. Together with equation (5.82), this implies the existence of an index $\hat{l}$ such that

$$d\alpha(x^0; \hat{r}^l) = \nabla\alpha^i(x^0)\hat{r}^l < 0$$

for all $l \geq \hat{l}$. This contradiction to the assumptions of the Theorem concludes the proof. □

PROOF OF THEOREM 5.29: From Theorem 5.24 we conclude both $\alpha(x^0) \leq \alpha_r(x^0)$ $\forall r$ and

$$d\alpha(x^0; r) \geq 0 \text{ for all } r \text{ satisfying } \alpha(x^0) = \alpha_r(x^0).$$

Adding the second equation to the foregoing inequality we get

$$\alpha(x^0) \leq \alpha_r(x^0) + d\alpha(x^0; r) \text{ for all } r \text{ with } \alpha(x^0) \geq \alpha_r(x^0).$$

This means $0 \in \partial_{rad}\alpha(x^0)$. □

PROOF OF THEOREM 5.30: From Theorem 5.26 it follows

$$\alpha(x^0) \leq \alpha_r(x^0) \ \forall r.$$

Obviously we have only to investigate r with $\|r\| = 1$. Since the interior of the radial subdifferential is an open set and $0 \in \text{int } \partial_{rad}\alpha(x^0)$ we get

$$\varepsilon r \in \partial_{rad}\alpha(x^0) \ \forall r$$

with some $\varepsilon > 0$. This means

$$\alpha(x^0) + \langle r, \varepsilon r\rangle = \alpha(x^0) + \varepsilon\|r\|^2 \leq \alpha_r(x^0) + d\alpha(x^0; r) \qquad (5.83)$$

for all r satisfying $\alpha(x^0) \geq \alpha_r(x^0), \|r\| = 1$. Thus, if for some $r \neq 0$ the equation $\alpha(x^0) = \alpha_r(x^0)$ is valid, then $d\alpha(x^0; r) > 0$ follows from (5.83). Hence, all assumptions of Theorem 5.25 are satisfied and we conclude $x^0 \in \Psi_{loc}$. □

PROOF OF THEOREM 5.31: We start with the necessary condition. Let there be an index $i \in \{1, \ldots, k\}$ with

$$f(x^i(y^0), y^0) = \varphi(y^0), \ F(x^0, y^0) = F(x^i(y^0), y^0) = \varphi_0(y^0)$$

and a direction $\hat{r}$ satisfying the conditions (5.45)–(5.47). Then, by the (ULR), there exists r^0 sufficiently close to $\hat{r}$ with

$$\nabla_y G_j(y^0)r^0 < 0 \text{ for all } j : G_j(y^0) = 0$$

while maintaining all the other inequalities in (5.46) and (5.47) as strong inequalities. Consider the point $\hat{y} = y^0 + \hat{t}r^0$ for sufficiently small $\hat{t} > 0$. Then,

$$G_j(\hat{y}) = G_j(y^0) + \hat{t}\nabla_y G_j(y^0)r^0 + o(\hat{t}) < 0 \text{ for all } j,$$

implying

$$G_j(\hat{y}) \le 0 \text{ for all } j,$$

$$f(x^i(\hat{y}), \hat{y}) - f(x^j(\hat{y}), \hat{y}) = f(x^i(y^0), y^0) - f(x^j(y^0), y^0) +$$
$$\hat{t}(\nabla_x f(x^i(y^0), y^0)x^{i'}(y^0; r^0) + \nabla_y f(x^i(y^0), y^0)r^0$$
$$-\nabla_x f(x^j(y^0), y^0)x^{j'}(y^0; r^0) - \nabla_y f(x^j(y^0), y^0)r^0) + o(\hat{t}) \le 0$$

for all r^0 sufficiently close to $\hat{r}$ by (5.47) and Lipschitz continuity of $x^i(\cdot), x^j(\cdot)$. This implies $f(x^j(y^0), y^0) > f(x^i(y^0), y^0)$. Since this is valid for all $j \in \{1, \ldots, k\}$ we derive that $f(x^i(\hat{y}), \hat{y}) = \varphi(\hat{y})$ for all $\hat{t} > 0$ sufficiently small. Considering the objective function value we derive

$$F(x^i(\hat{y}), \hat{y}) = F(x^i(y^0), y^0) +$$
$$\hat{t}\left(\nabla_x F(x^i(y^0), y^0)x^{i'}(y^0; r^0) + \nabla_y F(x^i(y^0), y^0)r^0\right)$$
$$+o(\hat{t}) < F(x^i(y^0), y^0) = F(x^0, y^0)$$

for sufficiently small $\hat{t} > 0$ by (5.46). Hence, (x^0, y^0) cannot be a local optimistic optimal solution.

Now assume that the point (x^0, y^0) is not a local optimistic optimal solution. Then there exists a sequence $\{y^t\}_{t=1}^{\infty}$ converging to y^0 with $\varphi_o(y^t) < \varphi_o(y^0)$ for all t. Let $\{x^t\}_{t=1}^{\infty}$ be a corresponding sequence of optimal solutions of the problems (5.4). By finiteness of the number of different optimal solutions in the lower level problem there is one function out of $\{x^i(y)\}_{i=1}^{k}$ such that $x^t = x^i(y^t)$ for infinitely many terms in $\{x^t\}_{t=1}^{\infty}$. Let without loss of generality the sequences $\{x^t\}_{t=1}^{\infty}$ and $\{x^i(y^t)\}_{t=1}^{\infty}$ coincide.

The sequence $\{(y^t - y^0)/\|y^t - y^0\|\}_{t=1}^{\infty}$ has at least one accumulation point. Let this sequence without loss of generality converge itself to the direction r^0. Then,

$$G_j(y^t) = G_j(y^0) + \|y^k - y^0\|\nabla_y G_j(y^0)\frac{y^k - y^0}{\|y^k - y^0\|} + o(\|y^k - y^0\|) \le 0$$

which implies $\nabla_y G_j(y^0)r^0 \leq 0$ for all j with $G_j(y^0) = 0$, i.e. condition (5.48) is satisfied by r^0.

By lower semicontinuity of the function $\varphi_o(\cdot)$ [17] and $\varphi_o(y^t) < \varphi_o(y^0) = F(x^0, y^0)$ we have that $F(x^0, y^0) = F(x^i(y^0), y^0)$ and by upper semicontinuity of the set of global optimal solutions in the lower level problem also $f(x^0, y^0) = f(x^i(y^0), y^0)$. Let $r^t = \frac{y^t - y^0}{\|y^t - y^0\|}$ for all t. Now, analogous differential considerations as for the functions G_j lead to:

$$f(x^i(y^t), y^t) = f(x^i(y^0), y^0) +$$
$$\|y^t - y^0\|(\nabla_x f(x^i(y^0), y^0)x^{i'}(y^0; r^t) + \nabla_y f(x^i(y^0), y^0)r^t)$$
$$+o(\|y^t - y^0\|) \leq f(x^j(y^t), y^t) = f(x^j(y^0), y^0) +$$
$$\|y^t - y^0\|(\nabla_x f(x^j(y^0), y^0)x^{j'}(y^0; r^t) + \nabla_y f(x^j(y^0), y^0)r^t)$$
$$+o(\|y^t - y^0\|)$$

for $j = 1, \ldots, k$ by $f(x^i(y^t), y^t) = \varphi_o(y^t)$. Passing to the limit for $t \to \infty$ shows that condition (5.50) is satisfied if $f(x^j(y^0), y^0) = \varphi_o(y^0)$.

Now we come to (5.49). Inserting $x^i(y^t)$ into the upper level objective function and using $\varphi_o(y^t) < \varphi_o(y^0) = F(x^i(y^0), y^0)$ we derive

$$F(x^i(y^0), y^0) > F(x^i(y^t), y^t) = F(x^i(y^0), y^0) +$$
$$\|y^t - y^0\|(\nabla_x F(x^i(y^0), y^0)x^{i'}(y^0; r^t) + \nabla_y F(x^i(y^0), y^0)r^t)$$
$$+o(\|y^t - y^0\|)$$

showing that r^0 satisfies also condition (5.49). Hence there exists at least one of the systems (5.48) - (5.50) having a nontrivial solution which contradicts our assumptions. Hence, the assertion is valid. $\square$

PROOF OF THEOREM 5.32: Since the second part of this proof can be shown repeating the proof of Theorem 5.31 we show only the first part. Let

$$K_{\{y:G(y)\leq 0\}}(y^0) \not\subseteq \bigcup_{i\in T} K_{\mathrm{Supp}(x,x^j)}(y^0).$$

This implies that there exists a direction $r^0 \in K_{\{y:G(y)\leq 0\}}(y^0)$ which does not belong to the contingent cones $K_{\mathrm{Supp}(x,x^j)}(y^0)$ for all $j \in T$. By the definition, there are sequences $\{y^k\}_{k=1}^{\infty}$ and $\{t_k\}_{k=1}^{\infty}$ with $G(y^k) \leq 0$ for all k, $\lim_{k\to\infty} y^k = y^0$, $\lim_{k\to\infty} t_k = 0$ and $\lim_{k\to\infty} t_k^{-1}[y^k - y^0] = r^0$. Closedness of the sets $\mathrm{Supp}(x, x^j)$ and the definition of the contingent cone implies that

$$y^k \notin \bigcup_{i\in T} \mathrm{Supp}(x, x^j) \tag{5.84}$$

for all sufficiently large k.

Let $x^i(y^k) \in \Psi(y^k)$ be given with $F(x^i(y^k), y^k) = \varphi_p(y^k)$ for all k and suppose without loss of generality that $x^i(y^0) = \lim_{k \to \infty} x^i(y^k)$ exists. Then, since (5.84) we have

$$F(x^i(y^0), y^0) < \varphi_p(y^0)$$

or, with other words, there exists $\varepsilon > 0$ such that $\varphi_p(y^k) \leq \varphi_p(y^0) - \varepsilon$ for all sufficiently large k. Hence, there exists $\delta > 0$ with

$$\inf\{\varphi_p(y) : G(y) \leq 0, \|y - y^0\| < \delta\} \leq \lim_{k \to \infty} \varphi_p(y^k) < \varphi_p(y^0).$$

This implies that the point (x^0, y^0) is not a local pessimistic optimal solution. $\square$

Chapter 6

SOLUTION ALGORITHMS

In this Chapter algorithms are given computing Bouligand or Clarke stationary points of problem (5.1), (5.2) in the case when the lower level problem (5.1) has a unique optimal solution for all values of the parameter. Having results of structural parametric optimization in mind [141, 142], such an assumption is in general only allowed for convex lower level problems. Whence we will add a convexity assumption to the lower level problem throughout this Chapter. We will describe different algorithms for bilevel programming problems: Section 6.1 contains the adaptation of a classical descent algorithm. We show that we need a certain generalized direction finding problem to achieve convergence to a Bouligand stationary point. Using the results in the monograph [224] and the paper [255] a bundle algorithm is given in Section 6.2. Bundle algorithms proved to be very robust and efficient minimization algorithms in nondifferentiable optimization (see also [150, 151, 225]). Using bundle algorithms Clarke stationary solutions can be computed. A trust region method developed in [269] can be used to compute a Bouligand stationary method. We will present this method in Section 6.4. After that we will derive penalty function approaches solving different reformulations (5.26), (5.27) or (5.36) of the bilevel programming problem. We close the presentations in this Chapter with smoothing methods.

It should be mentioned again that focus is mainly on the computation of (local) optimal solutions of the bilevel programming problem in the sense of Definition 5.1. Since all the approaches in this Chapter start with reformulations of the bilevel problem the computed solutions are (Clarke or Bouligand) stationary points of these reformulated problems only. For the relations between these points and (local) optimal solutions of the bilevel problem we refer to Theorem 5.15. It turns out that, if the lower level problem is a convex one having a strongly stable optimal solution, then (Clarke or Bouligand) stationary points of these reformulated problems have the same property also for the original problems. If the algorithms in this Chapter are applied to bilevel problems satisfying all the assumptions but lower level convexity, solutions will be computed which are in general not feasible for the bilevel problem since the lower level solution is in general only a local but not a global optimal solution.

Since algorithmic considerations do not belong to the main points in this book and since convergence proofs for them are often lengthy we will give the results in this Chapter without proofs.

6.1   A DESCENT ALGORITHM

We consider the bilevel programming problem with explicit upper level constraints in the form

$$\min_{y}\{F(x,y) : G(y) \leq 0,\ x \in \Psi(y)\}, \tag{6.1}$$

where $\Psi(y)$ is defined by (5.1) and is assumed to have a unique optimal solution for all values of the parameter y. Then, this problem reduces to the one–level problem

$$\min_{y}\{\mathcal{F}(y) = F(x(y), y) : G(y) \leq 0\}. \tag{6.2}$$

Assume that the lower level problem (5.1) is a convex parametric optimization problem and that the assumptions (MFCQ), (CRCQ) and (SSOC) are satisfied at all points (x,y) with $x \in \Psi(y)$, $G(y) \leq 0$. Then the unique optimal solution of this problem is strongly stable (Theorem 4.4), it is a PC^1 function (Theorem 4.10) and hence locally Lipschitz continuous (Theorem 4.6). Thus, the objective function of problem (6.2) is directionally differentiable (Theorem 4.11). This motivates the following prototype of a *descent algorithm*:

Descent algorithm for the bilevel problem:

   Input: Bilevel optimization problem (5.1), (6.1).

   Output: A Clarke stationary solution.

   1. Select y^0 satisfying $G(y^0) \leq 0$, set $k := 0$, choose $\varepsilon, \delta \in (0, 1)$.

   2. Compute a direction r^k, $\|r^k\| \leq 1$, satisfying
   $$\mathcal{F}'(y^k; r^k) \leq s^k, \nabla_y G_i(y^k) r^k \leq -G_i(y^k) + s^k, \quad i = 1, \ldots, l,$$
   and $s^k < 0$.

   3. Choose a step-size t^k such that
   $$\mathcal{F}(y^k + t^k r^k) \leq \mathcal{F}(y^k) + \varepsilon t^k s^k, G(y^k + t^k r^k) \leq 0.$$

   4. Set $y^{k+1} := y^k + t^k r^k$, compute $x^k \in \Psi(y^k)$, set $k := k + 1$.

   5. If a stopping criterion is satisfied stop, else goto step 2.

For computing the direction of descent we can exploit the necessary optimality condition for bilevel programming problems given in Theorem 5.4 together with the formulae for computing the directional derivative of the solution function of the lower level problem given in Theorem 4.11. This leads to the following problem to be solved for some index set $K^k \in \mathcal{I}(\lambda^k)$ and some vertex $\nu^k := (\lambda^k, \mu^k) \in E\Lambda(x^k, y^k)$, where $z^k := (x^k, y^k)$:

$$s \to \min_{d,r,\gamma,\eta,s}$$

$$\mathcal{F}'(y^k; r^k) := \nabla_x F(z^k)d + \nabla_y F(z^k)r \leq s$$

$$\nabla_y G_i(y^k)r \leq -G_i(y^k) + s, \ i = 1, \ldots, l$$

$$\nabla^2_{xx}L(z^k, \nu^k)d + \nabla^2_{xy}L(z^k, \nu^k)r + \nabla^T_x g(z^k)\gamma + \nabla^T_x h(z^k)\eta = 0$$

$$\nabla_x g_i(z^k)d + \nabla_y g_i(z^k)r \begin{cases} = 0, & i \in K^k \\ \leq -g_i(x^k, y^k) + s, & i \notin K^k \end{cases} \quad (6.3)$$

$$\nabla_x h_j(z^k)d + \nabla_y h_j(z^k)r = 0, j = 1, \ldots, q$$

$$\lambda_i + \gamma_i + s \geq 0, \ i \in K^k, \ \gamma_i = 0, \ i \notin K^k, \ \|r\| \leq 1.$$

Let $(d^k, r^k, \gamma^k, \eta^k, s^k)$ be a feasible solution of problem (6.3) with $s^k < 0$. Then, the verification if the computed direction r^k is such that $(\lambda^k, \mu^k) \in S(r^k)$ is not necessary since problem $QP(\lambda^k, \mu^k, r^k)$ has a feasible solution if and only if $(\lambda^k, \mu^k) \in S(r^k)$ by linear programming duality. Note that the (necessary and sufficient) optimality conditions of problem $QP(\lambda^k, \mu^k, r^k)$ are part of the constraints of problem (6.3) for suitably chosen index set $K^k \in \mathcal{I}(\lambda^k)$. Theorem 5.4 shows that if problem (6.3) has a negative optimal value for some index set $K^k \in \mathcal{I}(\lambda^k)$ and some vertex $\nu^k := (\lambda^k, \mu^k) \in E\Lambda(x^k, y^k)$, then the point (x^k, y^k) is not locally optimal. Note that the inactive inequalities both in the lower and the upper levels are treated in a way which is sensible only if the assumptions (ULR) and (FRR) are satisfied. If one of these conditions is violated a simple degenerate situation can force the objective function value to be zero without any implication for the bilevel problem. The use of this formulation of the direction finding problem is motivated by techniques which avoid zigzagging [31].

In place of a step–size we can use a kind of Armijo step-size rule [31], i.e. we select the largest number t^k in $\{\rho, \rho^2, \rho^3, \rho^4, \ldots\}$, where $\rho \in (0, 1)$, such that

$$\mathcal{F}(y^k + t^k r^k) \leq \mathcal{F}(y^k) + \varepsilon t^k s^k, \ \varepsilon \in (0, 1), \ \text{and} \ G(y^k + t^k r^k) \leq 0. \ (6.4)$$

Of course, this rule can be replaced by other step–size selection methods and can also be refined according to the ideas in [31, 123, 124].

Applying the resulting algorithm to the bilevel programming problem we get convergence to a Clarke–stationary point:

THEOREM 6.1 ([76]) *Consider problem (5.1), (6.1) with a convex parametric lower level problem under assumptions (C), (ULR), (FRR), (MFCQ), (CRCQ), and (SSOC) for all (x, y), $x \in \Psi(y)$, $G(y) \leq 0$. Then, for the sequence $\{(x(y^k), y^k, r^k, d^k, \lambda^k, \mu^k, \gamma^k, \eta^k, s^k, K^k)\}_{k=1}^{\infty}$ computed by the algorithm, the sequence $\{s^k\}_{k=1}^{\infty}$ has zero as the only accumulation point.*

It should be mentioned that, to achieve local convergence, the strong assumptions (MFCQ), (CRCQ), (SSOC), (FRR) are only needed at the limit point of the computed iterates.

COROLLARY 6.1 *Consider the bilevel programming problem (5.1), (6.1) and let all assumptions of Theorem 6.1 be valid. Then, each accumulation point (x^0, y^0) of the sequence of iterates is Clarke stationary.*

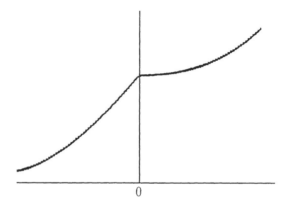

0

Figure 6.1.   Convergence to Clarke stationary point

The following example shows that the result of the above algorithm is in general not a Bouligand stationary point.

Example: In this example the following problem is solved:

$$\min\{(y+2)^2 + (x-2)^2 : x \in \Psi(y)\},$$

where

$$\Psi(y) = \operatorname*{Argmin}_{x} \{x^2 : x \geq (y-1)^2, \ x \geq (y+1)^2\}.$$

Then,

$$x(y) = \max\{(y-1)^2, (y+1)^2\} = \begin{cases} (y-1)^2 & \text{if } y \leq 0 \\ (y+1)^2 & \text{if } y \geq 0 \end{cases}$$

k	y^k	x^k	r^k	d^k	t^k	$\mathcal{F}(y^k; r^k)$	$\mathcal{F}(y^k)$
1	1	4	-1	-4	0.5	-50	13
2	0.5	2.25	-1	-3	0.25	-6.5	6.3125
3	0.25	1.56	-1	-2.5	0.125	-2.31	5.2561
4	0.125	1.27	-1	-2.25	0.0625	-0.94	5.048
5	0.0625	1.129	-1	-2.125	0.0312	-0.423	5.0125
6	0.0312	1.0635	-1	-2.06	0.0156	-0.199	5.003
7	0.0156	1.0315	-1	-2.03	0.0078	-0.097	5.0007
8	0.0078	1.0157	-1	-2.016	0.0039	-0.0008	5.0002
⋮	⋮	⋮	⋮	⋮	⋮	⋮	⋮

Table 6.1. Convergence to Clarke stationary point

Hence,

$$\mathcal{F}(y) = \begin{cases} (y+2)^2 + ((y-1)^2 - 2)^2 & \text{if } y \leq 0 \\ (y+2)^2 + ((y+1)^2 - 2)^2 & \text{if } y \leq 0 \end{cases}$$

which is equal to

$$\mathcal{F}(y) = \min\{(y+2)^2 + ((y-1)^2 - 2)^2, (y+2)^2 + ((y+1)^2 - 2)^2\}.$$

This function is plotted in Figure 6.1. Now, if the algorithm is started with small positive y and the selected step sizes are not too large, the algorithm converges to the point $y^0 = 0$. One possible sequence of iterates is given in Table 6.1. At all iteration points, the function $\mathcal{F}(y)$ is indeed differentiable and we have $\mathcal{F}'(y^k; r^k) = -\mathcal{F}'(y^k)$. Hence, it can be seen that $\{\mathcal{F}'(y^k)\}_{k=1}^\infty$ converges to zero and the limit point $y^0 = 0$ of $\{y^k\}_{k=1}^\infty$ is Clarke stationary. But this point is not Bouligand stationary since $r = -1$ is a descent direction for the function $\mathcal{F}(y)$ at y^0. The reason for this behavior is that a new strongly active selection function arises at the point $y^0 = 0$ for the PC^1 function $x(\cdot)$. □

The above algorithm is not able to "see" this new strongly active selection function if never a "large" step–size is tried. But in higher dimensions also the use of larger step–sizes can not prevent the algorithm from such a behavior. The only way out of this situation is to enlarge the family of index sets in the lower level beyond $\bigcup_{(\lambda,\mu)\in\Lambda(x^k,y^k)} \mathcal{I}(\lambda)$. To do that, take $\kappa > 0$ and let

$$I_\kappa(x, y) = \{j : -\kappa \leq g_j(x, y) \leq 0\}$$

denote the set of κ–active lower level inequalities. The idea of the following complete algorithm is to use the κ–active inequalities in the lower level to enable the algorithm to foresee that an additional selection function will get active at an accumulation point of the iterates without explicitly computing that limit point. Then, if we have also an approximation of the corresponding Lagrange multiplier at hand, we can use the approximation of the direction finding problem to compute an approximation of the descent direction at the limit point. If the limit point is Clarke but not Bouligand stationary this new descent direction must give a much larger descent than that at "late" iterates without this foreseeing possibility. This can be used to pass the accumulation point without computing it explicitly.

The computation of the approximate Lagrange multiplier can be done by solving the quadratic optimization problem

$$(\lambda^k, \mu^k) \in \underset{(\lambda,\mu)}{\text{Argmin}} \ \{\|\nabla_x L(x(y^k), y^k, \lambda, \mu)\|^2 : \lambda_j = 0, \ j \notin K^k\}$$

for some set $K^k \in I_\kappa(x^k, y^k)$ such that the gradients

$$\{\nabla_x g_i(x^k, y^k) : i \in K^k\} \cup \{\nabla_x h_j(x^k, y^k) : j = 1, \ldots, q\}$$

are linearly independent. The optimal solution of this problem is uniquely determined and converges to the true Lagrange multiplier vector (λ^0, μ^0) for $\lim_{k \to \infty} (x^k, y^k) = (x^0, y^0)$ if $K^k \in \mathcal{I}(\lambda^0)$. The correct selection of the set $K^k \in \mathcal{I}(\lambda^0)$ can be left to an oracle. We suggest to use trial and error to realize such a selection.

Due to the posed assumptions, problem (6.3) has a negative optimal objective function value if the accumulation point (x^0, y^0) is not Bouligand stationary and the correct values (λ^0, μ^0) are inserted. Due to $K^k \in \mathcal{I}(\lambda^0)$ and Theorem 4.3 the optimal value function of this problem depends continuously on the data, especially on the lower level Lagrange multipliers. Hence, if (λ^k, μ^k) is closely enough to (λ^0, μ^0) the optimal objective function value of the direction finding problem will be negative provided the correct set K^k is used. This verifies usefulness of the following *descent algorithm*:

Modified descent algorithm for the bilevel problem:

   Input: Bilevel optimization problem (5.1), (6.1).
   Output: A Bouligand stationary solution.
   1. Select y^0 solving $G(y^0) \leq 0$, choose a small $\varepsilon' > 0$, a sufficiently small $\kappa > 0$, $\varepsilon \in (0,1)$, a factor $\varrho \in (0,1)$, a $w < 0$, and set $k := 0$.
   2. Choose (K^k, λ^k, μ^k) with
   $$(\lambda^k, \mu^k) \in E\Lambda(x(y^k), y^k) \text{ and } K^k \in \mathcal{I}(\lambda^k)$$
   and compute an optimal solution $(d^k, r^k, \gamma^k, \eta^k, s^k)$ for problem (6.3).

If $s^k < w$ then goto step 3'. If $s^k \geq w$ and not all possible samples (λ^k, μ^k, K^k) are tried then continue with step 2.

If all (λ^k, μ^k) and all K^k tried to be used then set $w := w/2$.

If $|w| < \varepsilon'$ then goto step 2' else to step 2.

2'. Choose (K^k, λ^k, μ^k) satisfying
$$K^k \subseteq I_\kappa(x(y^k), y^k) \text{ and (C2) as well as}$$

$$(\lambda^k, \mu^k) \in \operatorname*{Argmin}_{(\lambda, \mu)} \{\|\nabla_x L(x(y^k), y^k, \lambda, \mu)\|^2 : \lambda_j = 0, j \notin K^k\},$$

and compute an optimal solution $(d^k, r^k, \gamma^k, \eta^k, s^k)$ for the problem (6.3).

If $s^k < w$ then goto step 3'. If $s^k \geq w$ and not all possible samples (λ^k, μ^k, K^k) are tried then continue with step 2'. If all K^k tried to be used then set $w := w/2$. If $|w| < \varepsilon'$ then stop.

3'. Select the largest number t^k in $\{\rho, \rho^2, \rho^3, \rho^4, \ldots\}$, such that
$$\mathcal{F}(y^k + t^k r^k) \leq \mathcal{F}(y^k) + \varepsilon t^k s^k, \text{ and } G(y^k + t^k r^k) \leq 0.$$

If $t^k < \varepsilon'$ then drop the actual set K^k and continue searching for a new set K^k in step 2 or 2'.

4. Set $y^{k+1} := y^k + t^k r^k$, $k := k + 1$. Goto Step 2.

The value ε' must be so small that the exit in step 3' can only be used if a set K^k is selected in the step 2' such that the problem (6.3) has a negative optimal value, but the corresponding direction r is a direction of ascent. This is obviously possible, if K^k is nowhere a set of active constraints locally around y^k.

The choice of κ seems to be a difficult task. But, on the first hand, there is a positive κ^0 such that $I_\kappa(x(y^0), y^0) = I(x(y^0), y^0)$ for each $0 < \kappa < \kappa^0$. Moreover, the set $I(x(y^0), y^0) \setminus I(x(y^k), y^k)$ should contain only a few elements (one or two) for sufficiently large k for most of the instances. On the one hand, searching for a direction of descent by use of the step 2' of the algorithm can result in a drastic increase of the numerical effort at least if κ is too large. Thus, we suggest to use the step 2' only in the case when the value of $\mathcal{F}(y^k; r^k)$ is sufficiently small and then only for small κ. On the other hand, if the step 2' successfully terminates with a useful direction r and with a set $K \not\subseteq I(x(y^k), y^k)$, then the calculated descent in the objective function value can be expected to be much larger than during the last iterations. Using this algorithm we can show convergence to a Bouligand stationary point:

THEOREM 6.2 ([76]) *Consider the bilevel programming problem (5.1), (6.1) where the lower level problem is assumed to be a convex parametric optimization problem. Let the assumptions (C), (ULR), (FRR), (MFCQ), (CRCQ), and (SSOC) be satisfied for all points (x, y), $x \in$*

$\Psi(y)$, $G(y) \leq 0$. *Take a sufficiently small and fixed parameter $\kappa > 0$. Let the sequence $\{(y^k, \lambda^k, \mu^k, d^k, r^k, s^k, t^k, \gamma^k, \eta^k, K^k)\}_{k=1}^{\infty}$ be computed by the modified descent algorithm. If y^0 is an accumulation point of $\{y^k\}_{k=1}^{\infty}$, then $(x(y^0), y^0)$ is a Bouligand stationary point of the bilevel programming problem.*

6.2   A BUNDLE ALGORITHM

Under the assumptions of Theorem 4.10 on the lower level problem, problem (6.2) is a problem of minimizing a Lipschitz continuous function on the feasible set $Y = \{y : G(y) \leq 0\}$. If assumption (FRR) is also satisfied, then

$$\partial \mathcal{F}(y^0) = \{\nabla_x F(x(y^0), y^0)\partial x(y^0) + \nabla_y F(x(y^0), y^0)\}, \qquad (6.5)$$

with

$$\partial x(y^0) = \text{conv } \{\nabla^\top x^I(y^0) : I \in \bigcup_{(\lambda,\mu) \in \Lambda(x(y^0),y^0)} \mathcal{I}(\lambda)\}$$

by Theorem 4.12. In the opposite case, if assumption (FRR) is not valid, formula (6.5) determines the pseudodifferential of the function $\mathcal{F}(\cdot)$ by Corollary 4.1. Hence, we can apply the bundle-trust region algorithm [160, 161, 223, 224]. A detailed description of this method can be found in [255]. We recall the formulation of this algorithm in the case when the constraints $G(y) \leq 0$ are absent. The inclusion of linear constraints is easy applying the ideas in [302] or using of a feasible directions approach in the direction finding problem (6.19) below. Nonlinear constraints $G(y) \leq 0$ can be treated using a feasible directions approach as in [150]. Denote a generalized gradient in the sense of Clarke for the function $\mathcal{F}(\cdot)$ at a point $\bar{y}$ by $v(\bar{y})$. The bundle method has its roots in cutting plane methods for minimizing convex functions. Let $\{y^i\}_{i=1}^k$, $\{z^i\}_{i=1}^k$ be trial points and iterates already computed. Then, the cutting plane method minimizes the function

$$\max_{1 \leq i \leq k} \{v(y^i)d + v(y^i)(z^k - y^i) + \mathcal{F}(y^i)\} \qquad (6.6)$$

with respect to d, where $d = y - z^k$. The model (6.6) is a piecewise affine linear lower approximate for $\mathcal{F}(y)$ if this function would be convex. A direction d minimizing this model is a descent direction if y is not a stationary point. Unfortunately, cutting plane methods using the model (6.6) for computing descent directions have a very slow convergence speed. Obviously, especially in nonconvex optimization, the model substituting the minimized function cannot give many information about the function far away from z^k. This was the reason for adding the quadratic regularization term $1/(2t_k)d^\mathsf{T}d$ in (6.6):

$$\max_{1 \leq i \leq k} \{v(y^i)d - \alpha_{ik}\} + \mathcal{F}(z^k) + \frac{1}{2t_k}d^\mathsf{T}d, \tag{6.7}$$

with

$$\alpha_{ik} = \mathcal{F}(z^k) - v(y^i)(z^k - y^i) - \mathcal{F}(y^i) \tag{6.8}$$

for all i, k and positive t_k. Using the model (6.7) as a substitute of the function $\mathcal{F}(y)$ a class of much more effective algorithms has been developed known as bundle algorithms and going back to [170, 293]. The optimal solution d of problem (6.7) is a descent direction for $\mathcal{F}(y)$ at $y = z^k$ provided the model approximates the function $\mathcal{F}(y)$ sufficiently good in a neighborhood of a nonstationary point z^k. Then a new iteration point $z^{k+1} = z^k + d$ is computed which gives a sufficient descent in the model function $\mathcal{F}(z^k + d) < \mathcal{F}(z^k)$. If, e.g. for nonsmooth $\mathcal{F}(y)$, the model approximates $\mathcal{F}(y)$ rather poor in a neighborhood of z^k, the direction d will not lead to a sufficiently good (if any) descent. In this case, no new iteration point is computed but the new trial point $y^{k+1} = z^k + d$ is used to improve the model by adding one generalized gradient from $\partial\mathcal{F}(y^{k+1})$. Schematically this leads to the following algorithm (see [224]):

Schematic step in the bundle trust region algorithm [224]
   Input: Sequences of iterates $\{z^i\}_{i=1}^k$ and trial points $\{y^i\}_{i=1}^k$, a regularization parameter t_k.
   Output: A new sufficiently better iteration point z^{k+1} or an improved model.
   1. Compute an optimal solution d^k of (6.7). Set $y^{k+1} = z^k + d^k$.
   2. If $\mathcal{F}(y^{k+1})$ is sufficiently smaller than $\mathcal{F}(z^k)$ then either
   a) enlarge t_k and go back to 1. or
   b) make a serious step: Set $z^{k+1} = y^{k+1}$.
   If $\mathcal{F}(y^{k+1})$ is not sufficiently smaller than $\mathcal{F}(z^k)$ then either
   c) reduce t_k and go back to 1. or
   d) make a null step: Set $z^{k+1} = z^k$, compute $v(y^{k+1}) \in \partial\mathcal{F}(y^{k+1})$.

This variant of the bundle algorithm has been implemented by H. Schramm and J. Zowe [255] and has shown to be a robust and very efficient solution algorithm for nonsmooth optimization problems in extensive test series including also real life problems and bilevel optimization problems (see e.g. [224]).
   The direction finding problem of minimizing the function (6.7) can equivalently be written as

$$u + \frac{1}{2t_k}d^\mathsf{T}d \to \min_{v,d} \tag{6.9}$$

$$-u + v(y^i)d \leq \alpha_{ik}, \ i = 1, \ldots, k.$$

Actually, computational complexity forces us to use only part of the sequence $\{y^i\}_{i=1}^k$ in the formulation of the direction finding problem (6.9). This implies that this problem has only some part of the given constraints. The correct selection of the elements $\{y^i\}_{i \in J_k}$ in $\{y^i\}_{i=1}^k$ is done in the outer iterations of the bundle algorithm.

From the optimality conditions of the strongly convex optimization problem (6.9) with linear constraints we get

LEMMA 6.1 ([255]) *Let (u^k, d^k) be the (unique) optimal solution of the problem (6.9). Then there exist numbers $\lambda_i^k \geq 0, \ i = 1, \ldots, k, \ \sum\limits_{i=1}^{k} \lambda_i^k = 1$ such that*

$$\lambda_i^k(u^k - v(y^i)d^k + \alpha_{ik}) = 0, \ i = 1, \ldots, k \qquad (6.10)$$

$$d^k = -t_k \sum_{i=1}^{k} \lambda_i^k v(y^i)^\top \qquad (6.11)$$

$$u^k = -t_k \left\| \sum_{i=1}^{k} \lambda_i^k v(y^i) \right\|^2 - \sum_{i=1}^{k} \lambda_i^k \alpha_{ik} \qquad (6.12)$$

$$= -\frac{1}{t_k} \|d^k\|^2 - \sum_{i=1}^{k} \lambda_i^k \alpha_{ik} \qquad (6.13)$$

The subgradient inequality

$$\mathcal{F}(y) \geq \mathcal{F}(y^i) + v(y^i)(y - y^i) \qquad (6.14)$$

for convex functions implies that $\alpha_{ik} \geq 0$ provided that the function $\mathcal{F}(\cdot)$ is convex. Adding (6.8) and (6.14) we get

$$v(y^i)(y - z^k) \leq \mathcal{F}(y) - \mathcal{F}(z^k) + \alpha_{ik} \qquad (6.15)$$

for all $y \in \mathbb{R}^m$ and all $i = 1, \ldots, k$. This implies that the inequality constraint corresponding to $v(y^i)$ is less bounding the larger α_{ik} is. Take $\lambda \geq 0$ with $\sum\limits_{i=1}^{k} \lambda_i = 1$, multiply conditions (6.15) by λ_i and sum up. Then we obtain

$$\sum_{i=1}^{k} \lambda_i v(y^i)(y - z^k) \leq \mathcal{F}(y) - \mathcal{F}(z^k) + \sum_{i=1}^{k} \lambda_i \alpha_{ik} \qquad (6.16)$$

which is valid for all $y \in \mathbb{R}^m$. The following Lemma is a direct consequence of (6.16):

LEMMA 6.2 ([255]) *Let $\varepsilon > 0$. If there exists a vector $\lambda \geq 0$, $\sum_{i=1}^{k} \lambda_i = 1$*

such that

$$\left\| \sum_{i=1}^{k} \lambda_i v(y^i) \right\| \leq \varepsilon \ \text{and} \ \sum_{i=1}^{k} \lambda_i \alpha_{ik} \leq \varepsilon$$

then

$$\mathcal{F}(z^k) \leq \mathcal{F}(y) + \varepsilon \|y - z^k\| + \varepsilon \ \forall y \in \mathbb{R}^m.$$

Setting $\varepsilon = 0$ and using (6.12) we see that $u^k = 0$ implies optimality of the point z^k as expected.

In [224, 255] refined rules for making serious vs. null steps are made which should now be given. Let numbers $0 < m_1 < m_2 < 1$, $0 < m_3 < 1$ be given as well as an upper bound T for t_k, some small $\nu > 0$ and a stopping parameter $\varepsilon \geq 0$. Let $\{z^i\}_{i=1}^{k}$ be a sequence of iterates and $\{y^i\}_{i=1}^{k}$ a sequence of trial points where elements of the generalized gradient $v(y^i) \in \partial \mathcal{F}(y^i)$ have been computed, $i = 1, \ldots, k$.

Refined inner iteration in the bundle algorithm

   Input: A function $\mathcal{F}(\cdot)$ to be minimized, an iteration point z^k and a sequence of trial points $\{y^i\}_{i=1}^{k}$.

   Output: Either a new iteration point z^{k+1} or a refined model (6.9).

1. Put $t^1 := t_{k-1}$, $t^* := 0$, $\bar{t} := T$, $j := 1$.
2. Compute the optimal solution (u^j, d^j) of (6.9). If

$$\frac{1}{t^j} \|d^j\| \leq \varepsilon \ \text{and} \ \frac{1}{t^j} \|d^j\|^2 + u^j \geq -\varepsilon$$

then stop: z^k is ε-optimal in the sense of Lemma 6.2. Otherwise put $w^j := z^k + d^j$ and compute $v(w^j) \in \partial \mathcal{F}(w^j)$.
3. If SS(i) and SS(ii) hold, then make a serious step: $z^{k+1} := y^{k+1} := w^j$ and stop.
4. If SS(i) holds but not SS(ii) then refine the parameter t^j: Set $t^* := t^j$, $t^{j+1} := 0.5(\bar{t} + t^*)$, $j := j + 1$ and goto 2.
5. If NS(i) and NS(ii) hold, make a null step: $z^{k+1} := z^k$, $y^{k+1} := w^j$ and stop.
6. If NS(i) holds but not NS(ii) make a refinement of the parameter t^j: Put $\bar{t} := t^j$, $t^{j+1} := 0.5(\bar{t} + t^*)$, $j := j + 1$ and goto Step 2.

Set

$$\varrho_k := \sum_{i=1}^{k} \lambda_i^k v(y^i), \ \sigma_k := \sum_{i=1}^{k} \lambda_i^k \alpha_{ik}. \tag{6.17}$$

Here, the conditions SS(i), SS(ii), NS(i), and NS(ii) are the following:

Condition

$$\text{SS(i):} \quad \mathcal{F}(w^j) - \mathcal{F}(z^k) < m_1 u^j$$

guarantees a decrease of at least $m_1 u^j$ which is at least m_1 times the directional derivative of the model function (6.6).

The second condition for a serious step is the following:

$$\text{SS(ii):} \quad v(w^j)^\top d^j \geq m_2 u^j \text{ or } t^j \geq T - \nu.$$

The first condition in SS(ii) checks if a substantial change in the model (6.9) is made. If this substantial change is not guaranteed and t^j is not large enough then a larger t^j is tried even if SS(i) is satisfied.

Condition NS(i) is the negation of SS(i):

$$\text{NS(i):} \quad \mathcal{F}(w^j) - \mathcal{F}(z^k) \geq m_1 u^j.$$

Then, either the model (6.6) is not adequate or the parameter t^j was too large. Hence, a smaller t^j can be used.

If the first condition in

$$\text{NS(ii):} \quad \mathcal{F}(z^k) - v(w^j)^\top (z^k - w^j) - \mathcal{F}(w^j) \leq m_3 \sigma_{k-1} \text{ or}$$
$$|\mathcal{F}(z^k) - \mathcal{F}(w^j)| \leq \|\varrho_{k-1}\| + \sigma_{k-1}$$

is satisfied, a null step leads to a significant change of the model function (6.6) and can be done. This is due to

$$v(y^{k+1})^\top (y - z^{k+1}) \leq \mathcal{F}(y) - \mathcal{F}(z^{k+1}) + \alpha_{i+1,k+1}$$

with $\alpha_{i+1,k+1} \leq m_3 \sigma_{k-1}$ and $0 < m_3 < 1$ after performing the null step. This can be interpreted as $v(w^j) = v(y^{i+1})$ being close to $\partial \mathcal{F}(z^k)$ and it makes sense to add the vector $v(w^j)$ to the bundle of all the elements of the generalized gradients $\{v(y^i)\}_{i=1}^k$. In this case, inequality NS(i) guarantees that the new inequality for (6.9) contributes non-redundant information:

$$v(y^{k+1})^\top (d^k) - \alpha_{k+1,k+1} = \mathcal{F}(y^{k+1}) - \mathcal{F}(z^{k+1})$$
$$\geq \quad m_1 u^k > u^k \geq v(y^i) d^k - \alpha_{ik} \tag{6.18}$$

for all $i = 1, \ldots, k$. The reason why a null step is made also if the first condition in NS(ii) is not satisfied but the second one is, comes from the proof of the bundle algorithm in [254, 255].

In [224, 255], the inner iteration of the bundle algorithm is embedded into the overall *bundle trust-region algorithm*. The main task in this outer algorithm besides fixing the initial conditions is it to update the bundle of subgradients.

Bundle algorithm for minimizing a convex function

Input: A convex function $\mathcal{F}(y)$ to be minimized.

Output: A Clarke stationary point.

1. Choose a starting point $z^1 \in \mathbb{R}^m$ and parameters $T > 0$, $0 < m_1 < m_2 < 1$, $0 < m_3 < 1$, $\nu > 0$, $\varepsilon > 0$ and an upper bound $j_{\max} \geq 3$ for the maximal number of subgradients in the bundle.

2. Compute $\mathcal{F}(z^1)$ and a subgradient $v(z^1) \in \partial \mathcal{F}(z^1)$. Set $k := 1$, $J_1 := \{1\}$.

3. Apply the inner iteration algorithm either to compute a new (trial or iteration) point z^{k+1} and a new subgradient $v(y^{k+1})$ or realize that z^k is ε-optimal.

4. If $|J_k| = j_{\max}$ reduce the bundle in Step 5. Else goto Step 6.

5. Choose a subset $J \subset J_k$ with $|J| \leq j_{\max} - 2$ and
$$\max\{i \in J_k : \alpha_{ik} = 0\} \in J.$$

Introduce some additional index $\tilde{k}$ and set
$$v(y^{\tilde{k}}) := \varrho_k, \quad \alpha_{\tilde{k}k} := \sigma_k, \quad J := J \cup \tilde{k}.$$

6. Compute $\alpha_{i,k+1}$ for all i, set $J_{k+1} := J \cup \{k+1\}$, $k := k+1$ and goto Step 2.

Some remarks to the algorithm are in order. First, the use of the subset J_k of subgradients, and hence the use of only a subset of the constraints in the problem (6.9), is needed due to to storage space limitations. The method for reducing J_k used in the above algorithm corresponds to the aggregate subgradient technique developed in [150]. Usually, the formula $v(y^{\tilde{k}}) = \varrho_k$ is not correct since there is in general no point $y^{\tilde{k}}$ satisfying this equation. In [255] this artificial notation has been used to keep with the previous notation. Also $\alpha_{\tilde{k}k} := \sigma_k$ has no interpretation as a linearization error as α_{ik} has. But it can easily be checked that the inequality

$$\mathcal{F}(x) - \mathcal{F}(z^k) + \alpha_{\tilde{k}k} \geq v(y^{\tilde{k}})(x - z^k)$$

remains valid for this setting. And this inequality is used for the subgradients in the convergence proof. Second, the method for reducing J_k is especially constructed such that all necessary information is maintained and information resulting from points which are fare away from the present iteration point is secured in an aggregated form.

For the bundle algorithm, convergence to optimal solutions of the problem of minimizing a convex function on the whole space can be shown provided the set of optimal solutions is not empty:

THEOREM 6.3 ([224, 255]) *Let the function* $\mathcal{F}(x)$ *be convex. If the set of optimal solutions* $\operatorname*{Argmin}_{x} \mathcal{F}(x) \neq \emptyset$, *then the bundle algorithm converges to some* $x^* \in \operatorname*{Argmin}_{x} \mathcal{F}(x)$ *for k tending to infinity, i.e.*

$$\lim_{k \to \infty} z^k \in \operatorname*{Argmin}_{x} \mathcal{F}(x).$$

If the set $\operatorname*{Argmin}_{x} \mathcal{F}(x) = \emptyset$, *then*

$$\lim_{k \to \infty} \mathcal{F}(z^k) = \inf_{x} \mathcal{F}(x).$$

Now we consider the nonconvex case. Note that $\alpha_{ik} \geq 0$ can be guaranteed only in the case of minimizing a convex function (i.e. $\mathcal{F}(\cdot)$ is accurately approximated by (6.7) generally only in the convex case). Since the function $\mathcal{F}(\cdot)$ as defined in (6.2) is in general not a convex function, the functions (6.6) can not be used to describe appropriate local approximations of $\mathcal{F}(\cdot)$. To overcome this difficulty, α_{ik} is replaced in [255] by

$$\beta_{ik} = \max\{\alpha_{ik}, c_0 \|z^k - y^i\|\}$$

for a small $c_0 > 0$ and we get the direction finding problem

$$\max_{1 \leq i \leq k} \{v(y^i)d - \beta_{ik}\} + \frac{1}{2t_k} d^\top d \to \min_{d} \qquad (6.19)$$

in place of (6.7). This change does not affect Lemma 6.1 and we will again introduce the vector ϱ_k and the number σ_k as in (6.17). Then, the conditions

$$\left\| \sum_{j \in J_k} \lambda_j^k v(y^j) \right\| \leq \varepsilon, \quad \sum_{j \in J_k} \lambda_j^k \beta_{jk} \leq \varepsilon$$

mean that zero lies up to ε in the convex hull of generalized gradients of the function $\mathcal{F}(y)$ at points which are sufficiently close to z^k. Since the function $\mathcal{F}(y)$ is locally Lipschitz continuous this approximates the Clarke stationarity of the point z^k.

Now consider the refined inner iteration in the bundle algorithm. The first part $v(y^j)^\top d^j \geq m_2 u^j$ of condition SS(ii) comes from convexity and has to be dropped now. This condition is also not needed for the convergence proof. But we have also to add one condition NS(iii) to the remaining rules

SS(i): $\mathcal{F}(y^j) - \mathcal{F}(z^k) \leq m_1 u^j$,
NS(i): $\mathcal{F}(y^j) - \mathcal{F}(z^k) \geq m_1 u^j$, and

NS(ii): $\beta_{jk} \leq m_3\sigma_{k-1}$ or $|\mathcal{F}(y^j) - \mathcal{F}(z^k)| \leq \|\varrho_{k-1}\| + \sigma_{k-1}$
NS(iii): $v(y^j)^\top d^j \geq m_2 u^j$.

The introduction of NS(iii) is due to nonconvexity. Assume that NS(i) is satisfied. For convex functions $\mathcal{F}(y)$ this implies (6.18) which corresponds to NS(iii). The result of (6.18) was a sufficient change of the function (6.6) locally modelling the function $\mathcal{F}(y)$ after a null step. In the convergence analysis the inequality (6.18) was essentially used. For nonconvex $\mathcal{F}(y)$ the conditions NS(i) and NS(ii) may hold without NS(iii) being valid. Hence we have to add this additional condition for a null step.

It remains to describe the ideas used in [224, 255] in the case when the conditions NS(i) and NS(ii) but not NS(iii) are satisfied. In this case a line search method is applied as an emergency exit. Such a line search for nonconvex functions $\mathcal{F}(y)$ has been investigated in detail in [123, 124, 169]. In [169] it has been shown that for weakly semismooth functions $\mathcal{F}(y)$ a line search can be constructed guaranteeing that, after finitely many steps, it stops with a step size $s_j \in (0,1)$ such that with $y^{j+1} = z^k + s_j d^j$, $v(y^{j+1}) \in \partial\mathcal{F}(y^{j+1})$ either one of the following conditions is satisfied:

$$\mathcal{F}(y^{j+1}) - \mathcal{F}(z^k) < m_1 s_j u^j, \quad v(y^{j+1})^\top d^j \geq m_2 u^j \qquad (6.20)$$

or

$$\mathcal{F}(y^{j+1}) - \mathcal{F}(z^k) \geq m_1 s_j u^j \quad \beta_{j+1,k} \leq m_3\sigma_{k-1}$$
$$v(y^{j+1})^\top d^j - \beta_{j+1,k} \geq m_2 u^j. \qquad (6.21)$$

According to the previous discussion the first case enables us to make a serious step whereas the second one implies that all conditions for a null step are satisfied. Hence, a short serious step is executed if (6.20) is valid. The new iteration point is then $z^{k+1} = z^k + s_j d^j$. In the other case, if (6.21) is true, a null step is done leading to $z^{k+1} = z^k$. In both cases one generalized gradient $v(z^k + s_j d_j) \in \partial\mathcal{F}(z^k + s_j d_j)$ is added to the bundle.

Summing up this leads to the following inner iteration of the *bundle algorithm*. The outer iteration remains unchanged.

Refined inner iteration in the bundle algorithm, nonconvex case

> **Input:** A locally Lipschitz function $\mathcal{F}(\cdot)$ to be minimized, an iteration point z^k and a sequence of trial points $\{y^i\}_{i=1}^k$.
> **Output:** Either a new iteration point z^{k+1} or a refined model (6.9).
> 1. Put $t^1 := t_{k-1}$, $t^* := 0$, $\bar{t} := T$, $j := 1$.

2. Compute the optimal solution (u^j, d^j) of (6.9) with α_{jk} replaced by β_{jk}. If

$$\frac{1}{t^j}\|d^j\| \leq \varepsilon \text{ and } \frac{1}{t^j}\|d^j\|^2 + u^j \geq -\varepsilon$$

then stop: z^k is "almost stationary". Otherwise put $w^j := z^k + d^j$ and compute $v(w^j) \in \partial \mathcal{F}(w^j)$.

3. If SS(i) hold, then make a serious step: $z^{k+1} := y^{k+1} := w^j$ and stop.

4. If NS(i), NS(ii) and NS(iii) hold, make a null step: $z^{k+1} := z^k, y^{k+1} := w^j$ and stop.

5. If NS(i), NS(ii) hold, but not NS(iii), then:

  a. If the second condition in NS(ii) holds, then make a line search along the line $z^k + sd^j, s \geq 0$.

  b. Otherwise refine t_j: $\bar{t} = t_j, t_{j+1} = (\bar{t} + t^*)/2$ and repeat 1.

6. If NS(i) holds but not NS(ii) make a refinement of the parameter t^j: Put $\bar{t} := t^j, t^{j+1} := 0.5(\bar{t} + t^*), j := j + 1$ and goto Step 2.

To show convergence of the algorithm a new assumption has to be posed:

$$\{z^k\}_{k=1}^{\infty} \text{ is bounded.} \tag{6.22}$$

This assumption can be satisfied if the process starts sufficiently close to a local optimum. If a convex function is minimized, this assumption is automatically satisfied.

THEOREM 6.4 ([254]) *If the function $\mathcal{F}(\cdot)$ is bounded below and the sequence $\{z^k\}_{k=1}^{\infty}$ computed by the above algorithm remains bounded, then there exists an accumulation point $\bar{z}$ of $\{z^k\}_{k=1}^{\infty}$ such that $0 \in \partial \mathcal{F}(\bar{z})$.*

Repeating the ideas in [254], it is easy to see that, if we are not sure to compute a generalized gradient of the function $\mathcal{F}(y)$ in all of the iteration points but compute elements of the pseudodifferential of $\mathcal{F}(y)$ in any case, then $0 \in \Gamma_{\mathcal{F}}(\bar{z})$, i.e. there is an accumulation point of the sequence of iterates being pseudostationary.

The following corollary is a simple consequence of the previous theorem and strong stability (cf. Theorem 4.4).

COROLLARY 6.2 *If the assumptions (C), (MFCQ), (CRCQ), (SSOC), and (FRR) are satisfied for the convex lower level problem at all points (x, y), $x \in \Psi(y)$, the sequence of iteration points $\{z^k\}_{k=1}^{\infty}$ remains bounded, then the bundle algorithm computes a sequence $\{z^k\}_{k=1}^{\infty}$ having at least one accumulation point $\bar{z} \in Y$ with $0 \in \partial \mathcal{F}(\bar{z})$. If assumption (FRR) is not satisfied, then the point $\bar{z}$ is pseudostationary.*

6.3 PENALTY METHODS

Penalty methods belong to the first attempts for solving bilevel programming problems [2, 3]. Different penalty functions have been used, some of them will be given below.

DEFINITION 6.1 *A penalty function for a set* $M \subseteq \mathbb{R}^p$ *is a continuous function* $v : \mathbb{R}^p \to \mathbb{R}$ *with*

$$v(z) \begin{cases} = 0, & z \in M \\ > 0, & z \notin M. \end{cases}$$

A penalty function v *(for a set* M*) is called exact (at a point* $\overline{z} \in M$*) if for every smooth objective function* $u : M \to \mathbb{R}$ *there exists a real number* τ *such that* $\overline{z}$ *is a local minimizer of the function* $u(z) + \tau v(z)$ *provided that* $\overline{z}$ *is local optimal solution of the problem* $\min\{u(z) : z \in M\}$.

Penalty functions have widely been used in nonlinear optimization and can be found in almost all text books on Nonlinear Optimization (see e.g. [31, 40, 300]). Exact penalty functions have been applied to nonlinear programming problems e.g. in [51, 89, 115, 261, 299]. Many results on exact penalty functions for Mathematical Problems with Equilibrium Constraints can be found in the monograph [188] and in the papers [196, 252]. The various approaches differ in the reformulation of the bilevel programming problem used as a basis for the penalty function. In the first result the problem (5.5) is replaced by (5.26). Note that all these problems are, even in the case of a convex lower level problem, in general not equivalent to the optimistic bilevel programming problem (5.1), (5.5). Theorem 5.15 has shown that all local optimistic optimal solutions of the bilevel programming problem are also local optimal solutions of the reformulation (5.27), a result which similarly is true also for (5.18) and (5.36).

THEOREM 6.5 ([188]) *Let* F, G, f, g, h *be analytic (vector-valued) functions. Let the lower level problem (5.1) be convex and assume that the assumption (MFCQ) is satisfied for this problem for all* $y \in \mathbb{R}^m$ *with* $G(y) \leq 0, \Psi(y) \neq \emptyset$. *Let the feasible set of the bilevel programming problem be not empty and bounded. Take any* $c > 0$ *such that* $\{(x, y) : G(y) \leq 0, x \in \Psi(y)\} \subseteq V_c(0)$. *Then, there exist a scalar* $\alpha^* > 0$ *and an integer* N^* *such that for all* $\alpha \geq \alpha^*, N \geq N^*$, *any vector* (x^*, y^*) *solves (5.5) if and only if for some* (λ^*, μ^*) *the tuple* $(x^*, y^*, \lambda^*, \mu^*)$ *solves the following problem*

$$F(x, y) + \alpha r(x, y, \lambda, \mu)^{1/N} \to \min_{x, y, \lambda, \mu}$$
$$G(y) \leq 0, \ \|(x, y)\| \leq c, \lambda \geq 0,$$

where

$$r(x, y, \lambda, \mu) = \|L(x, y, \lambda, \mu)\| + \|h(x, y)\|$$
$$+ \sum_{i=1}^{p} (\max\{0, g_i(x, y)\} + |\lambda_i g_i(x, y)|).$$

This Theorem uses a standard exact penalty function applied to problem (5.26). The difficulty which makes the fractional exponent necessary results from the complementarity condition in (5.26).

If the bilevel programming problem is replaced by the nonsmooth one-level programming problem (5.27), the following function serves as a penalty function:

$$\Gamma(x, y, \lambda, \mu) = \max\{|\min\{-g(x, y), \lambda\}|, |\nabla_x L(x, y, \lambda, \mu)|, |h(x, y)|, G(y)\} \tag{6.23}$$

Here, $|a|$ is understood component wise: $|a| = (|a_i|)_{i=1}^{p}$ for $a \in \mathbb{R}^p$. The following theorem applies the penalty function (6.23) to bilevel programming problems. It says that a local optimistic optimal solution of the bilevel programming problem with a convex lower level problem (5.1) is a Bouligand stationary point of an exact penalty function for (5.27). This opens the possibility to apply algorithms solving nonsmooth optimization problems to the bilevel programming problem.

THEOREM 6.6 ([249]) *A feasible point* $(\overline{x}, \overline{y}, \overline{\lambda}, \overline{\mu})$ *of (5.27) is Bouligand stationary for* $F(x, y) + \tau\Gamma(x, y, \lambda, \mu)$ *for sufficiently large* τ *if and only if* $\nabla F(\overline{x}, \overline{y})(d\, r)^\top \geq 0$ *for all* $(d, r, \zeta, \xi)\top \in \mathbb{R}^n \times \mathbb{R}^m \times \mathbb{R}^p \times \mathbb{R}^q$ *satisfying*

$$\nabla(\nabla_x L(\overline{x}, \overline{y}, \overline{\lambda}, \overline{\mu}))(d\, r\, \zeta\, \xi)^\top = 0$$
$$\min\{-\nabla g_i(\overline{x}, \overline{y})(d\, r)^\top, \zeta_i\} = 0 \quad \forall i: g_i(\overline{x}, \overline{y}) = \overline{\lambda}_i = 0$$
$$\nabla g_i(\overline{x}, \overline{y})(d\, r)^\top = 0 \quad \forall i: \overline{\lambda}_i > 0$$
$$\nabla h(\overline{x}, \overline{y})(d\, r)^\top = 0 \tag{6.24}$$
$$\nabla G_i(\overline{y})r \leq 0 \quad \forall i: G_i(\overline{y}) = 0$$
$$\zeta_i = 0 \quad \forall i: g_i(\overline{x}, \overline{y}) < 0$$

i.e. it is Bouligand stationary for the problem (5.27).

A third reformulation of the bilevel programming problem is given by (5.36). Here the optimal value function of the lower level problem is used to replace this problem. Again, the exact penalty function can be minimized using algorithms for nonsmooth optimization since the function $\varphi(\cdot)$ is locally Lipschitz continuous under the assumptions used (see Theorem 4.14).

THEOREM 6.7 ([297]) *Consider the problem (5.36). Assume that the assumptions (C) and (MFCQ) are satisfied for the convex lower level* *problem (5.1) for all y with $G(y) \leq 0$, $\Psi(y) \neq \emptyset$. Let (x^0, y^0) be a local optimistic optimal solution of the bilevel problem. Then, problem (5.36) is partially calm at this point if and only if there exists $\kappa > 0$ such that (x^0, y^0) is also a local minimizer of the problem*

$$F(x, y) + \kappa(f(x, y) - \varphi(y)) \to \min_{x, y}$$

$$g(x, y) \leq 0$$

$$h(x, y) = 0$$

$$G(y) \leq 0$$

In [262] this penalty function has been used to construct an algorithm computing a global optimistic optimal solution. Using the sufficient conditions for partial calmness in Section 5.6 this result implies the penalty function approach for linear bilevel programming problems in [7].

6.4   A TRUST REGION METHOD

As in bundle methods, a local model $\Delta_{\tau,k}(s)$ of the objective function is minimized in each step of a trust region method [252, 269]. To introduce this local model function, we start with the nonsmooth problem (5.27). By Theorem 6.6 the task of finding a Bouligand stationary solution for problem (5.27) can be solved by minimizing the function

$$H_\tau(x, y, \lambda, \mu) := F(x, y) + \tau \Gamma(x, y, \lambda, \mu) \text{ for sufficiently large } \tau.$$

Fix some iterate $(x^k, y^k, \lambda^k, \mu^k)$ and consider the function $\Delta_{\tau,k}(s)$ which is obtained from $H_\tau(x, y, \lambda, \mu)$ if each of the functions F, G, L, g, h, λ is replaced by its linearization at the point $z^k := (x^k, y^k)$, $\nu^k := (\lambda^k, \mu^k)$ in direction $s = (s_z, s_\nu)^\top$ with $s_z = (s_x, s_y)^\top$, $s_\nu = (s_\lambda, s_\mu)^\top$:

$$\Delta_{\tau,k}(s) = F(z^k) + \nabla F(z^k) s_z$$
$$+ \quad \tau \max\{| \min\{-g(z^k) - \nabla g(z^k) s_z, \lambda^k + s_\lambda\}|,$$
$$|\nabla_x L(z^k, \nu^k) + \nabla \nabla_x L(z^k, \nu^k)^\top s_z + \nabla_x g(z^k)^\top s_\lambda + \nabla_x h(z^k)^\top s_\mu|,$$
$$|h(z^k) + \nabla h(z^k) s_z|, G(y^k) + \nabla G(y^k) s_y\} + 0.5 s^\top B_k s.$$

Since it is assumed that this model $\Delta_{\tau,k}(s)$ approximates the penalty function in a certain neighborhood of the current iterate $(x^k, y^k, \lambda^k, \mu^k)$, the model $\Delta_{\tau,k}(s)$ is minimized on a an appropriate closed neighborhood of zero. If the resulting change in the function value is not less than some fixed percentage of the change predicted by the model function, a new iterate $(x^k, y^k, \lambda^k, \mu^k) + s$ is computed. In the opposite case, the trust region radius is updated. The *trust region algorithm* roughly works as follows [269]:

Trust region algorithm for the bilevel problem:

　　Input: An instance of problem (5.27).

　　Output: A Bouligand stationary point.

　　0. Choose two fixed parameters $0 < c_0 < c_1 < 1$, a starting point $(x^0, y^0, \lambda^0, \mu^0)$, an initial symmetric matrix B_0 of appropriate dimension, a starting trust region radius δ_0 and a minimal radius $\delta_{\min}$. Set $k := 0$.

　　1. Compute a vector s^k with $\|s^k\| \le \delta_k$ and $\Delta_{\tau,k}(s^k) < \Delta_{\tau,k}(0)$. If no such vector exists, terminate the algorithm. Otherwise compute

$$r_k = \frac{H_\tau(x^k, y^k, \lambda^k, \mu^k) - H_\tau(x^k + s_x^k, y^k + s_y^k, \lambda^k + s_\lambda^k, \mu^k + s_\mu^k)}{\Delta_{\tau,k}(0) - \Delta_{\tau,k}(s^k)}.$$

　　2. Set

$$(z^{k+1}, \nu^{k+1}) = \begin{cases} (z^k, \nu^k) & \text{if } r_k \le c_0 \\ (z^k + s_z^k, \nu^k + s_\nu^k) & \text{if } r_k > c_0 \end{cases}$$

and adjust the trust region radius

$$\delta_{k+1} = \begin{cases} \text{Reduce}(\delta_k) & \text{if } r_k \le c_0 \\ \max\{\delta_{\min}, \delta_k\} & \text{if } c_0 < r_k < c_1 \\ \text{Increase}(\delta_k) & \text{if } r_k > c_1 \end{cases}.$$

Terminate the method if a prescribed rule is satisfied, otherwise set $k := k + 1$, select a new symmetric matrix B_k and return to Step 1.

In the following convergence theorem we need some assumptions:

(S1) If $(x^*, y^*, \lambda^*, \mu^*)$ is an accumulation point of the trust region sequence $\{(x^k, y^k, \lambda^k, \mu^k)\}_{k=1}^\infty$ which is not a Bouligand stationary point of the function $H_\tau(x, y, \lambda, \mu)$ then there exists a positive number α and a normalized descent direction $\|s\| = 1$ of $H_\tau(x, y, \lambda, \mu)$ such that: for all $\delta > 0$ there exists an open neighborhood V of $(x^*, y^*, \lambda^*, \mu^*)$ such that for all k with $(x^k, y^k, \lambda^k, \mu^k) \in V$ and all $\delta_k > \delta$ the direction s^k determined in Step 1 of the algorithm satisfies

$$\Delta_{\tau,k}(0) - \Delta_{\tau,k}(s^k) \ge \max_{\delta \le t \le \delta_k} \alpha[\Delta_{\tau,k}(0) - \Delta_{\tau,k}(ts)].$$

　　The essence of this condition is that the search direction is a tradeoff between the global minimum of the model function and a pure descent direction of this function at the current iterate. The computation of a pure descent direction is not sufficient for convergence due to the non-convexity of the penalty function. This assumption can be satisfied if the linear independence constraint qualification is satisfied for problem (5.30) and appropriate Lagrange multipliers are taken [269]. Obviously, this assumption is also valid if it would be possible to compute a direction s^k with

$$\Delta_{\tau,k}(0) - \Delta_{\tau,k}(s^k) \geq \max_{\|s\| \leq \delta_k} \alpha[\Delta_{\tau,k}(0) - \Delta_{\tau,k}(s)]$$

in each iteration of the algorithm. Unfortunately, this problem is $\mathcal{NP}$-hard.

Next two conditions are given which are conditions on the parameters of the algorithm and can easily be satisfied.

(S2) For each δ_0 the assignment $\delta_k = \text{Reduce}(\delta_{k-1})$ generates a strictly decreasing positive sequence converging to zero and there exists a constant $c_R \in (0,1)$ such that $\text{Reduce}(\delta) \geq c_R\delta$ is satisfied. For each δ_0 the setting $\delta_k = \text{Increase}(\delta_{k-1})$ generates a strictly increasing sequence and there exist a number $c_I > 1$ and $0 \leq \delta_{\min} < \delta_{\max}$ such that $\text{Increase}(\delta) \geq c_I\delta$ for all $\delta \leq \delta_{\max}$ and $\text{Increase}(\delta) \geq \max\{c_I\delta, \delta_{\min}\}$ if $\delta < \delta_{\min}$.

(S3) The sequence of the matrices $\{B_k\}_{k=1}^{\infty}$ is bounded.

The following theorem says that this algorithm computes a sequence of iterates each accumulation point of which is a Bouligand stationary point:

THEOREM 6.8 ([269]) *Let $\{x^k, y^k, \lambda^k, \mu^k\}_{k=1}^{\infty}$ be an infinite sequence of iterates computed with the trust region algorithm and let $(x^*, y^*, \lambda^*, \mu^*)$ be an accumulation point of this sequence. Then, if the assumptions (S1)–(S3) are satisfied and if $\delta_{\min} > 0$, the point $(x^*, y^*, \lambda^*, \mu^*)$ is Bouligand stationary for the function $H_\tau(x, y, \lambda, \mu)$.*

Theorem 6.8 implies, that any accumulation point of the sequence of iterates is a Bouligand stationary point of the penalty function $H_\tau(x, y, \lambda, \mu)$. Theorem 6.6 then helps to verify that this point is a Bouligand stationary point of the problem (5.27) provided that the penalty parameter is sufficiently large. This implies that we have to find an update rule for the parameter. In [269] an update rule from [298] has been applied to reach that goal. This update rule constructs two sequences $\{\tau_k\}_{k=1}^{\infty}$ and $\{\beta_k\}_{k=1}^{\infty}$ as follows: Let α_0, β_0 be constants with $0 < \beta_0 < 1 < \alpha_0$. If

$$\Delta_{\tau_k,k}(0) - \Delta_{\tau_k,k}(s^k) < \beta_k\tau_k \min\{\delta_k, \Gamma(x^k, y^k, \lambda^k, \mu^k)\}$$

then $\tau_{k+1} = \alpha_0\tau_k$, $\beta_{k+1} = \beta_0\beta_k$ else both τ_k and β_k remain unchanged. One more assumption is needed to investigate the convergence properties of the trust region algorithm with added penalty update rule.

(S4) If $(x^*, y^*, \lambda^*, \mu^*)$ is an accumulation point of the sequence of iterates computed by the trust region algorithm which is a Bouligand stationary point then there exists an open neighborhood V of $(x^*, y^*, \lambda^*, \mu^*)$ such that the following is valid: If $(x^k, y^k, \lambda^k, \mu^k) \in V$ and if the kth iteration is not successful, i.e. if $(x^{k+1}, y^{k+1}, \lambda^{k+1}, \mu^{k+1}) = (x^k, y^k, \lambda^k, \mu^k)$ and $\delta_{k+1} \geq c_R \delta_k$ then there exists τ_1 independent of k such that

$$\Delta_{\tau_k, k}(0) - \Delta_{\tau_k, k}(s^k) \geq \tau_1 [\Delta_{\tau_{k+1}, k+1}(0) - \Delta_{\tau_{k+1}, k+1}(s^{k+1})].$$

This assumption essentially means that the solution technique used in the subproblem does not delude us into believing that significantly better points exist if the iteration point is sufficiently close to the accumulation point, the point itself remains the same but the trust region radius is increased. Then we have

THEOREM 6.9 ([269]) *Let* $\{(x^k, y^k, \lambda^k, \mu^k)\}_{k=1}^{\infty}$ *be the sequence of iterates computed with the trust region algorithm completed with the penalty parameter update rule and suppose that the assumptions (S1)–(S4) are satisfied and* $\delta_{\min} > 0$. *Then, if* $\lim_{k \to \infty} \tau_k < \infty$ *every accumulation point of the sequence* $\{(x^k, y^k, \lambda^k, \mu^k)\}_{k=1}^{\infty}$ *is Bouligand stationary of the function* $H_\tau(x, y, \lambda, \mu)$ *and* $\Gamma(x, y, \lambda, \mu) = 0$.

If we replace the assumption (S1) with a similar condition using the function $\Gamma(x, y, \lambda, \mu)$ in place of the local model function, then we get also a result for unbounded sequence $\{\tau_k\}_{k=1}^{\infty}$.

(S1') If $(x^*, y^*, \lambda^*, \mu^*)$ is an accumulation point of the trust region sequence $\{(x^k, y^k, \lambda^k, \mu^k)\}_{k=1}^{\infty}$ which is not a Bouligand stationary point of the function $\Gamma(x, y, \lambda, \mu)$ then there exists positive numbers τ_0, ζ and a normalized descent direction $\|s\| = 1$ of $\Gamma(x, y, \lambda, \mu)$ such that: for all $\delta > 0$ there exists an open neighborhood V of $(x^*, y^*, \lambda^*, \mu^*)$ such that for all k with $(x^k, y^k, \lambda^k, \mu^k) \in V$ and all $\tau_k \geq \tau_0$, $\delta_k > \delta$ the direction s^k determined in Step 1 of the algorithm satisfies

$$\Delta_{\tau, k}(0) - \Delta_{\tau, k}(s^k) \geq \max_{\delta \leq t \leq \delta_k} \zeta [\Delta_{\tau, k}(0) - \Delta_{\tau, k}(ts)].$$

THEOREM 6.10 ([269]) *Let* $\{(x^k, y^k, \lambda^k, \mu^k)\}_{k=1}^{\infty}$ *be the sequence of iterates computed with the trust region algorithm completed with the penalty parameter update rule and suppose that the assumptions (S1'), (S2), (S3) are satisfied and* $\delta_{\min} > 0$. *Then, if* $\lim_{k \to \infty} \tau_k = \infty$ *every accumulation point of the sequence* $\{(x^k, y^k, \lambda^k, \mu^k)\}_{k=1}^{\infty}$ *is Bouligand stationary of the function* $\Gamma(x, y, \lambda, \mu)$.

6.5 SMOOTHING METHODS

In [101] the ideas of the perturbed Fischer-Burmeister function [98, 143] have been used to construct a smooth approximation of the KKT-transformation of the bilevel programming problem.

DEFINITION 6.2 *The Fischer-Burmeister function is* $\Phi : \mathbb{R}^2 \to \mathbb{R}$ *defined by*

$$\Phi(a, b) = a + b - \sqrt{a^2 + b^2}.$$

The perturbed Fischer-Burmeister function is $\Phi : \mathbb{R}^3 \to \mathbb{R}$ *given by*

$$\Phi(a, b, \varepsilon) = a + b - \sqrt{a^2 + b^2 + \varepsilon}.$$

The Fischer-Burmeister function has the property

$$\Phi(a, b) = 0 \text{ if and only if } a \geq 0, \ b \geq 0, \ ab = 0$$

but it is nondifferentiable at $a = b = 0$. Its perturbed variant satisfies

$$\Phi(a, b, \varepsilon) = 0 \text{ if and only if } a > 0, \ b > 0, \ ab = \varepsilon/2$$

for $\varepsilon > 0$. This function is smooth with respect to a, b for $\varepsilon > 0$. Using this function, the problem (5.26) can be approximated by

$$
\begin{aligned}
F(x, y) &\to \min_{x, y} \\
G(y) &\leq 0 \\
\nabla_x L(x, y, \lambda, \mu) &= 0 \\
\Phi(-g_i(x, y), \lambda_i, \varepsilon) &= 0, \ i = 1, \ldots, p \\
h(x, y) &= 0.
\end{aligned}
\tag{6.25}
$$

Let

$$
\begin{aligned}
\mathcal{L}^\varepsilon(x, y, \lambda, \mu, \kappa, \zeta, \tau, \omega) &= F(x, y) + \kappa^\top G(y) + \nabla_x L(x, y, \lambda, \mu)\omega \\
&+ \zeta^\top \Phi(-g(x, y), \lambda, \varepsilon) + \tau^\top h(x, y)
\end{aligned}
$$

denote the Lagrange function of the problem (6.25).

Then, the usual necessary optimality conditions of first and second orders can be defined for problem (6.25): The Karush-Kuhn-Tucker conditions state that there exist vectors $\kappa, \zeta, \tau, \omega$ satisfying

$$
\begin{aligned}
\nabla_x \mathcal{L}^\varepsilon(x, y, \lambda, \mu, \kappa, \zeta, \tau, \omega) &= 0 \\
\nabla_y \mathcal{L}^\varepsilon(x, y, \lambda, \mu, \kappa, \zeta, \tau, \omega) &= 0 \\
\nabla_\lambda \mathcal{L}^\varepsilon(x, y, \lambda, \mu, \kappa, \zeta, \tau, \omega) &= 0 \\
\nabla_x g(x, y)\omega &= 0 \\
\nabla_x h(x, y)\omega &= 0 \\
h(x, y) &= 0 \\
\Phi(-g_i(x, y), \lambda_i, \varepsilon) &= 0, i = 1, \ldots, p \\
\nabla_x L(x, y, \lambda, \mu) &= 0 \\
G(y) \leq 0, \gamma \geq 0, \qquad \gamma^\top G(y) &= 0
\end{aligned}
\tag{6.26}
$$

(cf. [101]).

The Clarke derivative of the function $\Phi(-g_i(x, y), \lambda_i)$ exists and is contained in the set

$$C_i(x, y, \lambda) = \{r = -\xi_i(\nabla g_i(x, y), 0) + \chi_i(0, 0, 1) : (1 - \xi_i)^2 + (1 - \chi_i)^2 \leq 1\}.$$

Let $(x^\varepsilon, y^\varepsilon, \lambda^\varepsilon, \mu^\varepsilon)$ be solutions of (6.25) and assume that this sequence has an accumulation point $(\overline{x}, \overline{y}, \overline{\lambda}, \overline{\mu})$ for $\varepsilon \to 0+$. It is then easy to see that, for each i such that $g_i(\overline{x}, \overline{y}) = \overline{\lambda}_i = 0$ any accumulation point of the sequence

$$\{(\nabla_x \Phi(-g_i(z^\varepsilon), \lambda_i^\varepsilon, \varepsilon), \nabla_y \Phi(-g_i(z^\varepsilon), \lambda_i^\varepsilon, \varepsilon), \nabla_\lambda \Phi(-g_i(z^\varepsilon), \lambda_i^\varepsilon, \varepsilon))\}$$

belongs to $C_i(x, y, \lambda)$, hence is of the form

$$r = -\xi_i(\nabla g_i(x, y), 0) + \chi_i(0, 0, 1).$$

Then, it is said that the sequence $\{(x^\varepsilon, y^\varepsilon, \lambda^\varepsilon, \mu^\varepsilon)\}$ is asymptotically weakly nondegenerate, if in this formula neither ξ_i nor χ_i vanishes for any accumulation point of $\{(x^\varepsilon, y^\varepsilon, \lambda^\varepsilon, \mu^\varepsilon)\}$. Roughly speaking this means that both $g_i(x^\varepsilon, y^\varepsilon)$ and λ_i^ε approach zero in the same order of magnitude [101].

THEOREM 6.11 ([101]) *Let* $\{(x^\varepsilon, y^\varepsilon, \lambda^\varepsilon, \mu^\varepsilon)\}$ *satisfy the necessary optimality conditions of second order for problem (6.25). Suppose that the sequence* $\{(x^\varepsilon, y^\varepsilon, \lambda^\varepsilon, \mu^\varepsilon)\}$ *converges to some* $(\overline{x}, \overline{y}, \overline{\lambda}, \overline{\mu})$ *for* $\varepsilon \to 0+$. *If the (PLICQ) holds at the limit point and the sequence* $\{(x^\varepsilon, y^\varepsilon, \lambda^\varepsilon, \mu^\varepsilon)\}$ *is asymptotically weakly nondegenerate, then* $(\overline{x}, \overline{y}, \overline{\lambda}, \overline{\mu})$ *is a Bouligand stationary solution for problem (5.26).*

Similar ideas have been used in [136] for the construction of several optimization algorithms for optimization problems with complementarity constraints which can also be used to compute optimistic optimal solutions for bilevel programming problems.

Chapter 7

NONUNIQUE LOWER LEVEL SOLUTION

If the lower level parametric optimization problem (5.1) can have non–unique optimal solutions for some values of the parameter the bilevel programming problem is in general not stable. This is due to the possible lack of lower semicontinuity of the solution set mapping of the follower's problem in this case. This has at least two consequences:

1. If the leader has computed one solution he cannot be sure that this solution can be realized. On the first hand this led to the formulation of the optimistic and the pessimistic bilevel programming problems (5.3) and (5.6). But both the optimistic and the pessimistic solutions can in general not be assumed to be good approximations of the realized solutions in practice.

2. Small changes in the problem data can result in drastic changes of the (lower level optimal) solution. Hence, if the leader has not solved the true problem but only an approximation of it his computed solution can be far from its practical realization.

To circumvent these difficulties it could be preferable for the leader not to take the (best possible or computed) solution but a small perturbation of this solution guaranteeing that the follower has a uniquely determined solution which is strongly stable. We will give some results guaranteeing that such a solution can be computed in the first section 7.1. The question of approximability of bilevel programming problems is investigated in Chapter 7.2. Focus in the last section of this chapter is on solution algorithms for problems with non–unique lower level problems.

7.1   APPROXIMATION OF DESIRED SOLUTIONS

As indicated at the beginning of this chapter the possibilities for the leader to get more safety in the realization of his computed solution (and the corresponding objective function value) by a subsequent perturbation of his selection are investigated. Assume that the leader has selected a feasible solution y^0 of the upper level problem (5.2) and that $x^0 \in \Psi(y^0)$ is one optimal solution of the follower's problem (5.1) the leader used in his computations. Then, the leader hopes that the realization of this solution will lead to an upper level objective function value close to $F(x^0, y^0)$. Due to the reasons given above this need not to be the case.

One way out of this dilemma is to perturb y^0 such that the set of optimal decisions of the follower for the perturbed problem is a small set in a neighborhood of x^0. If this is possible, the set of possible values for the leader's objective function is also small and, in the best case, cut down to a unique value. The idea for applying subsequent perturbations comes from assertions that many optimization problems have generically a unique optimal solution.

THEOREM 7.1 ([220]) *Consider a linear optimization problem*

$$\min\{\langle y, x \rangle : Ax = b, \ x \geq 0\}$$

having a parametric objective function. Then, for almost all $y \in \mathbb{R}^n$ this problem has a unique optimal solution.

THEOREM 7.2 *Consider the convex parametric optimization problem (5.1) and let the assumptions (C) and (MFCQ) be satisfied at some point (x, y^0) with $x \in \Psi(y)$. Then, for all positive definite matrices M and for all y sufficiently close to y^0, the problem*

$$\min\{f(x, y) + x^\top M x : g(x, y) \leq 0, \ h(x, y) = 0\} \qquad (7.1)$$

has a unique optimal solution.

More general genericity results can be found in the papers [62, 128]. A related result has been obtained in [246]: If problem (5.1) with $y \in \mathbb{R}^1$ is considered then it can be shown that for almost all vectors a, b, c and for almost all matrices A, B, C of apropriate dimensions the problem

$$\begin{aligned}
f(x, y) + a^\top x + 0.5 x^\top A a &\to \min \\
g(x, y) + b + Bx &\leq 0 \\
h(x, y) + c + Cx &= 0
\end{aligned} \qquad (7.2)$$

belongs to the so-called class of generic optimization problems $\mathcal{F}^{**}$ [113]. The elements of this class have the property that the family of all their stationary solutions which are not strongly stable in the sense of Kojima (cf. Theorem 4.4) forms a discrete point set [113]. This implies that for almost all y the optimal solution of problem (7.2) is locally unique and strongly stable.

In [147], the following genericity result for the pessimistic bilevel programming problem (5.6) has been given.

DEFINITION 7.1 *Let $F^* = \inf\{\varphi_p(y) : y \in Y\}$ denote the optimal value of the pessimistic problem (5.6) and let for simplicity $Y = \mathbb{R}^m$. A sequence $\{y^k\}_{k=1}^{\infty} \subseteq Y$ is called minimizing sequence if $\lim_{k\to\infty} \varphi_p(y^k) = F^*$. Problem (5.6) is called well-posed if every minimizing sequence $\{y^k\}_{k=1}^{\infty}$ converges itself to an optimal solution.*

It is easy to see that well–posedness implies that the problem has a unique optimal solution.

DEFINITION 7.2 *Let E be a subset of the set of all continuous functions $F : \mathbb{R}^n \times \mathbb{R}^m \to \mathbb{R}$. Then, the set E is called generically well–posed if there exists a dense subset (w.r.t. the maximum norm) $E' \subseteq E$ such that problem (5.6) is well posed for every $F \in E'$.*

Let $\Phi(F) = \{y : \varphi_p(y) = F^*\}$ denote the solution set mapping of problem (5.6) with respect to variations of the upper level objective function F.

THEOREM 7.3 ([147]) *Let $E = C(\mathbb{R}^n \times \mathbb{R}^m, \mathbb{R})$ be the set of all continuous functions mapping $\mathbb{R}^n \times \mathbb{R}^m$ to $\mathbb{R}$. Then the following conditions are equivalent:*

1. *Problem (5.6) is well–posed.*

2. *$\Phi : E \to \mathbb{R}^m$ is upper semicontinuous and single–valued at F.*

For a proof of this and the next theorem the reader is referred to the original paper. This result shows that small perturbations of the upper level objective function can not result in large changes of the optimal solution if and only if each minimizing sequence of the original problem converges to a pessimistic optimal solution.

THEOREM 7.4 ([147]) *Let $\Psi(y)$ be an upper semicontinuous mapping and let assumption (C) be satisfied. Then E is generically well–posed.*

Both results are generalized in [147] also to the case that pairs (x, y) of lower and upper variables are considered as (feasible resp. optimal) solutions.

Unfortunately, these ideas are not easily applicable to the case treated here since they are valid only if general perturbations are allowed. But we can try to implement the following idea:

Method for approximation of a computed solution

Input: A solution (x^0, y^0) of e.g. one of the problems (5.3) or (5.6).

Output: A perturbed point $(\tilde{x}, \tilde{y})$ such that $\|(\tilde{x}, \tilde{y}) - (x^0, y^0)\|$ is small, $\tilde{x} \in \Psi(\tilde{y})$ and $\Psi(\tilde{y})$ is "small".

1. Compute a direction r^0 such that:
$$\forall \{r^k\}_{k=1}^{\infty} : \lim_{k \to \infty} r^k = r^0, \forall \{t^k\}_{k=1}^{\infty} : \lim_{k \to \infty} t^k = 0, t^k > 0, \forall \{x^k\}_{k=1}^{\infty} :$$
$x^k \in \Psi(y^0 + t^k r^k)$ we have $\lim_{k \to \infty} x^k = x^0$.

2. Choose a step-size $t > 0$ sufficiently small and compute a perturbed solution $\tilde{y} = y^0 + tr^0, \tilde{x} \in \Psi(\tilde{y})$.

Consider first the case that each optimal solution of the lower level problem (5.1) is strongly stable in the sense of Kojima [157]. Then, if assumption (C) is satisfied, $\Psi(y^0) = \{x^1(y^0), \ldots, x^k(y^0)\}$ for some finite index k. Let assumptions (MFCQ), (SSOC), and (CRCQ) be satisfied at all points (x^i, y^0) with $x^i \in \Psi(y^0)$. Then, there exist k PC^1–functions $x^i(\cdot)$ with $\{x^1(y), \ldots, x^k(y)\} = \Psi_{loc}(y)$ for all y sufficiently close to y^0. This implies that, for each $i = 1, \ldots, k$ there exist j_i strongly active selection functions $x^{ij}(\cdot)$ with

$$x^{i'}(y^0; r) \in \{\nabla x^{i1}(y^0)r, \ldots, \nabla x^{ij_i}(y^0)r\}$$

by Corollary 4.1. Assume w.l.o.g. that the leader has computed the solution (x^1, y^0) and denote

$$A := \{z = \nabla_x f(x^i, y^0) \nabla x^{ij}(y^0) + \nabla_y f(x^i, y^0) : i > 1, j = 1, \ldots, j_i\}$$

and

$$B := \{z = \nabla_x f(x^1, y^0) \nabla x^{1j}(y^0) + \nabla_y f(x^1, y^0) : j = 1, \ldots, j_1\}.$$

THEOREM 7.5 *Consider the problem (5.1) and let the assumptions (C), (MFCQ), (CRCQ), and (SSOC) be satisfied for all global optimal solutions. If* conv $A \cap$ conv $B = \emptyset$ *then there exists a direction r^0 satisfying the conditions in Step 1 of the above algorithm.*

The proof of this theorem uses a hyperplane separating the sets conv A and conv B. The computation of the separating hyperplane can be done by minimizing the distance between the sets conv A and conv B. This reduces to a quadratic optimization problem. The optimal solution of this problem can be used to compute the direction r^0 desired for the implementation of Step 1 of the above algorithm. But it should be noticed that this is not as easy as it seems since the sets A and B are in general not known. To compute them we need to compute all globally optimal solutions of a nonconvex optimization problem.

This approach is not possible if there are global optimal solutions in the lower level not being strongly stable. Then, if the solution set mapping of the lower level problem is not lower semicontinuous, the desired perturbation cannot exist in general. This follows since a lack of lower semicontinuity implies that the set of reachable points

$$\mathcal{RP}(y^0) := \{x \in \mathbb{R}^n : \exists\{(x^k, y^k)\}_{k=1}^{\infty}, \ y^k \neq y^0, \ x^k \in \Psi(y^k), \ \forall k,$$
$$\lim_{k \to \infty} (x^k, y^k) = (x, y^0)\}$$

(7.3)

is in general a proper subset of $\Psi(y^0)$. It can be difficult to compute elements of the reachable set. The following theorems give approximations of that set:

THEOREM 7.6 ([256]) *Consider problem (5.1) at a point $y = y^0$ and let the assumptions (C) and (MFCQ) at all points $x \in \Psi(y^0)$ be satisfied. Then,*

$$\mathcal{RP}(y^0) \subseteq \bigcup_{\|r\|=1} \{x \in \Psi(y^0) : \min_{(\lambda,\mu) \in \Lambda(x,y^0)} \nabla_y L(x, y^0, \lambda, \mu) r \leq D^+ \varphi(y^0; r)\}.$$

Using the directional differentiability of the optimal value function of problem (5.1) in the case that the stronger linear independence constraint qualification (LICQ) is satisfied for all optimal solutions $x \in \Psi(y^0)$ (cf. Corollary 4.4) we get an even stronger result not using the upper Dini derivative:

COROLLARY 7.1 *Consider problem (5.1) and let the assumptions (C) and (LICQ) be satisfied at all points $x \in \Psi(y^0)$. Then,*

$$\mathcal{RP}(y^0) \subseteq \bigcup_{\|r\|=1} \operatorname*{Argmin}_{x \in \Psi(y^0)} \nabla_y L(x, y^0, \lambda, \mu) r.$$

Define a directional version of the reachable set by

$$\mathcal{RP}(y^0; r) := \{x \in \mathbb{R}^n : \exists\{(x^k, t^k)\}_{k=1}^{\infty}, \ t^k > 0, \ x^k \in \Psi(y^0 + t^k r), \ \forall k,$$
$$\lim_{k \to \infty} (x^k, t^k) = (x, 0)\}.$$

Then, the result of the above corollary reads as

$$\mathcal{RP}(y^0; r) \subseteq \operatorname*{Argmin}_{x \in \Psi(y^0)} \nabla_y L(x, y^0, \lambda, \mu) r \ \forall r.$$

(7.4)

To find a direction which can be used in Step 2 of the Method for approximation of a computed solution we can try to determine a direction r such that the right hand side of (7.4) reduces to a singleton. Provided that we are able to compute all globally optimal solutions of the nonconvex optimization problem (5.1) for $y = y^0$ this is an easy task if assumption (SSOC) is added to the assumptions of the last corollary. Then, the number of globally optimal solutions is finite and the right-hand side problem in (7.4) is a linear optimization problem parametrized in r. If (x^0, y^0) is to be approximated and $\nabla_y L(x^0, y^0, \lambda^0, \mu^0)$ is a vertex of the set

$$\mathrm{conv}\ \{\nabla_y L(x, y^0, \lambda, \mu) : \{(\lambda, \mu)\} = \Lambda(x, y^0), x \in \Psi(y^0)\}$$

then a direction r needed in Step 2 can be computed using results of parametric linear optimization [178].

THEOREM 7.7 ([256]) *Consider the convex parametric optimization problem (5.1) and let the assumptions (C) and (MFCQ) be satisfied at* $y = y^0$. *Then,*

$$\mathcal{RP}(y^0) \subseteq \bigcup_{\|r\|=1} \mathcal{T}(y^0; r), \qquad (7.5)$$

where

$$\mathcal{T}(y^0; r) = \underset{x \in \Psi(y^0)}{\mathrm{Argmin}} \ \underset{(\lambda, \mu) \in \Lambda(x, y^0)}{\max} \ \nabla_y L(x, y^0, \lambda, \mu) r. \qquad (7.6)$$

The proof of this theorem follows directly from the proof of Theorems 4.16 and 7.6. The following simple example shows that the inclusion in the last theorem can be strict:

Example: Consider the problem

$$\min\{(x_1 - x_2)^2 - x_1 + x_2 : y_1 x_1 - (y_2 + (1 - y_2)^2) x_2 \le 0, \ x_1, x_2 \in [0, 1]\}$$

at $y^0 = (1, 1)^\top$. The feasible set and the set of free minimizers of the objective function in this problem are plotted in Figure 7.1. Then,

$$\Psi(y^0) = \{(x_1, x_2)^\top : x_1 = x_2 \in [0, 1]\}$$

and the set of Lagrange multipliers $\Lambda(x, y^0) = \{\lambda^0 = (1, 0, 0, 0, 0)\}$ reduces to a singleton, $\nabla_y L(x, y^0, \lambda^0) = (x_1, -x_2)$. Thus,

$$\underset{x \in \Psi(y^0)}{\mathrm{Argmin}} \ \nabla_y L(x, y^0, \lambda^0) r = \underset{x_1 \in [0,1]}{\mathrm{Argmin}} \ x_1(r_1 - r_2) = \begin{cases} 0, & r_1 > r_2 \\ r_1 - r_2, & r_1 < r_2 \\ \Psi(y^0), & r_1 = r_2 \end{cases}$$

Thus, $\displaystyle\bigcup_{\|r\|=1} \mathcal{T}(y^0, r) = \Psi(y^0)$. Let $\omega = y_2 + (1 - y_2)^2$. Then,

$$
\Psi(y) = \begin{cases} \operatorname{conv}\left\{\left(\dfrac{\omega}{2(\omega - y_1)}, \dfrac{y_1}{2(\omega - y_1)}\right)^\mathsf{T}, (1, 0.5)^\mathsf{T}\right\}, & y_1 \leq 0.5\omega \\[3mm] \left\{\left(1, \dfrac{y_1}{\omega}\right)^\mathsf{T}\right\}, & y_1 \in [0.5\omega, \omega] \\[3mm] \{(0, 0)^\mathsf{T}\}, & \omega \leq y_1 \end{cases}
$$

for $y \neq (1, 1)^\mathsf{T}$. The corresponding Lagrange multipliers are

$$
\Lambda(x, y) = \begin{cases} \left(1/y_1, 0, 0, 1 - \dfrac{\omega}{y_1}, 0\right)^\mathsf{T}, & \text{if } \omega < y_1 \\[3mm] \left(\dfrac{2x_2 - 1}{\omega}, 0, 2x_2 - 1 - \lambda_1 y_1, 0, 0\right)^\mathsf{T}, & \text{if } 0.5\omega \leq y_1 \leq \omega \\[3mm] (0, 0, 0, 0, 0)^\mathsf{T}, & \text{if } 0.5\omega \geq y_1 \end{cases}
$$

Hence, $\mathcal{RP}(y^0) = \{(1, 1)^\mathsf{T}, (0, 0)^\mathsf{T}\}$. □

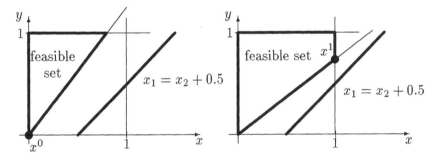

Figure 7.1. Feasible set and free minimizers in Example on page 222 for $\omega < y_1$ and $\omega > y_1$

If we fix a direction r, then we get a directional version of Theorem 7.7:

$$
\emptyset \neq \mathcal{RP}(y^0; r) \subseteq \mathcal{T}(y^0; r).
$$

Hence, in order to compute an element of $\mathcal{RP}(y^0)$ we can try to determine a direction r such that $\mathcal{T}(y^0; r)$ reduces to a singleton. Then, since $\mathcal{RP}(y^0; r) \neq \emptyset$, we have also computed an element of $\mathcal{RP}(y^0)$. For that we will use the following equivalent setting of the right-hand side problem in equation (7.6) which is a simple implication of linear programming duality together with validity of the saddle point inequality.

THEOREM 7.8 ([30]) *Let the assumptions (C) and (MFCQ) be satisfied for the parametric convex optimization problem (5.1) at a point y^0. Take*

$x^0 \in \Psi(y^0)$, $(\lambda^0, \mu^0) \in \Lambda(x^0, y^0)$. Then, $\varphi'(y^0; r)$ is equal to the optimal value (and the elements of $\mathcal{T}(y^0; r)$ are the x-parts of the optimal solutions) of the following optimization problem $H(x^0, \lambda^0, \mu^0, r)$:

$$
\begin{aligned}
\nabla_x f(x^0, y^0) d \;+\; & \nabla_y f(x, y^0) r && \to && \min_{d, x} \\
\nabla_x g_i(x^0, y^0) d \;+\; & \nabla_y g_i(x, y^0) r && \leq 0, && i \in I(x^0, y^0), \\
\nabla_x h_j(x^0, y^0) d \;+\; & \nabla_y h_j(x, y^0) r && = 0, && j = 1, \ldots, q, \\
& \nabla_x L(x, y^0, \lambda^0, \mu^0) && = 0, && \\
& g_i(x, y^0) && \leq 0, && i \notin J(\lambda^0), \\
& g_i(x, y^0) && = 0, && i \in J(\lambda^0), \\
& h(x, y^0) && = 0. &&
\end{aligned}
$$

In general, problem $H(x^0, \lambda^0, \mu^0, r)$ is a non-convex optimization problem. It will be convenient for us to use the following assumption which together with the convexity of $\Psi(y^0)$ guarantees convexity of this problem:

(CG) The gradients w.r.t. the parameter y of the constraint functions $\nabla_y g_i(\cdot, y)$, $i = 1, \ldots, p$, are convex and $\nabla_y h_j(\cdot, y)$, $j = 1, \ldots, q$ are affine for each fixed y.

If, in addition to the assumptions in Theorem 7.8 these conditions are also satisfied then problem $H(x^0, \lambda^0, \mu^0, r)$ is a convex optimization problem.

In the following we will need a tangent cone to $\Psi(y^0)$ which is defined by

$$
\begin{aligned}
K^0_{\Psi(y^0)}(x^0, \lambda^0, \mu^0) := \{ u : \;\; & \nabla^2_{xx} L(x^0, y^0, \lambda^0, \mu^0) u = 0, \\
& \nabla_x g_i(x^0, y^0) u = 0, i \in J(\lambda^0), \\
& \nabla_x g_i(x^0, y^0) u \leq 0, i \in I(x^0, y^0) \setminus J(\lambda^0), \\
& \nabla_x h_j(x^0, y^0) u = 0, j = 1, \ldots, q \}.
\end{aligned}
$$

(7.7)

Using assumption (CG) we derive the following sufficient condition for $\mathcal{T}(y^0; r)$ reducing to a singleton:

THEOREM 7.9 *Consider the convex parametric optimization problem (5.1) and let the assumptions (C), (CG), and (MFCQ) be satisfied at $y = y^0$. Let $x^0 \in \mathcal{T}(y^0; r)$ for some $r \in \mathbb{R}^m$. Then, if $\mathcal{T}(y^0; r)$ does not reduce to a singleton, the following system of linear equations and inequalities has a nontrivial solution u for arbitrary $(\lambda^0, \mu^0) \in S(r)$ (where $S(r)$ is defined in (4.22)):*

$$
u^\top \nabla^2_{xy} L(x^0, y^0, \lambda^0, \mu^0) r \leq 0, \tag{7.8}
$$
$$
u \in K^0_{\Psi(y^0)}(x^0, \lambda^0, \mu^0)
$$

Note that the linear system (7.8) is homogeneous, thus having the trivial feasible solution in any case. This implies that if the following optimization problem has a unique optimal solution then the set $T(y^0; r)$ reduces to a singleton:

$$\min_u \{u^\top \nabla^2_{xy} L(x^0, y^0, \lambda^0, \mu^0) r : u \in K^0_{\Psi(y^0)}(x^0, \lambda^0, \mu^0)\}. \tag{7.9}$$

THEOREM 7.10 *Consider the parametric convex problem (5.1) under (C), (MFCQ) at some point (x^0, y^0) with $x^0 \in \Psi(y^0)$. Let $r^0 \in \mathbb{R}^m$ be fixed and $(\lambda^0, \mu^0) \in S(r^0)$. If problem (7.9) with $r = r^0$ has an optimal solution, then $x^0 \in T(y^0; r^0)$.*

The following example continuing the example on page 222 gives some illustration of Theorem 7.9.

Example: For the problem in the example on page 222 we obtain: $K^0_{\Psi(y^0)}(x, \lambda^0, \mu^0) = D$ with

$$D = \begin{cases} \{u : 2(u_1 - u_2) = 0, u_1 - u_2 = 0\}, & 0 < x_1 = x_2 < 1, \\ \{u : 2(u_1 - u_2) = 0, u_1 - u_2 = 0, u_1, u_2 \le 0\}, & x_1 = x_2 = 1, \\ \{u : 2(u_1 - u_2) = 0, u_1 - u_2 = 0, u_1, u_2 \ge 0\}, & x_1 = x_2 = 0 \end{cases}$$

and $u^\top \nabla^2_{xy} L(x, y^0, \lambda^0, \mu^0) r = r_1 u_1 - r_2 u_2$. Thus, for $0 < x_1 = x_2 < 1$, system (7.8) has a nontrivial solution for each r. For the vertices $x = (0, 0)^\top$ and $x = (1, 1)^\top$ of $\Psi(y^0)$ problem (7.9) has at most one solution. Hence, the sets $T(y^0; r)$ as well as $\mathcal{RP}(y^0; r)$ reduce to singletons for certain directions. For $x_1 = x_2 = 1$, one such direction is each r with $r_1 < r_2$. In the last case, r satisfying $r_1 > r_2$ is a desired direction. The usefulness of these values can be confirmed considering the general solution of this problem outlined in the example on page 222.   □

It should be noticed that, even if the set $T(y^0; r)$ consists of only one point it is not guaranteed that $\Psi(y^0 + tr)$ also reduces to a singleton for small positive t. But, $\Psi(y^0 + tr)$ is small and has a small distance to $T(y^0; r)$. Hence, the leader can select the unique point in $T(y^0; r)$ as (approximate) substitute for the real choice of the follower if he decides to take $y = y^0 + tr$ as his solution.

The last theorem in this section gives a sufficient condition for the existence of the desired direction:

THEOREM 7.11 *Consider the problem (5.1) at a point $y = y^0$, let $x^0 \in T(y^0; r)$ for some $r \in \mathbb{R}^m$. Assume that (C) and (MFCQ) as well as (CG) are satisfied. Let (λ^0, μ^0) be a vertex of $S(r)$ such that*

1. $K^0_{\Psi(y^0)}(x^0, \lambda^0, \mu^0) \cap (-K^0_{\Psi(y^0)}(x^0, \lambda^0, \mu^0)) = \{0\}$, *and,*

2. *setting* $I^0 = I(x^0, y^0)$, *the matrix*

$$\begin{pmatrix} 0 & \nabla^2_{xy}L(x^0, y^0, \lambda^0, \mu^0) \\ \nabla_x g_{I^0}(x^0, y^0) & \nabla_y g_{I^0}(x^0, y^0) \\ \nabla_x h(x^0, y^0) & \nabla_y h(x^0, y^0) \end{pmatrix}$$

has full row rank.

Then, there exists a direction r^0 *such that* $\{x^0\} = \mathcal{T}(y^0; r^0)$.

For some more results about this approach the reader is referred to [69].

7.2   STABILITY OF BILEVEL PROBLEMS

The lack of lower semicontinuity of the solution set mapping of the lower level problem (5.1) can result in an unstable behavior of the bilevel programming problem (5.2) with respect to small perturbations as well of the parameter (which means an insufficient exactness in the leader's choice) as of the data of the lower level problem. This is demonstrated in the following two examples:

Example: Consider the bilevel problem

$$\min_y \{(x - y)^2 + y^2 : -20 \le y \le 20, \ x \in \Psi(y)\},$$

where

$$\Psi(y) = \operatorname*{Argmin}_x \{xy : -y - 1 \le x \le -y + 1\}.$$

Then,

$$\Psi(y) = \begin{cases} [-1, 1], & \text{if } y = 0 \\ \{-y - 1\}, & \text{if } y > 0 \\ \{-y + 1\}, & \text{if } y < 0. \end{cases}$$

Let $F(x, y) = (x - y)^2 + y^2$. Then, inserting the optimal solution of the lower level problem into this function (in the cases, when this solution is uniquely determined), we get

$$F(x(y), y) = \begin{cases} (-2y - 1)^2 + y^2, & \text{if } y > 0 \\ (-2y + 1)^2 + y^2, & \text{if } y < 0. \end{cases}$$

On the regions where these functions are defined, both take their infima for y tending to zero with $\lim_{y \to 0} F(x(y), y) = 1$. Figure 7.2 shows the graph of the point-to-set mapping $y \mapsto F(\Psi(y), y)$. This can be used to confirm that $(x^0, y^0) = (0, 0)$ is the (unique) optimistic optimal solution

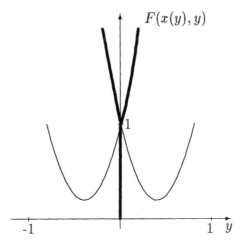

Figure 7.2.   Upper level objective function values in Example on page 226

of the problem in this example. Now, if the leader is not exactly enough in choosing his solution, then the real outcome of the problem has an objective function value above 1 which is far away from the optimistic optimal value zero.   □

Example: Consider the bilevel programming problem

$$\min_{y}\{-x + y^2 : -0.5 \le y \le 0.5, \ x \in \Psi(y)\}$$

where

$$\Psi(y) = \ \text{Argmin}_{x} \ \{xy^2 : -1 \le x \le 1\}.$$

Since $\Psi(y) = \{-1\}$ for $y \ne 0$, and $\Psi(0) = [-1, 1]$, the unique optimistic optimal solution of the bilevel problem is $(x, y) = (1, 0)$. The optimistic optimal function value is -1.

Now assume that the follower's problem is perturbed:

$$\Psi\alpha(y) = \ \text{Argmin}_{x} \ \{xy^2 + \alpha x^2 : -1 \le x \le 1\}$$

for small $\alpha > 0$. Then,

$$x_\alpha(y) = \begin{cases} -1 & \text{if } y^2 > 2\alpha \\ -y^2/(2\alpha) & \text{if } y^2 \le 2\alpha \end{cases}$$

Inserting this function into the leader's objective function gives

$$F(x_\alpha(y), y) = \begin{cases} y^2 + 1 & \text{if } y^2 > 2\alpha \\ y^2 + y^2/(2\alpha) & \text{if } y^2 \le 2\alpha \end{cases}$$

to be minimized on $[-0.5, 0.5]$. The (unique) optimal solution of this problem is $y_\alpha = 0$ for all $\alpha > 0$, with $f(x_\alpha(0), 0) = 0$. For $\alpha \to 0$ the leader's objective function value tends to 0 which is not the optimistic optimal objective function value. $\square$

THEOREM 7.12 *Consider the bilevel programming problem (5.1), (5.2) and let the assumptions (C) and (MFCQ) for all y with $G(y) \leq 0$, be satisfied. Let us consider sequences of sufficiently smooth functions $\{f_k, g^k, h^k, F_k, G^k\}_{k=1}^\infty$ converging pointwise to (f, g, h, F, G). Let*

$$\Psi_k(y) = \operatorname*{Argmin}_x \{f_k(x, y) : g^k(x, y) \leq 0, h^k(x, y) = 0\} \qquad (7.10)$$

and consider the sequence of perturbed optimistic bilevel programming problems

$$\min_{x,y}\{F_k(x, y) : G^k(y) \leq 0, x \in \Psi_k(y)\}. \qquad (7.11)$$

Let (x^k, y^k) be an optimal solution of the problem (7.11) for $k = 1, 2, \ldots$. Then, we have:

1. *The sequences $\{x^k, y^k\}_{k=1}^\infty$ have accumulation points, and for each accumulation point $(\overline{x}, \overline{y})$ we have $\overline{x} \in \Psi(\overline{y})$.*

2. *$(\overline{x}, \overline{y})$ is a lower optimal solution of (5.2).*

It has been shown in the papers [175, 179, 180] that this result holds even in a much more general setting. This is also true if the optimistic bilevel problem is replaced by the pessimistic one.

Under stronger assumptions even piecewise smoothness of the solution of the optimistic bilevel programming problem can be achieved.

THEOREM 7.13 *Consider the bilevel programming problem (5.1), (5.2) with a convex lower level problem. Let this problem depend on an additional parameter z which means that smooth (at least twice continuously differentiable) functions $F, G, f, g, h : \mathbb{R}^n \times \mathbb{R}^m \times \mathbb{R}^l$ are given with $F(x, y, 0) = F(x, y), G(y, 0) = G(y), f(x, y, 0) = f(x, y), g(x, y, 0) = g(x, y), h(x, y, 0) = h(x, y)$. Denote by $\Lambda(x, y, z)$ the set of Lagrange multipliers for the perturbed lower level problem*

$$\min_x\{f(x, y, z) : g(x, y, z) \leq 0, h(x, y, z) = 0\}.$$

Let (x^0, y^0) be a local optimistic optimal solution of problem (5.1), (5.2) and assume:

1. *assumptions (C) and (LICQ) are satisfied for the lower level problem at (x^0, y^0), which is a convex parametric optimization problem for all z,*

2. *for the unique* $(\lambda^0, \mu^0) \in \Lambda(x^0, y^0, 0)$ *suppose that the (LICQ) and the (SSOC) are satisfied for each of the following problems at the local optimal solution* $(x^0, y^0, \lambda^0, \mu^0)$

$$F(x, y, z) \to \min_{x,y,\lambda,\mu}$$

$$G(y, z) \leq 0$$

$$\nabla_x L(x, y, \lambda, \mu, z) = 0$$

$$h(x, y, z) = 0 \qquad (7.12)$$

$$g_i(x, y, z) = 0, i \in I$$

$$g_i(x, y, z) \leq 0, \ i \notin I$$

$$\lambda_i = 0, \ i \notin I$$

$$\lambda_i \geq 0, \ i \in I$$

for arbitrary $I \in \mathcal{I}(\lambda^0)$ *and* $z = 0$ *(cf. problem (5.28))*

Then, there exists a PC^1 *function* $(x(\cdot), y(\cdot))$ *being a local optimal solution of the problem*

$$\min_{x,y}\{F(x, y, z) : G(y, z) \leq 0, x \in \Psi(y, z)\} \qquad (7.13)$$

with

$$\Psi(y, z) = \operatorname*{Argmin}_{x} \ \{f(x, y, z).g(x, y, z) \leq 0, h(x, y, z) = 0\}$$

for all z *in some open neighborhood* $V_\varepsilon(0)$.

Related stability results for the optimal solution of the KKT reformulation of the bilevel programming problems can be found in [249].

The investigation of unstable optimization problems is difficult both from a theoretical and from a numerical points of view. To avoid this difficult property of bilevel programming problems we have also another possibility: we can enlarge the solution set mapping of the lower level problem such that a continuous point-to-set mapping $\Psi' : \mathbb{R}^m \to 2^{\mathbb{R}^n}$ arises. Then, using the results in the monograph [17], the resulting problem reacts smoothly on smooth perturbations of the problem functions as well as on changes of the values of the variables. One possibility is to replace $\Psi(y)$ by the set of ε-optimal solutions of (5.1) for $\varepsilon > 0$. The point-to-set mapping

$$y \mapsto \Psi_\varepsilon(y) := \{x \in M(y) : f(x, y) \leq \varphi(y) + \varepsilon\}$$

is locally Lipschitz continuous under presumably weak assumptions (cf. Theorem 4.3). Then, we have the following relations between the original bilevel problem (5.1), (5.2) and the relaxed problem

$$\text{``}\min_{y}\text{''}\{F(x,y) : G(y) \le 0, \ x \in \Psi_\varepsilon(y)\} \tag{7.14}$$

where

$$\Psi_\varepsilon(y) = \{x \in M(y) : f(x,y) \le \varphi(y) + \varepsilon\}. \tag{7.15}$$

THEOREM 7.14 *Consider the optimistic bilevel programming problem related to problem (7.14):*

$$\min_{x,y}\{F(x,y) : G(y) \le 0, \ x \in \Psi_\varepsilon(y)\} \tag{7.16}$$

and let $\{F_k, f_k\}_{k=1}^{\infty}$ be sequences of continuous functions converging pointwise to (F, f). Let the assumptions (C) and (MFCQ) be satisfied for the lower level problem (5.1) at all feasible points $(x,y), x \in M(y), \ G(y) \le 0$, assume that (ULR) is valid for the upper level problem. Let $\varepsilon > 0$ and define

$$\varphi_k(y) = \min_{x}\{f_k(x,y) : x \in M(y)\},$$

$$\Psi_\varepsilon^k(y) = \{x \in M(y) : f_k(x,y) \le \varphi_k(y) + \varepsilon\},$$

$$\Phi_\varepsilon^k = \min_{x,y}\{F_k(x,y) : G(y) \le 0, \ x \in \Psi_\varepsilon^k(y)\}, \tag{7.17}$$

$$\Phi_\varepsilon = \min_{x,y}\{F(x,y) : G(y) \le 0, \ x \in \Psi_\varepsilon(y)\}, \tag{7.18}$$

$$\Phi = \min_{x,y}\{F(x,y) : G(y) \le 0, \ x \in \Psi(y)\}. \tag{7.19}$$

Then,

1. $\liminf_{k\to\infty} \Phi_\varepsilon^k \ge \Phi_\varepsilon$,

2. $\limsup_{\varepsilon\to 0+} \Phi_\varepsilon = \Phi$.

Comprehensive related results both for the optimistic and pessimistic bilevel problems even under much weaker presumptions can be found e.g. in [173, 179, 180, 181, 184]. Stability of Stackelberg problems depending on an additional parameter have been investigated in [174].

7.3   SPECIAL PERTURBATIONS
7.3.1   LINEAR PERTURBATIONS

In [301] the objective function of the lower level problem is linearly perturbed and it is shown that this leads to an equivalent problem if the perturbation is used in the upper level objective function, too. To formulate this result we need

DEFINITION 7.3 *Let $C \subset \mathbb{R}^n$. A function $f : \mathbb{R}^n \to \mathbb{R}$ is said to satisfy a semi-Lipschitz condition relative to C if there exists a constant $L \geq 0$ such that*

$$f(x) - f(x') \leq L\|x - x'\| \ \forall x \in C, \ x' \notin C.$$

A function satisfying a semi-Lipschitz continuity condition relative to C need not be Lipschitz continuous on C, although a (globally) Lipschitz continuous function clearly is also semi-Lipschitz continuous relative to any set.

Now consider the problem of computing an optimistic optimal solution of the bilevel programming problem (5.1), (5.2). Let $r \in \mathbb{R}^n$ be arbitrarily chosen and consider the perturbed problem

$$\min_{x,y,v}\{F(x, y) + r\|v\| : (x, y) \in D, \ x \in \Psi^v(y)\}, \qquad (7.20)$$

where

$$\Psi^v(y) = \operatorname*{Argmin}_{x} \ \{f(x, y) - \langle v, x\rangle : g(x, y) \leq 0, h(x, y) = 0\}$$

and $D \subseteq \mathbb{R}^n \times \mathbb{R}^m$ is a closed set. Then, we have the following result:

THEOREM 7.15 ([301]) *Suppose that the point-to-set mapping $\mathcal{F} \cap D : v \mapsto \operatorname{grph}\Psi^v \cap D$ is upper Lipschitz continuous at $v = 0$ with a constant κ and that the upper level objective function satisfies a semi-Lipschitz condition relative to $\operatorname{grph}\Psi \cap D$ with modulus M. Then, for any $r > \kappa M$ the problems (5.1), (5.2) and (7.20) are equivalent in the following sense: If $\overline{z} = (\overline{x}, \overline{y})$ is an optimistic optimal solution of (5.1), (5.2), then $(\overline{z}, 0)$ solves (7.20) and, if $(\overline{z}, \overline{v})$ is an optimal solution of (7.20) then $\overline{v} = 0$ and $\overline{z}$ is a solution of (5.1), (5.2).*

7.3.2 TYKHONOV REGULARIZATION

If the convex lower level problem (4.1) has not a unique optimal solution but the upper level objective function $F(x, y)$ is strongly convex with respect to x, then the regularized problem

$$\min_{x}\{f(x, y) + \alpha F(x, y) : g(x, y) \leq 0 \ h(x, y) = 0\} \qquad (7.21)$$

has a uniquely determined optimal solution for each $\alpha > 0$ where the feasible set is not empty. The idea of using such type of a regularization is due to [277]. Clearly, if the assumptions of Theorem 4.3 are satisfied, then the solution set mapping

$$\Psi_\alpha(y) = \operatorname*{Argmin}_{x} \ \{f(x, y) + \alpha F(x, y) : g(x, y) \leq 0 \ h(x, y) = 0\}$$

of the problem (7.21) is upper semicontinuous, i.e. for each sequence $\{(x^k, y^k, \alpha^k)\}_{k=1}^{\infty}$ with $\lim_{k\to\infty} y^k = y^0$, $\lim_{k\to\infty} \alpha^k = 0+$ and $x^k \in \Psi_{\alpha^k}(y^k)$ for all k each accumulation point of the sequence $\{x^k\}_{k=1}^{\infty}$ belongs to $\Psi_0(y^0) := \Psi(y^0)$. It is even possible to get the following stronger result:

THEOREM 7.16 *Consider the convex parametric optimization problem (4.1) and let the assumptions (C) and (MFCQ) be satisfied at* $y = y^0$. *Let* $\nabla^2_{xx} F(x, y)$ *be positive definite at each* (x, y). *Then,*

1. *the unique optimal solution* $x_\alpha(y) \in \Psi_\alpha(y)$ *is strongly stable at each* $\alpha > 0$ *and directionally differentiable,*

2. *for each* $\alpha > 0$ *we have the following two inequalities:*

$$F(x_\alpha(y), y) \leq \min_x \{F(x, y) : x \in \Psi(y)\}, \ f(x_\alpha(y), y) \geq \varphi(y),$$

3. *for fixed* y, $\lim_{\alpha\to 0+} \{x_\alpha(y)\} = \text{Argmin}_x \ \{F(x, y) : x \in \Psi(y)\}.$

If we do not consider globally optimal solutions in the lower level problem but ε-optimal ones, then following the pattern of [1] we can obtain a quantitative version of the last theorem. For doing so, we need the notion of ε-optimal solutions of the lower level problem defined as

$$\Psi_{\alpha,\varepsilon}(y) := \{x \in M(y) : f(x, y) + \alpha F(x, y) \leq \varphi_\alpha(y) + \varepsilon\},$$

where $M(y)$ again denotes the feasible set of the lower level problem and

$$\varphi_\alpha(y) := \min_x \{f(x, y) + \alpha F(x, y) : x \in M(y)\}.$$

THEOREM 7.17 *Let the assumption (C) be satisfied for the lower level problem (5.1). Let* $\alpha > 0$, $\varepsilon \geq 0$. *Then there exists a constant* c *such that*

$$\Psi_{\alpha,\varepsilon}(y) \subseteq \Psi_{0,\varepsilon+\alpha c}(y)$$

for each y *with* $G(y) \leq 0$.

This can now be used to derive some quantitative bounds for the deviation of the optimal function values of the original and the relaxed bilevel problems.

THEOREM 7.18 *Let the assumption of Theorem 7.17 be satisfied. Then, for each* $x \in \Psi_{\alpha,\varepsilon}(y)$, $G(y) \leq 0$, *we have*

$$\min_x\{F(x, y) : x \in \Psi(y)\} + \frac{\varepsilon}{\alpha} \geq F(x, y) \geq \min_x\{F(x, y) : x \in \Psi_{0,\varepsilon+\alpha c}(y)\}.$$

This implies

$$\min_{x,y}\{F(x,y) : x \in \Psi(y),\ G(y) \le 0\} + \frac{\varepsilon}{\alpha}$$
$$\ge\ \min_{y}\{F(x_{\alpha,\varepsilon}(y),y) : G(y) \le 0\}$$
$$\ge\ \min_{x,y}\{F(x,y) : x \in \Psi_{0,\varepsilon+\alpha c}(y),\ G(y) \le 0\},$$

where $x_{\alpha,\varepsilon}(y) \in \Psi_{\alpha,\varepsilon}(y)$ for each y.

COROLLARY 7.2 *Under the assumptions of Theorem 7.17, if $\alpha \to 0+$, $\varepsilon \to 0+$ such that $\frac{\varepsilon}{\alpha} \to 0$, then,*

$$\min_{x}\{F(x,y) : x \in \Psi_{\alpha,\varepsilon}(y)\} \to \min_{x}\{F(x,y) : x \in \Psi(y)\}$$

as well as

$$\min_{x,y}\{F(x,y) : x \in \Psi_{\alpha,\varepsilon}(y),\ G(y) \le 0\} \to \Phi$$

with Φ from equation (7.19).

Note that in its original version in [1], the last theorems are given with respect to the regularized lower level problem

$$\varphi_\alpha^0(y) := \min_{y}\{f(x,y) - \alpha F(x,y) : x \in M(y)\}.$$

In this case, this idea is used to approximate the pessimistic and not the optimistic approach to bilevel programming with non-unique lower level solutions. It should be noted that the assertion of Theorem 7.16 cannot be generalized to the case when both parameters α and y converge to $0+$ and y^0, respectively. This can be seen in the following example taken from [250]:

Example: Consider the bilevel problem

$$\min_{y}\{(x-y)^2 + y^2 : -20 \le y \le 20,\ x \in \Psi(y)\},$$

where

$$\Psi(y) = \underset{x}{\text{Argmin}} \ \{xy : -y - 1 \le x \le -y + 1\}.$$

Then, for $0 < \alpha < 0.25$,

$$x_\alpha(y) = \begin{cases} -y - 1 & \text{if } y > -\frac{2\alpha}{4\alpha-1} \\ y(1 - \frac{1}{2\alpha}) & \text{if } \frac{2\alpha}{4\alpha-1} \le y \le -\frac{2\alpha}{4\alpha-1} \\ -y + 1 & \text{if } y < \frac{2\alpha}{4\alpha-1} \end{cases}$$

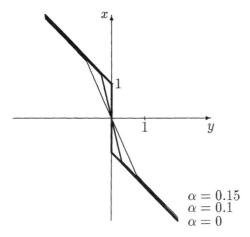

Figure 7.3.   The optimal solution of the regularized lower level problem in the example on page 233

The mapping $\Psi(\cdot)$ $(\alpha = 0)$ as well as the function $x_\alpha(\cdot)$ for $\alpha = 0.15$ and $\alpha = 0.1$ are shown in figure 7.3.

Then, if we take $y = 2\alpha$ and let α converge to zero, we get

$$\lim_{y \to 0} x_\alpha(y) = \lim_{y \to 0}(y - 1) = -1 \notin \underset{x}{\text{Argmin}} \ \{x^2 : x \in \Psi(0)\}.$$

$\square$

Nevertheless, the optimal solution of the regularized lower level problem has better regularity properties than that of the original one. Moreover, the above theorems can be used as a motivation for using this regularization approach when solving bilevel programming problems. This has been done by the first time in [213].

In [76] the algorithm described in Section 6.1 has been applied to the solution of bilevel programming problems with non-unique optimal solutions in the lower level using the Tykhonov regularization. We will describe this algorithm next. It solves the sequence of problems

$$\min_{x,y}\{F(x, y) : G(y) \leq 0, x \in \Psi_\alpha(y)\} \tag{7.22}$$

for $\alpha \searrow 0$. Assume that the lower level problem is a convex parametric optimization problem and that the upper level objective function is strongly convex with repect to x. Then, by strong convexity of $f(\cdot, y) + \alpha F(\cdot, y)$ on R^n for each fixed y, problem (7.22) is equivalent to the following nondifferentiable optimization problem with an implicitly defined objective function:

$$\min_y \{\mathcal{F}_\alpha(y) \mid G(y) \le 0\}, \tag{7.23}$$

where $\mathcal{F}(y) := F(x_\alpha(y), y)$, $x_\alpha(y) \in \Psi_\alpha(y) \; \forall \; y$. Using the idea of the descent algorithm in Section 6.1 this problem can be solved with the following prototype algorithm:

Prototype descent algorithm

  Input: Bilevel programming problem with strongly convex upper level objective function

  Output: A Bouligand stationary point

  1: Select y^0 solving $G(y^0) \le 0$, set $k := 0$, $\alpha^0 > 0$, $\varepsilon, \delta \in (0, 1)$.

  2: Compute a direction $r^k, \|r^k\| \le 1$, satisfying the following inequalities for $s^k < 0$:

  $$\mathcal{F}_\alpha'(y^k; r^k) \le s^k, \nabla_y G_i(y^k) r^k \le -G_i(y^k) + s^k, \quad i = 1, \ldots, l,$$

  3: Choose a step-size t^k such that
  $$\mathcal{F}_\alpha(y^k + t^k r^k) \le \mathcal{F}_\alpha(y^k) + \varepsilon t^k s^k, G(y^k + t^k r^k) \le 0.$$

  4: Set $y^{k+1} := y^k + t^k r^k$, $\quad k := k + 1$.

  5: If a stopping criterion is satisfied: if α is sufficiently small, then stop; else set $\alpha := \delta\alpha$ and compute $x_\alpha(y^k)$. Goto step 2.

To inspire life into the algorithm a more detailed description of steps 2 and 3 is necessary. This has already been done in Section 6.1. We will not repeat this here but only mention the main steps.

As in the descent method in Section 6.1 the necessary optimality condition from Theorem 5.4 can be used for computing the direction of descent. Using a smoothing operation to avoid zigzagging we again come up with the direction finding problem (6.3). In Step 2 we are searching for a set $K \in \mathcal{I}(\lambda)$ with $(\lambda, \mu) \in \Lambda(x_\alpha(y), y)$ such that this direction finding problem has a negative optimal value.

As in Section 6.1, in Step 3 of the prototype of the descent algorithm we use a kind of Armijo step-size rule: take the largest number t^k in $\{\rho, \rho^2, \rho^3, \rho^4, \ldots\}$, where $\rho \in (0, 1)$, such that

$$\mathcal{F}_\alpha(y^k + t^k r^k) \le \mathcal{F}_\alpha(y^k) + \varepsilon t^k s^k, \; \varepsilon \in (0, 1), \quad \text{and } G(y^k + t^k r^k) \le 0. \tag{7.24}$$

Then, convergence of the algorithm for fixed $\alpha > 0$ to a Clarke stationary point and of the modified algorithm to a Bouligand stationary point is mentioned in Corollary 6.1 and Theorem 6.2. The main point in this modification was the introduction of relaxed index sets of active constraints needed to teach the algorithm to foresee the index set of active constraints in the lower level problem at an accumulation point of the iterates.

Now, in order to get an algorithm for solving the sequence of bilevel problems (7.22) for $\alpha \searrow 0$ we add the update role for the regularity parameter α from the Prototype algorithm to the Modified descent algorithm for the bilevel problem in Section 6.1. Then we get convergence of the algorithm to a Bouligand stationary point. Note that the additional strong sufficient optimality condition of second order at the accumulation point of the iterates guarantees that the lower level optimal solution of the original problem is strongly stable and directional differentiable thus making the use of Bouligand stationarity possible.

THEOREM 7.19 ([76]) *Consider the optimistic bilevel problem (5.1), (5.5) and let the assumptions (C), (FRR), (MFCQ), (ULR), and (CRCQ) be satisfied at all points* $(x, y), G(y) \leq 0, x \in M(y)$. *Take a sufficiently small and fixed parameter* $\kappa > 0$. *Let the sequence of iterates* $\{(y^k, \lambda^k, \mu^k, d^k, r^k, s^k, t^k, \nu^k, \omega^k, K^k)\}_{k=1}^{\infty}$ *be computed by the modified descent algorithm where* $\alpha^k \searrow 0$. *Thus, if* (x^0, y^0) *is an accumulation point of the sequence* $\{(x^k, y^k)\}_{k=1}^{\infty}$ *satisfying (SSOC), then* (x^0, y^0) *is a Bouligand stationary point of problem (5.1), (5.5).*

For a proof the reader is referred to the original paper [76], we will not include it here.

7.3.3   LEAST-NORM REGULARIZATION

If the function $F(x, y)$ does not have the appropriate properties such that the problem (7.21) can be used for approximating the lower level problem, then another kind of regularization can be applied. This has been done in [183] by applying an idea originally used in [266], namely the approximation of the lower level solution by the least norm element in the set $\Psi(y)$:

$$\Psi^0(y) = \operatorname*{Argmin}_{x} \{\|x\| : x \in \Psi(y)\}. \tag{7.25}$$

Unfortunately, by use of this approach it is not possible to overcome all difficulties when computing optimistic respectively pessimistic optimal solutions, namely the solution set mapping of the such regularized lower level problem is in general not lower semicontinuous. This can be seen in the following example:

Example: [183, 186] Consider the problem

$$\min\{(y - 0.5)x : 0 \leq x \leq 1\}.$$

Then

$$\Psi(y) = \begin{cases} \{0\}, & \text{if } y > 0.5 \\ [0,1] & \text{if } y = 0.5 \\ \{1\}, & \text{if } y < 0.5 \end{cases}, \Psi^0(y) = \begin{cases} \{0\}, & \text{if } y \geq 0.5 \\ \{1\}, & \text{if } y < 0.5 \end{cases}$$

Hence, if we take the function $F(x, y) = -xy$, then

$$F(x(y), y) = \begin{cases} 0, & \text{if } y \geq 0.5, \\ -y, & \text{if } y < 0.5, \end{cases}$$

where $\{x(y)\} = \Psi^0(y)$. Consequently, the infimum of the function $F(x(y), y)$ over $[0, 1]$ is equal to -0.5 and there is no $y \in [0, 1]$ such that $F(x(y), y) = -0.5$.   □

For overcoming this difficulty again the idea of using ε-optimal solutions in the lower level problem can be used: Consider a parametric lower level problem (5.1) and let $x_\varepsilon(y)$ be an optimal solution of the regularized problem

$$\Psi_\varepsilon^0(y) = \operatorname*{Argmin}_x \{\|x\| : x \in \Psi_\varepsilon(y)\}. \tag{7.26}$$

If problem (5.1) is convex, then $x_\varepsilon(y)$ is uniquely determined provided that $\Psi(y) \neq \emptyset$. Moreover, we have the following result:

THEOREM 7.20 ([183]) *Consider the convex parametric optimization problem (5.1) at $y = y^0$ and let the assumptions (C) and (MFCQ) be satisfied at all points (x, y^0) with $x \in M(y^0)$. Then,*

$$\lim_{\varepsilon \to +0} x_\varepsilon(y) = x_0(y).$$

This result is no longer true if the parameter y also converges to y^0. This can be seen in the following

Example: [183] We continue the last example on page 236. Then,

$$\Psi_\varepsilon(y) = \begin{cases} [0, 1] & \text{if } 0.5 - \varepsilon < y < \varepsilon + 0.5 \\ [0, 2\varepsilon/(2y - 1)] & \text{if } \varepsilon + 0.5 < y \\ [1 - \frac{2\varepsilon}{1 - 2y}, 1] & \text{if } y \leq 0.5 - \varepsilon \end{cases}$$

and

$$x_\varepsilon(y) = \begin{cases} 0, & \text{if } y > 0.5 - \varepsilon \\ 1 - \frac{2\varepsilon}{1 - 2y}, & \text{else} \end{cases}$$

The upper and lower bounds of the sets $\Psi_\varepsilon(y)$ as well as three examples of this set for fixed y (left: $y \leq 0.5 - \varepsilon$, middle: $0.5 - \varepsilon < y < \varepsilon + 0.5$, right: $\varepsilon + 0.5 < y$) are depicted in Fig. 7.4. Now, if $\{y^k\}_{k=1}^\infty \subset (-\infty, 0.5)$

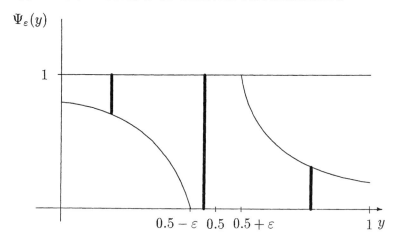

Figure 7.4.   Set of ε optimal solutions in the example on page 237

is a sequence converging to 0.5 and $\varepsilon^k = (0.5 - y^k)^2$ for each k, then $x_{\varepsilon^k}(y^k) = 1 - \frac{2\varepsilon^k}{1-2y^k}$ converging to 1 for $k \to \infty$ and $x_0(0.5) = 0$.   □

Nevertheless, by (Lipschitz) continuity of the point-to-set mapping $y \mapsto \Psi_\varepsilon(y)$ for each $\varepsilon > 0$ (cf. Corollary 4.3) the mapping $y \mapsto \Psi_\varepsilon^0(y)$ is continuous for convex lower level problems satisfying the assumptions of Theorem 7.20. Hence, an optimal solution of the regularized bilevel programming problem

$$\min_y \{F(x,y) : x \in \Psi_\varepsilon^0(y),\ G(y) \leq 0\} \tag{7.27}$$

exists for each positive ε.

THEOREM 7.21 ([183]) *Consider the bilevel problem (5.1), (5.2), let the lower level problem be a convex parametric one and let the assumptions (C) and (MFCQ) be satisfied for all (x,y), $x \in M(y)$, $G(y) \leq 0$. Take any sequence $\{\varepsilon^k\}_{k=1}^\infty$ converging to $0+$. Denote by y^k a solution of the problem (7.27) for $\varepsilon = \varepsilon^k$. Then, any accumulation point of the sequence $\{(x_{\varepsilon^k}^0(y^k), y^k) : x_{\varepsilon^k}^0(y^k) \in \Psi_{\varepsilon^k}^0(y^k)\}_{k=1}^\infty$ is a lower optimal solution, i.e. it belongs to the set S defined in (5.8).*

We will close this Chapter with a second regularization approach which can be used in the cases when the upper level objective function does not have the properties making the approach in Subsection 7.3.2 possible. Then we can replace the lower level problem (5.1) by

$$\begin{cases} f(x,y) + \alpha\|x\|^2 \to \min_x \\ g(x,y) \leq 0, \\ h(x,y) = 0 \end{cases} \tag{7.28}$$

for $\alpha > 0$. Let $\Psi_\alpha(y)$ denote the set of optimal solutions for this problem. Some relations between the regularized and the original bilevel problems have been investigated in [183]. Then, for fixed $\alpha > 0$, and under presumably not too restrictive assumptions, the optimal solution of problem (7.28) is uniquely determined by a locally Lipschitz continuous function $x_\alpha(y)$ with respect to α, y. Hence, the regularized problem

$$\min_y \{\mathcal{F}_\alpha(y) := F(x_\alpha(y), y) : G(y) \leq 0\} \qquad (7.29)$$

is a Lipschitz optimization problem for $\alpha > 0$ which again can be solved by means of nondifferentiable minimization techniques as e.g. the bundle – trust region algorithm [151, 255]. This will lead to a modified bundle algorithm which, in general, is only an approximation algorithm to the bilevel problem.

The following results are obvious implications of upper semicontinuity (cf. Theorem 4.3).

THEOREM 7.22 *Consider the parametric problems (5.1) and (7.28) and let the assumptions (C) and (MFCQ) be satisfied. Then*

1. *For each sequences $\{y^k\}_{k=1}^\infty$, with $G(y^k) \leq 0$ for all k and $\{\alpha^k\}_{k=1}^\infty \subseteq \mathbb{R}_+$ converging to $\overline{y}, \overline{\alpha}$, resp., and for each sequence $\{x^k\}_{k=1}^\infty$ satisfying $x^k \in \Psi_{\alpha^k}(y^k)$ $\forall k$ the sequence $\{x^k\}_{k=1}^\infty$ has accumulation points $\overline{x}$ and all these points satisfy $\overline{x} \in \Psi_{\overline{\alpha}}(\overline{y})$.*

2. *For $\overline{\alpha} = 0$ we have*

$$\lim_{\substack{y^k \to \overline{y} \\ \alpha^k \searrow 0}} x_{\alpha^k}(y^k) = x(\overline{y})$$

provided that $\Psi(\overline{y}) = \{x(\overline{y})\}$.

It should be noted that in general we do not have

$$\lim_{\substack{y^k \to \overline{y} \\ \alpha^k \searrow 0}} x_{\alpha^k}(y^k) \in \operatorname*{Argmin}_x \{\|x\| : x \in \Psi(\overline{y})\}$$

without the assumption that $\Psi(\overline{y})$ reduces to a singleton, even if this limit exists. This can be seen in

Example: Consider the problem

$$\min_x \{xy : x \in [-1, 1]\}.$$

Then,

$$\Psi(y) = \begin{cases} [-1,1], & \text{if } y = 0, \\ \{1\}, & \text{if } y < 0, \\ \{-1\}, & \text{if } y > 0, \end{cases}$$

and

$$\Psi_\alpha(y) = \begin{cases} \{-y/(2\alpha)\} & \text{if } -y/(2\alpha) \in [-1,1], \\ \{1\}, & \text{if } -y/(2\alpha) \geq 1, \\ \{-1\}, & \text{if } -y/(2\alpha) \leq -1. \end{cases}$$

for $y \in [-2\alpha, 2\alpha]$. Hence, depending on the limit of the sequence $-y^k/\alpha^k$, the sequence $x_{\alpha^k}(y^k)$ can have any limit point in $[-1,1]$ for $y^k \to 0$, $\alpha^k \to 0$. □

Let (5.1) be a convex parametric optimization problem. For a fixed value $\alpha > 0$, the optimal solution of problem (7.28) is uniquely determined and the function $x_\alpha(\cdot)$ of optimal solutions for these problems is continuous. In the paper [74] the bundle algorithm in [151] has been used to solve problem (7.29) with $G(y) \equiv 0$. To do that the bundle algorithm is applied to a sequence of problems (7.29) with smaller and smaller values of the regularization parameter α. This will result in the application of the bundle algorithm to a sequence of optimization problems. Clearly, in the s-th application the algorithm is started with the last point y^{s-1} computed in the previous application and using a smaller value of the regularization parameter $0 < \alpha^s < \alpha^{s-1}$. The value of α can also be decreased within one application of the bundle algorithm. In [74] this is done in each serious step. But this cannot guarantee that α converges to zero within one application of the bundle algorithm. Then, this again results in the application of the bundle algorithm to a sequence of problems with smaller and smaller regularization parameter.

Summing up, a sequence of iterates $\{(x_{\alpha^s}(y^s), y^s, \alpha^s)\}_{s=1}^\infty$ is computed where both the sequences within one application of the bundle algorithm and within the sequence of applications are denoted by this sequence.

We will not include the overall algorithm here but refer the interested reader to the original papers [74, 151]. The following results, whose proofs are again not included, can then be shown.

THEOREM 7.23 ([74]) *Consider the bilevel programming problem (5.1), (5.2) and let the assumptions (C), (CRCQ), (FRR), and (MFCQ) be satisfied for the problems (7.28). Assume that the lower level problem is a convex one. Let $\{x_{\alpha^s}(y^s), y^s, \alpha^s\}_{s=1}^\infty$ be the sequence computed by the modified bundle algorithm with $\{\alpha^s\}_{s=1}^\infty$ being bounded from below by some $\alpha^0 > 0$. Then, every accumulation point $(x_{\overline{\alpha}}(\overline{y}), \overline{y}, \overline{\alpha})$ of this sequence satisfies*

$$0 \in \{\nabla_x F(x_{\overline{\alpha}}(\overline{y}), \overline{y})d + \nabla_y F(x_{\overline{\alpha}}(\overline{y}), \overline{y}) : d \in \partial_y x_{\overline{\alpha}}(\overline{y})\}.$$

If the sequence $\{\alpha^s\}_{s=1}^{\infty}$ converges to zero then the strong sufficient optimality condition of second order at the accumulation points of the sequence $\{(x_{\alpha^s}(y^s), y^s)\}_{s=1}^{\infty}$ is necessary in order to guarantee that the bundle algorithm is able to compute all the data necessary (see Section 6.2 and Theorem 4.12). If this sufficient optimality condition is satisfied then the above theorem shows that the accumulation points are Clarke stationary. Hence, we get

COROLLARY 7.3 ([74]) *Under the assumptions of Theorem 7.23 let* $\{\alpha^s\}_{s=1}^{\infty}$ *converge to zero. Let* $\overline{y}$ *be the limit point of the sequence* $\{y^s\}_{s=1}^{\infty}$ *and let* $x_0(\overline{y}) \in \Psi(\overline{y})$. *If the additional assumption (SSOC) is satisfied at* $(x_0(\overline{y}), \overline{y})$ *then*

$$0 \in \{\nabla_x F(x_0(\overline{y}), \overline{y})d + \nabla_y F(x_0(\overline{y}), \overline{y}) : d \in \partial_y x_0(\overline{y})\}.$$

In this Corollary, the restrictive assumption (SSOC) has been used. Together with the assumptions (MFCQ) and (CRCQ) this assumption guarantees that the solution function $x(\cdot)$ of the original lower level problem (5.1) is locally Lipschitz continuous at $\overline{y}$ (see Theorem 4.10). This implies also that α^s can theoretically tend to zero since we could minimize the function $\mathcal{F}_0(\cdot)$ itself by use of the bundle algorithm. In the other case, if the function $\mathcal{F}_0(\cdot)$ is not locally Lipschitz continuous, numerical difficulties make the decrease of α^s to zero impossible.

Now we come back to the more practical case that we need a sequence of applications of the bundle algorithm. To realize the convergence of α^s to zero during the bundle algorithms one more assumption is needed. Recall that $x_0(y) \in \Psi(y)$ is an arbitrary optimal solution of the problem (5.1). Consider the set

$$\widehat{Y} = \{y \in \mathbb{R}^m : \text{(SSOC) is satisfied at } (x_0(y), y)\}.$$

In general, the set $\widehat{Y}$ is neither open nor closed nor connected, but $x_0(\cdot)$ is locally Lipschitz continuous on $\widehat{Y}$. The following considerations are only useful if the set $\widehat{Y}$ has a suitable structure which is the main assumption in what follows. Let $\theta > 0$ be a (small) constant and consider a set D such that $D + 2\theta\mathcal{B}^m \subseteq \widehat{Y}$. Subsequently, we assume that $D \neq \emptyset$ exists. This is a weaker assumption than supposing (SSOC) throughout $\mathbb{R}^m$.

In the following theorem the resulting algorithm is investigated and we consider two cases: First the case when the bundle algorithm is restarted finitely many times. Then, the overall algorithm can be considered as being equal to one run of the bundle algorithm (namely the last one). Due to (C) and the special calculations in the modified bundle algorithm the sequence $\{y^s\}_{s=1}^{\infty}$ itself converges to some $\overline{y}$ in that case. Hence, also $\{(x_{\alpha^s}(z^s), z^s, \alpha^s)\}_{s=1}^{\infty}$ converges to $(x_{\overline{\alpha}}(\overline{y}^m), \overline{y}^m, \overline{\alpha}^m)$. Second the case when the bundle algorithm is restarted infinitely often. The sequence of all iteration points computed during the infinite number of applications of the modified bundle algorithm is then being considered in the second case. One accumulation point of this sequence is also an accumulation point of the sequence $\{(x_{\overline{\alpha}}(\overline{y}^m), \overline{y}^m, \overline{\alpha}^m)\}_{m \in N}$.

Now we are able to state the main convergence theorem for the modified bundle algorithm:

THEOREM 7.24 *Consider the regularized bilevel problem (7.29) and let the assumptions (C), (MFCQ), (CRCQ) and (FRR) be satisfied for all $y \in \mathbb{R}^m, \alpha > 0$. Let there exist $\alpha^* > 0$, $\widehat{\varepsilon} > 0$, and $\overline{F}$ such that for all $0 < \alpha < \alpha^*$ all stationary points $\widetilde{y}_\alpha$ of the functions $F(x_\alpha(y), y)$ have $F(x_\alpha(\widetilde{y}_\alpha), \widetilde{y}_\alpha) \leq \overline{F}$ and*

$$\{y : F(x_\alpha(y), y) \leq \overline{F} + \widehat{\varepsilon}\} \subseteq D \ \forall \ 0 < \alpha < \alpha^*,$$

where the set D is defined as above stated and bounded. Then we have the following:

1. *if only a finite number of restarts is needed to get convergence of $\{\alpha^s\}_{s=1}^\infty$ to zero, then the limit point $(x_0(\overline{y}), \overline{y})$ of the sequence computed by the algorithm satisfies*

$$0 \in \{\nabla_x F(x_0(\overline{y}), \overline{y})d + \nabla_y F(x_0(\overline{y}), \overline{y}) : d \in \partial_y x_0(\overline{y})\}.$$

2. *if an infinite number of restarts is necessary then there exist accumulation points $(x_{\overline{\alpha}}(\overline{y}), \overline{y}, \overline{\alpha})$ of the sequence computed by the algorithm with $\overline{\alpha} = 0$ satisfying*

$$0 \in \{\nabla_x F(x_{\overline{\alpha}}(\overline{y}), \overline{y})d + \nabla_y F(x_{\overline{\alpha}}(\overline{y}), \overline{y}) : d \in \partial_y x_{\overline{\alpha}}(\overline{y})\}.$$

Moreover, the sequence computed by the algorithm has accumulation points $(x_{\overline{\alpha}}(\overline{y}), \overline{y}, \overline{\alpha})$ with $\overline{\alpha} = 0$.

We close this Chapter with a last regularization idea originating from game theory.

REMARK 7.1 *In mathematical game theory, the use of mixed strategies allows to regularize games in which no equilibrium strategy exists. Here, a mixed strategy can be considered as a probability measure on the set of admissible strategies and, in the equilibrium solution for the mixed problem, the different strategies are taken with the probabilities given by the measure in the equilibrium strategy. The same idea is applied to Stackelberg games (which are bilevel programming problems where the sets of feasible solutions for each player do not depend on the decisions of the other one) in [191]. Then, the lower level problem*

$$\Psi_S(y) := \min_x \{f(x,y) : x \in X\}$$

is replaced by the problem of minimizing the average value of the objective function

$$\Psi_S^a := \min_\mu \{\langle \mu, f(x,y)\rangle : \mu \in \Omega(X)\},$$

where $\Omega(X)$ denotes the set of all Radon probability measures on X. Using this idea, the bilevel programming problem

$$\varphi_S := \text{``}\min_y\text{''}\{F(x,y) : y \in Y, \ x \in \Psi_S(y)\}$$

is replaced by the mixed problem (which is taken here in the pessimistic sense)

$$\min_y \max_\mu \{F(x,y) : y \in Y, \ x \in \Psi_S^a(y)\}.$$

Again, by use of ε-optimal solutions in the lower level of the mixed problem

$$\varphi_S^a(\varepsilon) := \min_y \max_\mu \{F(x,y) : y \in Y, \ x \in \Psi_{S\varepsilon}^a(y)\},$$

convergence of the optimal value of the regularized problem $\varphi_S^a(\varepsilon)$ to the optimal value of the pessimistic bilevel programming problem can be shown under suitable assumptions [191].

7.4   PROOFS

PROOF OF THEOREM 7.1: Let $x^0 \in \Psi(y)$ and let B be a corresponding basic matrix. Denote by $\Delta = (y_B^\top B^{-1} A)^\top - y$ the reduced cost vector. If, for all basic matrices, we have $\Delta_i < 0$ for all non-basic variables, then the optimal solution is unique. If this is not the case then, for some non-basic variable x_{i_0} and some basic matrix B the objective function coefficient y_{i_0} is determined by $y_{i_0} = y_B^\top B^{-1} A^{i_0}$. Since this is an additional linear constraint on the objective function coefficients, the set of all objective functions satisfying this additional condition is of measure zero. Due to the finite number of different basic matrices and of the family of all index sets of non-basic variables, the assertion of the theorem follows. □

PROOF OF THEOREM 7.2: For convex problems, (MFCQ) at one feasible point is equivalent with Slater's condition and with fulfillment of (MFCQ) at all feasible points. This implies by Theorem 4.3 that the solution set mapping $\Psi(y)$ is upper semicontinuous at y^0. By compactness of the feasible set and continuity of the objective function, problems

(5.1) and (7.1) have solutions for y near y^0. Since the objective function of problem (7.1) is strongly convex, this problem has a unique optimal solution. $\square$

PROOF OF THEOREM 7.5: Since conv A and conv B are convex compact sets, conv $A \cap$ conv $B = \emptyset$ implies that these sets can be strongly separated. Hence there exists a direction r^0 such that

$$\nabla_x f(x^i, y^0) \nabla x^{ij}(y^0) r^0 + \nabla_y f(x^i, y^0) r^0$$
$$> \nabla_x f(x^1, y^0) \nabla x^{1s}(y^0) r^0 + \nabla_y f(x^1, y^0) r^0$$

for all $i > 1, j = 1, \ldots, j_i, s = 1, \ldots, j_1$. Due to Corollary 4.1 this implies

$$\nabla_x f(x^i, y^0) x^{i'}(y^0; r^0) + \nabla_y f(x^i, y^0) r^0$$
$$> \nabla_x f(x^1, y^0) x^{1'}(y^0; r^0) + \nabla_y f(x^1, y^0) r^0 \qquad (7.30)$$

for all $i > 1$. By (MFCQ), (SSOC), (CRCQ) and Theorem 4.3, $\Psi(y^0 + tr^0) \neq \emptyset$ for sufficiently small $t > 0$ and $x^{i_0}(y^0 + tr^0) \in \Psi(y^0 + tr^0)$ for some $i_0 \in \{1, \ldots, k\}$. Let $x^1(y^0 + tr^0) \notin \Psi(y^0 + tr^0)$. By $x^1(y^0) \in \Psi(y^0)$ this implies that

$$t^{-1}[f(x^{i_0}(y^0 + tr^0), y^0 + tr^0) - f(x^{i_0}(y^0), y^0)]$$
$$< t^{-1}[f(x^1(y^0 + tr^0), y^0 + tr^0) - f(x^1(y^0), y^0)]$$

or by directional differentiability

$$\nabla_x f(x^{i_0}, y^0) x_{i_0}'(y^0; r^0) + \nabla_y f(x^{i_0}, y^0) r^0$$
$$\leq \nabla_x f(x^1, y^0) x^{1'}(y^0; r^0) + \nabla_y f(x^1, y^0) r^0$$

contradicting (7.30). Hence, the theorem is true. $\square$

PROOF OF THEOREM 7.6: Let $\|r\| = 1$ and $x^0 \in \mathcal{RP}(y^0; r)$, where

$$\mathcal{RP}(y^0; r) := \{x \in \mathbb{R}^n : \exists \{(x^k, t^k)\}_{k=1}^\infty, \ t^k > 0, \ x^k \in \Psi(y^0 + t^k r), \ \forall k,$$
$$\lim_{k \to \infty} (x^k, t^k) = (x, 0)\}.$$

Let $\{t_k, x^k\}_{k=1}^\infty$ be such that $t_k > 0$, $x^k \in \Psi(y^0 + t_k r)$ for all k and $\lim_{k \to \infty} (x^k, t_k) = (x^0, 0)$. Since (MFCQ) is satisfied at (x^0, y^0) by our assumptions, the feasible set mapping $M(\cdot)$ is pseudo Lipschitz continuous at (x^0, y^0) [244]. This implies that there exists a sequence $\{v^k\}_{k=1}^\infty \subseteq M(y^0)$ such that

$$\|x^k - v^k\| \leq Lt_k \ \forall \ k \qquad (7.31)$$

and some Lipschitz constant $L < \infty$ [244]. Moreover, v^k can be selected in such a way that $g_i(v^k, y^0) = 0$ for all $i \in J(\lambda^k)$ and some vertex $(\lambda^k, \mu^k) \in E\Lambda(x^0, y^0)$. Since the vertex set of the compact convex polyhedral set $\Lambda(x^0, y^0)$ (cf. Theorem 4.1) is finite, we can assume that $(\lambda^k, \mu^k) \equiv (\lambda^0, \mu^0)$ for all k. Then,

$$\varphi(y^0) \leq f(v^k, y^0) = L(v^k, y^0, \lambda^0, \mu^0)$$

by the complementarity conditions and

$$\varphi(y^0 + t_k r) = f(x^k, y^0 + t_k r) \geq L(x^k, y^0 + t_k r, \lambda^0, \mu^0)$$

for all k. Applying the mean value theorem this implies

$$\varphi(y^0 + t_k r) - \varphi(y^0) \geq L(x^k, y^0 + t_k r, \lambda^0, \mu^0) - L(v^k, y^0, \lambda^0, \mu^0)$$
$$= \nabla_x L(v^*, y^*, \lambda^0, \mu^0)(x^k - v^k) + t_k \nabla_y L(v^*, y^*, \lambda^0, \mu^0) r,$$

where (v^*, y^*) is a point in the interval between (v^k, y^0) and $(x^k, y^0 + t_k r)$. Using the Lipschitz condition (7.31), dividing both sides by t_k and using $\lim_{k \to \infty} \nabla_x L(v^*, y^*, \lambda^0, \mu^0) = 0$ we obtain

$$t_k^{-1}[\varphi(y^0 + t_k r) - \varphi(y^0)] \geq \nabla_y L(v^*, y^*, \lambda^0, \mu^0) r + o(t_k)/t_k$$

for all k and some $(\lambda^0, \mu^0) \in E\Lambda(x^0, y^0)$. Passing to the limit for $k \to \infty$ and using Theorem 4.15 we derive the desired result. $\qquad \square$

PROOF OF THEOREM 7.8: Due to the convexity and regularity assumptions we have $\Lambda(x^0, y^0) = \Lambda(x, y^0)$ for all $x \in \Psi(y^0)$, i.e. the set of Lagrange multipliers does not depend on the optimal solution. This is a direct consequence of the saddle point inequality and its equivalence to optimality in convex regular optimization and has been shown e.g. in [228]. In accordance with Theorem 4.16 we have

$$\varphi(y^0; r) = \min_{x \in \Psi(y^0)} \max_{(\lambda, \mu) \in \Lambda(x, y^0)} \nabla_y L(x, y^0, \lambda, \mu) r.$$

Consider the inner problem with $x = x^0$ in the constraints which is an equivalent formulation by $\Lambda(x^0, y^0) = \Lambda(x, y^0)$:

$$\nabla_y f(x, y^0) r + \sum_{i \in I(x^0, y^0)} \lambda_i \nabla_y g_i(x, y^0) r + \sum_{j=1}^{q} \mu \nabla_y h_j(x, y^0) r \to \max_{\lambda, \mu}$$

$$\nabla_x f(x^0, y^0) + \sum_{i \in I(x^0, y^0)} \lambda_i \nabla_x g_i(x^0, y^0) + \sum_{j=1}^{q} \mu \nabla_x h_j(x^0, y^0) = 0$$

$$\lambda_i \geq 0, i \in I(x^0, y^0).$$

Dualizing this linear optimization problem we get the problem

$$\nabla_y f(x, y^0) r - \nabla_x f(x^0, y^0) d \to \min_d$$

$$\nabla_x g_i(x^0, y^0) d \geq \nabla_y g_i(x, y^0) r, i \in I(x^0, y^0)$$
$$\nabla_x h_j(x^0, y^0) d = \nabla_y h_j(x, y^0) r, j = 1, \ldots, q$$

having the same optimal value. Now, setting $d := -d$ we get

$$\nabla_y f(x, y^0) r + \nabla_x f(x^0, y^0) d \to \min_d$$

$$\nabla_x g_i(x^0, y^0) d + \nabla_y g_i(x, y^0) r \leq 0, i \in I(x, y^0)$$
$$\nabla_x h_j(x^0, y^0) d + \nabla_y h_j(x, y^0) r = 0, j = 1, \ldots, q.$$

Inserting this dual problem into the formula for computing the directional derivative of the function $\varphi(\cdot)$ and taking into consideration

$$\Psi(y^0) = \{x : \nabla_x L(x, y^0, \lambda^0, \mu^0) = 0,$$
$$h(x, y^0) = 0, g(x, y^0) \leq 0, \lambda^{0\top} g(x, y^0) = 0\}$$

we get the desired result. Remember that the problem $H(x^0, \lambda^0, \mu^0, r)$ has a solution by assumptions (C) and (MFCQ) and, hence, minimization w.r.t. both (x, d) and the iterated minimization are equivalent [187]. □

PROOF OF THEOREM 7.9: Let (x^1, d^1), (x^2, d^2) be optimal solutions of problem $H(x^0, \lambda^0, \mu^0, r)$ with $x^1 \neq x^2$. Then, by convexity of the solution set of problem $H(x^0, \lambda^0, \mu^0, r)$, we have that the set

$$\{(x, d) : (x, d) = \alpha(x^2, d^2) + (1 - \alpha)(x^1, d^1), \ \alpha \in [0, 1]\}$$

is contained in the set of optimal solutions of this problem. Thus, by means of Taylor's expansion formulae up to first order we easily derive that

$$(u, w) = (x^2 - x^1, d^2 - d^1)$$

solves the following system of equations and inequalities (where the abbreviations $z^0 = (x^0, y^0)$ and $z^1 = (x^1, y^0)$ are used):

$$
\begin{aligned}
u^\top \nabla_{xy}^2 f(z^1) r + \nabla_x f(z^0) w &= 0 \\
\nabla_{xx}^2 L(z^1, \lambda^0, \mu^0) u &= 0 \\
u^\top \nabla_{xy}^2 g_i(z^1) r + \nabla_x g_i(z^0) w &\leq 0, \ \forall i : \nabla_x g_i(z^0) d^1 + \nabla_y g_i(z^1) r = 0, \\
u^\top \nabla_{xy}^2 h_j(z^1) r + \nabla_x h_j(z^0) w &= 0, \ j = 1, \ldots q \\
\nabla_x g_i(z^1) u &= 0, \ i \in J(\lambda^0), \\
\nabla_x g_i(z^1) u &\leq 0 \ \forall i \in I(x^1, y^0) \setminus J(\lambda^0) \\
\nabla_x h_j(z^1) u &= 0, \ j = 1, \ldots, q.
\end{aligned}
$$

$$(7.32)$$

Now, take $(\lambda^1, \mu^1) \in \operatorname*{Argmax}_{(\lambda, \mu) \in \Lambda(z^1)} \nabla_y L(z^1, \lambda, \mu) r$. By independence of $\Lambda(x, y^0)$ on $x \in \Psi(y^0)$, we have also $(\lambda^1, \mu^1) \in \operatorname*{Argmax}_{(\lambda, \mu) \in \Lambda(z^0)} \nabla_y L(z^1, \lambda, \mu) r$ which, by complementary slackness, implies

$$\nabla_x g_i(x^0, y^0) d^1 + \nabla_y g_i(x^1, y^0) r = 0, \text{ for each } i \in J(\lambda^1)$$

since d^1 is the dual solution to (λ^1, μ^1) in the inner problem of (7.6). Then,

$$\{ (u, w) : u^\top \nabla_{xy}^2 f(z^1) r + \nabla_x f(z^0) w = 0,$$
$$u^\top \nabla_{xy}^2 g_i(z^1) r + \nabla_x g_i(z^0) w \leq 0, \forall i : \nabla_x g_i(z^0) d^1 + \nabla_y g_i(z^1) r = 0$$
$$u^\top \nabla_{xy}^2 h_j(z^1) r + \nabla_x h_j(z^0) w = 0, \ j = 1, \ldots, q \}$$

$$\subseteq \{ (u, w) : u^\top \nabla_{xy}^2 L(z^1, \lambda^1, \mu^1) r + \nabla_x L(z^0, \lambda^1, \mu^1) w \leq 0 \}.$$

Hence, by $\lambda^1 \in \Lambda(x^0, y^0)$ we obtain that $u \neq 0$ is also a solution of

$$\left.\begin{array}{rcl}
u^\top \nabla_{xy}^2 L(x^1, y^0, \lambda^1, \mu^1) r & \leq & 0 \\
\nabla_{xx}^2 L(x^1, y^0, \lambda^0, \mu^0) u & = & 0 \\
\nabla_x g_i(x^1, y^0) u & = & 0, i \in J(\lambda^0), \\
\nabla_x g_i(x^1, y^0) u & \leq & 0, i \in I(x^1, y^0) \setminus J(\lambda^0), \\
\nabla_x h_j(x^1, y^0) u & = & 0, j = 1, \ldots, q.
\end{array}\right\} \quad (7.33)$$

Now substituting (x^0, λ^0) for (x^1, λ^1) (which is possible since $x^0 \in \mathcal{T}(y^0; r)$) we conclude the desired proof. □

PROOF OF THEOREM 7.10: Consider the Karush-Kuhn-Tucker conditions for the problem $H(x^0, \lambda^0, \mu^0, r^0)$ at an optimal solution (x, d): there exist vectors $\eta, \delta, \xi, \theta, \kappa$ such that the following system is satisfied:

$$\nabla_x L(x^0, y^0, \eta, \kappa) = 0, \quad (7.34)$$
$$\nabla_{xy}^2 L(x, y^0, \eta, \kappa) r^0 + \nabla_{xx}^2 L(x, y^0, \lambda^0, \mu^0) \delta +$$
$$\sum_{i=1}^p \xi_i \nabla_x^\top g_i(x, y^0) + \sum_{j=1}^q \theta_j \nabla_x^\top h_j(x, y^0) = 0 \quad (7.35)$$
$$\nabla_x g_i(x^0, y^0) d + \nabla_y g_i(x, y^0) r^0 \leq 0, \ i \in I(x^0, y^0), \quad (7.36)$$
$$\nabla_x h_j(x^0, y^0) d + \nabla_y h_j(x, y^0) r^0 \leq 0, \ j = 1, \ldots, q, \quad (7.37)$$
$$\eta_i \geq 0, \ \eta_i [\nabla_x g_i(x^0) d + \nabla_y g_i(x, y^0) r^0] = 0, \ i \in I(x^0, y^0), (7.38)$$
$$g_i(x, y^0) \leq 0, \ i \notin J(\lambda^0), \quad (7.39)$$
$$g_i(x, y^0) = 0, \ i \in J(\lambda^0), \quad (7.40)$$
$$h(x, y^0) = 0, \quad (7.41)$$

$$\xi_i \geq 0, \; i \notin J(\lambda^0), \tag{7.42}$$

$$\xi_i g_i(x, y^0) = 0, \; i = 1, \ldots, p, \tag{7.43}$$

$$\nabla_x L(x, y^0, \lambda^0, \mu^0) = 0. \tag{7.44}$$

For $(x, d) = (x^0, d^0)$, conditions (7.34), (7.36)–(7.38) are the necessary and sufficient optimality conditions for $(\eta, \kappa) \in S(r^0)$. So, they are satisfied by $(\eta, \kappa) = (\lambda^0, \mu^0)$. Let problem (7.9) have an optimal solution. Then, $u = 0$ is an optimal solution and by linear programming duality there exist vectors δ, ξ, θ such that equation (7.35) is satisfied for $(\eta, \kappa) = (\lambda^0, \mu^0)$ and we have

$$\xi_i \geq 0, \; i \in I(x^0, y^0) \setminus J(\lambda^0)$$

and $\xi_i \nabla_x g_i(x^0, y^0)0 = 0$. Hence, condition (7.43) is also valid. Consequently, by $x^0 \in \Psi(y^0)$, $(\lambda^0, \mu^0) \in \Lambda(x^0, y^0)$, relations (7.34)–(7.44) are satisfied for $x = x^0$, i.e. the problem in Theorem 7.8 has (x^0, d^0) as an optimal solution. This implies $x^0 \in \mathcal{T}(y^0; r^0)$. □

PROOF OF THEOREM 7.11: By our assumptions on the tangent cones we have $K^0_{\Psi(y^0)}(x^0, \lambda^0, \mu^0) \cap (-K^0_{\Psi(y^0)}(x^0, \lambda^0, \mu^0)) = \{0\}$. Hence, by [156], there exists a vector $w \in R^n$ satisfying

$$w^\top \overline{u} > 0 \text{ for each } \overline{u} \in K^0_{\Psi(y^0)}(x^0, \lambda^0, \mu^0) \setminus \{0\}. \tag{7.45}$$

By the presumptions, there exist vectors $\hat{r}$, $\hat{d}$ satisfying the system of linear (in-)equalities

$$\nabla^2_{xy} L(x^0, y^0, \lambda^0, \mu^0)\hat{r} = w \tag{7.46}$$

$$\nabla_x g_i(x^0, y^0)\hat{d} + \nabla_y g_i(x^0, y^0)\hat{r} = 0, i \in J(\lambda^0), \tag{7.47}$$

$$\nabla_x g_i(x^0, y^0)\hat{d} + \nabla_y g_i(x^0, y^0)\hat{r} \leq 0, i \in I(x^0, y^0) \setminus J(\lambda^0), \tag{7.48}$$

$$\nabla_x h_j(x^0, y^0)\hat{d} + \nabla_y h_j(x^0, y^0)\hat{r} = 0, j = 1, \ldots, q. \tag{7.49}$$

By linear programming duality and $(\lambda^0, \mu^0) \in \Lambda(x^0, y^0)$, (7.47)–(7.49) imply $(\lambda^0, \mu^0) \in S(\hat{r})$. Condition (7.46) together with (7.45) is equivalent to zero being the unique optimal solution of problem (7.9) for $r = \hat{r}$. Theorem 7.10 now implies $x^0 \in \mathcal{T}(y^0; \hat{r})$ and Theorem 7.9 that $\{x^0\} = \mathcal{T}(y^0; \hat{r})$. □

PROOF OF THEOREM 7.12: The first result follows directly from upper semicontinuity of the solution set mapping of perturbed optimization problems and is analogous to Theorem 4.3. This implies $F(\overline{x}, \overline{y}) \leq \varphi_p(\overline{y})$. By upper semicontinuity of the solution set mapping of the parametric problem (7.10) this inequality holds true also for all y with $G(y) \leq$

0: for all sequences $\{u^k, v^k\}_{k=1}^\infty$ with $G(v^k) \leq 0, u^k \in \Psi(v^k)$ and each accumulation point (u, y) of this sequence we have $F(u, y) \leq \varphi_p(y)$. By assumption (C) and the continuity assumptions, $v^* = \inf_y \varphi_p(y) > -\infty$ even if no pessimistic optimal solution exists. Suppose now that the second assertion is not satisfied. Then,

$$v^* < F(\overline{x}, \overline{y}) \leq \varphi_p(\overline{y}). \tag{7.50}$$

Let $0 < \delta < (F(\overline{x}, \overline{y}) - v^*)/3$. Then, there exists $k_1 < \infty$ such that

$$F_k(x^k, y^k) > v^* + 2\delta \text{ for all } k \geq k_1.$$

By the properties of the infimum there exists a sequence $\{z^t\}_{t=1}^\infty$ such that $\lim_{t \to \infty} \varphi_p(z^t) = v^*$. Thus, there exists $t_1 < \infty$ such that

$$\varphi_p(z^t) \leq v^* + \delta \; \forall t \geq t_1.$$

Fix any $t \geq t_1$. Let $\{v^k\}_{k=1}^\infty$ be a sequence converging to z^t and take a corresponding sequence of feasible points $\{u^k\}_{k=1}^\infty$ with $g^k(u^k, v^k) \leq 0, h^k(u^k, v^k) = 0, x^k \in \Psi_k(v^k)$. Then, by upper semicontinuity of the solution set mapping, any accumulation point of $\{u^k\}_{k=1}^\infty$ belongs to $\Psi(z^t)$ and, thus, there exists $k_2 = k_2(t)$ such that

$$F_k(u^k, v^k) \leq \varphi_p(z^t) + \delta \; \forall k \geq k_2.$$

Putting the last three inequalities together we get

$$F_k(u^k, v^k) \leq \varphi_p(z^t) + \delta \leq v^* + 2\delta < F_k(x^k, y^k)$$

for $t \geq t_1, k \geq \max\{k_1, k_2\}$. On the other hand, by definition of the sequence $\{(x^k, y^k)\}_{k=1}^\infty$ we have $F_k(x^k, y^k) \leq F_k(u^k, v^k)$ which thus implies

$$F_k(u^k, v^k) < F_k(u^k, v^k).$$

Since this is impossible, $v^* \geq F(\overline{x}, \overline{y})$.   □

PROOF OF THEOREM 7.13: The assumptions to the lower level problem guarantee that, for each fixed (y, z) in some open neighborhood of $(y^0, 0)$, the lower level programming problem can locally be reduced to one of the following problems:

$$f(x, y, z) \to \min_x$$
$$g_i(x, y, z) = 0, \; i \in I$$
$$g_i(x, y, z) \leq 0, \; i \notin I$$
$$h(x, y, z) = 0$$

for some $I \in \mathcal{I}(\lambda^0)$ and $(\lambda^0, \mu^0) \in \Lambda(x^0, y^0, 0)$. By Theorem 5.15, an optimistic local optimal solution of problem (5.1), (5.2) is also a local optimal solution of problem (5.27) and hence for each of the problems (7.12) for $z = 0$. Due to (LICQ) and (SSOC), there exist PC^1 functions $(x^I(\cdot), y^I(\cdot), \lambda^I(\cdot), \mu^I(\cdot))$ defined on some open neighborhood $V^I(0)$ composing optimal solutions for the problem (7.12) for all $z \in V^I(0)$. Now, the local optimal solution function $(x(z), y(z), \lambda(z), \mu(z))$ of the KKT reformulation of problem (7.13) is obtained as a continuous selection of the continuously differentiable local solution functions $(x^I(\cdot), y^I(\cdot), \lambda^I(\cdot), \mu^I(\cdot))$ on the open neighborhood $V(0) = \bigcap\limits_I V^I(0)$. We have to show that $(x(z), y(z))$ is a local optimal solution of the problem (7.13). First, by convexity, $x(z) \in \Psi(y(z), z)$ for $z \in V(0)$. Second, to verify our aim, we argue from contradiction, i.e. assume that there exists a sequence $\{z^k\}_{k=1}^{\infty}$ converging to zero and there are sequences $\{\widehat{x}(z^k), \widehat{y}(z^k)\}_{k=1}^{\infty}$ converging to (x^0, y^0) with

$$G(\widehat{y}(y^k)) \leq 0, \ \widehat{x}(z^k) \in \Psi(\widehat{y}(z^k), z^k)$$

and

$$F(\widehat{x}(z^k), \widehat{y}(z^k), z^k) < F(x(z^k), y(z^k), z^k)$$

for all k. By the assumptions there are Lagrange multipliers

$$(\widehat{\lambda}(z^k), \widehat{\mu}(z^k)) \in \Lambda(\widehat{x}(z^k), \widehat{y}(z^k), z^k)$$

converging to some Lagrange multiplier $(\widehat{\lambda}, \widehat{\mu}) \in \Lambda(x^0, y^0, 0)$. By shrinking to a subsequence if necessary we can without loss of generality assume that $J(\widehat{\lambda}(z^k)) \equiv \widehat{I}$ for some $\widehat{I} \in \mathcal{I}(\lambda^0)$. Now, since each of the problems (7.12) with $I \supseteq J(\lambda)$, $I \in \mathcal{I}(\lambda)$ is equivalent to the problem with $I = J(\lambda)$, the point $(\widehat{x}(z^k), \widehat{y}(z^k), z^k, \widehat{\lambda}(z^k), \widehat{\mu}(z^k))$ is feasible and hence optimal for the corresponding problem (7.12) with $I = \widehat{I}$. Since (SSOC) is satisfied for problem (7.13) with $I = \widehat{I} \cup J(\lambda(z^k)) \in \mathcal{I}(\lambda^0)$ too, the optimal solution of this problem is locally unique, which implies $\widehat{x}(z^k) = x(z^k)$ and $\widehat{y}(z^k) = y(z^k)$ contradicting our assumption. This completes the proof. □

PROOF OF THEOREM 7.14: The assuptions imply that the lower level problem is stable, i.e. that for fixed $\varepsilon > 0$ we have

$$\lim_{k\to\infty} \varphi_k(y) = \varphi(y) \text{ and } \limsup_{k\to\infty} \Psi_\varepsilon^k(y) \subseteq \Psi_\varepsilon(y)$$

[17]. Let $\{x^k, y^k\}_{k=1}^{\infty}$ be a sequence of optimistic optimal solutions of the relaxed bilevel programming problem (7.17). Then, this sequence

has accumulation points $(\overline{x}, \overline{y})$ by (C) and each accumulation point is feasible for (7.16). This implies the first result.

Now, let (x^0, y^0) be an optimistic optimal solution of the bilevel programming problem (5.1), (5.2). Then, by (ULR) there exists a sequence $\{y^k\}_{k=1}^{\infty}$ converging to y^0 such that $G(y^k) \leq 0$ for all k. Now, by (MFCQ) we can find a corresponding sequence $\{x^k\}_{k=1}^{\infty}$ such that $x^k \in M(y^k)$ for all k and $\lim_{k \to \infty} x^k = x^0$. This implies that for each fixed $\varepsilon > 0$ there is $k = k(\varepsilon)$ such that

$$f_k(x^k, y^k) \leq \varphi_k(y^k) + \varepsilon \ \forall k \geq k(\varepsilon)$$

by $\lim_{k \to \infty} f_k(x^k, y^k) = \varphi(y^0) = \lim_{k \to \infty} \varphi(y^k)$. Hence, $x^k \in \Psi_\varepsilon^k(y^k)$ for all sufficiently large k. This implies

$$\Phi_\varepsilon \leq \liminf_{k \to \infty} \Phi_\varepsilon^k \leq \limsup_{k \to \infty} \Phi_\varepsilon^k \leq \lim_{k \to \infty} F(x^k, y^k) = \Phi.$$

The opposite inclusion follows from $\limsup_{\varepsilon \to 0+} \Psi_\varepsilon(y) \subseteq \Psi(y)$ for all y. $\quad\square$

PROOF OF THEOREM 7.15: Let $\overline{z} = (\overline{x}, \overline{y})$ be an optimistic optimal solution of (5.1), (5.2). Then,

$$F(\overline{x}, \overline{y}) \leq F(x', y')$$

for all $(x', y') \in D$, $x' \in \Psi(y')$. For any $v \in \mathbb{R}^n$ and $(x, y) \in D$, $x \in \Psi^v(y)$ we get:

If $x \in \Psi(y)$, then

$$F(\overline{x}, \overline{y}) \leq F(x, y) \leq F(x, y) + r\|v\|.$$

If $x \notin \Psi(y)$, then by upper Lipschitz continuity of the mapping $\mathcal{F} \cap D$ at $v = 0$, there exists $(x^0, y^0) \in D$, $x^0 \in \Psi(y^0)$ with $\|(x, y) - (x^0, y^0)\| \leq \kappa\|v\|$, and by semi-Lipschitz continuity of F we derive

$$F(\overline{x}, \overline{y}) \leq F(x^0, y^0) \leq F(x, y) + M\|(x, y) - (x^0, y^0)\| \leq F(x, y) + r\|v\|.$$

This implies that $(\overline{z}, 0)$ solves (7.20).

Conversely, if $(\overline{z}, \overline{v})$ solves (7.20) with $r > \kappa M$, then

$$(\overline{x}, \overline{y}) \in D, \overline{x} \in \Psi^v(\overline{y}) \text{ and } F(\overline{x}, \overline{y}) + r\|\overline{v}\| \leq F(x, y) + r\|v\|$$

for all $(x, y) \in D$, $x \in \Psi^v(y)$, $v \in V$, where V is some closed set. In particular,

$$F(\overline{x}, \overline{y}) + r\|\overline{v}\| \leq F(x, y) \ \forall \ (x, y) \in D, x \in \Psi(y). \tag{7.51}$$

If $\overline{x} \in \Psi(\overline{y})$, we get $\overline{v} = 0$ by minimizing the left-hand side. Otherwise, if $\overline{x} \notin \Psi(\overline{y})$, by the semi-Lipschitz continuity of F and the upper Lipschitz continuity of $F \cap D$ we find $(x^0, y^0) \in D$ with $x^0 \in \Psi(y^0)$ such that

$$r\|\overline{v}\| \leq F(x^0, y^0) - F(\overline{x}, \overline{y}) \leq M\|(x^0, y^0) - (\overline{x}, \overline{y})\| \leq \kappa M\|\overline{v}\|$$

by (7.51). Due to $r > \kappa M$ this yields $\overline{v} = 0$ and $(\overline{x}, \overline{y})$ solves (5.1), (5.2).
$\square$

PROOF OF THEOREM 7.16: The first assertion of the theorem follows immediately from Theorem 4.4. The second inequality of the second assertion is due to feasibility which then implies the first inequality to be satisfied. The last assertion is an obvious implication of the second one for fixed y.
$\square$

PROOF OF THEOREM 7.17: By (C) and the continuity of the function F, there exist numbers b, c satisfying

$$b \leq F(x, y) \leq b + c \ \forall \ (x, y) \text{ with } G(y) \leq 0, x \in M(y).$$

This implies

$$f(x, y) + \alpha F(x, y) \geq \varphi(y) + \alpha b \ \forall \ (x, y) \text{ with } G(y) \leq 0, x \in M(y),$$

where $\varphi(y) = \varphi_0(y)$. Let $x \in \Psi_{\alpha, \varepsilon}(y)$, $G(y) \leq 0$. Then,

$$
\begin{aligned}
f(x, y) + \alpha b &\leq f(x, y) + \alpha F(x, y) \\
&\leq \min_x \{f(x, y) + \alpha F(x, y) : x \in M(y)\} + \varepsilon \\
&\leq \min_x \{f(x, y) + \alpha(b + c) : x \in M(y)\} + \varepsilon = \varphi(y) + \alpha(b + c) + \varepsilon.
\end{aligned}
$$

Thus,

$$f(x, y) \leq \varphi(y) + \alpha c + \varepsilon \text{ for } x \in \Psi_{\alpha, \varepsilon}(y).$$

$\square$

PROOF OF THEOREM 7.18: Let $G(y) \leq 0$, $x \in \Psi_{\alpha, \varepsilon}(y)$. Then,

$$F(x, y) \geq \min_x \{F(x, y) : x \in \Psi_{\alpha, \varepsilon}(y)\} \geq \min_x \{F(x, y) : x \in \Psi_{0, \varepsilon + \alpha c}(y)\}$$

by Theorem 7.17. Moreover, by definition,

$$
\begin{aligned}
f(x, y) + \alpha F(x, y) &\leq \varphi_\alpha(y) + \varepsilon = \\
\min_x \{f(x, y) + \alpha F(x, y) : x \in M(y)\} + \varepsilon \\
&\leq \min_x \{f(x, y) + \alpha F(x, y) : x \in \Psi(y)\} + \varepsilon \\
&= \min_x \{\varphi(y) + \alpha F(x, y) : x \in \Psi(y)\} + \varepsilon \\
&= \varphi(y) + \min_x \{\alpha F(x, y) : x \in \Psi(y)\} + \varepsilon \\
&\leq f(x, y) + \alpha \min_x \{F(x, y) : x \in \Psi(y)\} + \varepsilon
\end{aligned}
$$

by feasibility of x. This implies the first sequence of inequalities. The second one follows by taking the minimum in each term. □

PROOF OF THEOREM 7.20: First, $\|x_\varepsilon(y)\| \le \|x_0(y)\| < \infty$ for all $\varepsilon > 0$ since $x_0(y) \in \Psi_\varepsilon(y) \ne \emptyset$ by our assumptions. Hence, the sequence $\{x_\varepsilon(y)\}_{\varepsilon>0}$ is bounded and has accumulation points $\overline{x}$. Let without loss of generality $\lim_{\varepsilon \to 0+} x_\varepsilon(y) = \overline{x}$ (take a convergent subsequence if necessary). Then, $x_\varepsilon(y) \in M(y)$ and $f(x_\varepsilon(y), y) \le \varphi(y) + \varepsilon$ for all $\varepsilon > 0$ which, for $\varepsilon \to 0+$, implies $\overline{x} \in \Psi(y)$. Now,

$$\|\overline{x}\| = \lim_{\varepsilon \to 0+} \|x_\varepsilon(y)\| \le \|x_0(y)\|$$

implies $\overline{x} = x_0(y)$ due to the uniqueness of the minimal norm element in the convex set $\Psi(y)$. □

PROOF OF THEOREM 7.21: Let $(\overline{x}, \overline{y})$ be an accumulation point of the sequence $\{(x^0_{\varepsilon^k}(y^k), y^k)\}_{k=1}^\infty$ which exists by (C). Without loss of generality, let $\lim_{k \to \infty} (x^0_{\varepsilon^k}(y^k), y^k) = (\overline{x}, \overline{y})$. First, due to Theorem 7.20 we obtain $\overline{x} \in \Psi(\overline{y})$ and, hence,

$$F(\overline{x}, \overline{y}) \le \max_x \{F(x, \overline{y}) : x \in \Psi(\overline{y})\}.$$

Assume that

$$F(\overline{x}, \overline{y}) > v^* := \inf_y \{\varphi_p(y) : G(y) \le 0\}.$$

Take δ such that

$$0 < \delta < (F(\overline{x}, \overline{y}) - v^*)/3.$$

Then, there is a $k_1 < \infty$ such that

$$F(x^0_{\varepsilon^k}(y^k), y^k) > v^* + 2\delta \; \forall \; k \ge k_1.$$

Since the infimum v^* must be finite due to (C) there exists a sequence $\{z^t\}_{t=1}^\infty$ such that $\lim_{t \to \infty} \varphi_p(z^t) = v^*$. Hence, there exists t_1 such that

$$\varphi_p(z^t) \le v^* + \delta \; \forall \; t \ge t_1.$$

Fix $t \ge t_1$. Then we can find a sequence $\{v^k\}_{k=1}^\infty$ converging to z^t and a corresponding sequence $\{u^k\}_{k=1}^\infty$ such that $u^k \in \Psi^0_{\varepsilon^k}(v^k)$ for all k. Then, due to Theorem 7.20 each accumulation point of $\{u^k\}_{k=1}^\infty$ belongs to $\Psi(z^t)$. Hence, there exists an index k_2 such that

$$F(u^k, v^k) \le \varphi_p(z^t) + \delta \; \forall \; k \ge k_2.$$

Putting the last three inequalities together we get for $t \geq t_1, k \geq \max\{k_1, k_2\}$

$$F(u^k, v^k) \leq \varphi_p(z^t) + \delta \leq v^* + 2\delta < F(x^0_{\varepsilon^k}(y^k), y^k).$$

This resulting inequality contradicts the rule for selecting the point $(x^0_{\varepsilon^k}(y^k), y^k)$ which proves the theorem. $\square$

Chapter 8

DISCRETE BILEVEL PROBLEMS

Bilevel programming problems containing integrality conditions have not yet encountered that attention that they should from the point of view of potential applications. We will start with some examples explaining the additional difficulties arising from the indivisibility conditions. After that we will add two ideas for attacking such problems. The first approach is based on cutting planes applied to the lower level problem. Cutting plane algorithms have found large attention in solving discrete optimization problems in the past either as stand-alone algorithms or in connection with enumeration methods as branch-and-cut algorithms. We will apply them to attack bilevel programming problems with lower level problems having a parameter in the objective function only. The second approach explicitly uses the structure of the solution set mapping of the discrete lower level optimization problem. We know now much about the structure of that mapping [18] but this has not often been used to solve bilevel programming problems with discreteness conditions. Even in continuous optimization the numerical description of that mapping is expensive and difficult and this, of course, is also true for discrete problems. Using the 0-1 knapsack problem in the lower level we will show that, in special cases, it is possible to find a way to successfully apply this knowledge to solve discrete bilevel programming problems.

We will not repeat solution algorithms for discrete bilevel programming problems based on enumeration principles. These can be found e.g. in the monograph [26] and in the papers [28, 87, 86, 215].

8.1 TWO INTRODUCTORY EXAMPLES

Example: This example stems from [279] and shows that the difficulties in investigating the discrete bilevel programming problem are

255

especially related with the position of the integrality conditions. Let the lower level problem be

$$\Psi_D(x) = \underset{x}{\text{Argmin}} \ \{x : x+y \leq 2, x-y \leq 2, -4x+5y \leq 10, -4x-5y \leq 10\}.$$

Picture 8.1 shows the feasible set of the corresponding bilevel programming problem with no integrality constraints (part a)), with additional integrality conditions on both the upper and the lower levels (picture b)), and with discreteness conditions only on the upper (part c)) respectively the lower levels (picture d)).   □

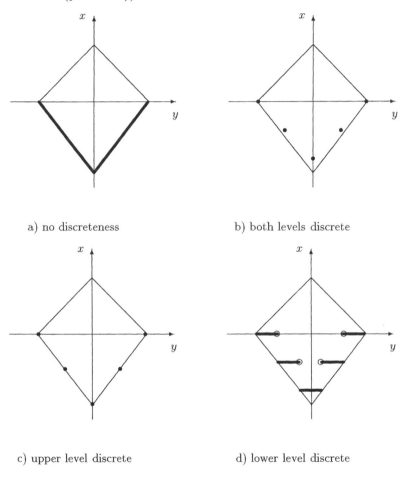

a) no discreteness   b) both levels discrete

c) upper level discrete   d) lower level discrete

Figure 8.1.   The linear discrete bilevel programming problem

Concerning the existence of optimistic optimal solutions we have the same situation as for linear bilevel programming problems in the case when the upper level problem is a discrete optimization problem but the lower one is linear continuous: An optimistic optimal solution exists since the solution set mapping of the lower level problem is upper semicontinuous provided it is bounded. The existence of a pessimistic optimal solution cannot be guaranteed in general if the lower level problem is a continuous one. If both the lower and the upper level problems are discrete optimization problems, the upper level objective is minimized over a discrete set in both the optimistic and the pessimistic problems and optimal solutions exist if the feasible set is bounded.

The situation is much more difficult in the case when the lower level programming problem is discrete and the upper level problem is a continuous one. Then, the feasible set of the discrete linear bilevel programming problem is the union (of a finite number under a boundedness condition) of sets whose closures are polyhedra [279]. These sets are in general neither open nor closed. Hence, the solution set mapping is not closed and we cannot guarantee even the existence of optimal optimistic solutions. The existence of nonunique lower level solutions for general discrete optimization problems can result also in the unsolvability of the pessimistic problem even if the feasible set is bounded.

Another unpleasant property of discrete bilevel programming problems in which both the upper and the lower level problems are discrete optimization problems is illustrated in an example in [215]:

Example: Consider the problem

$$\min_{y}\{-10x - y : x \in \Psi_D(y), y \text{ integer}\},$$

where the lower level problem is

$$\Psi_D(y) = \operatorname*{Argmin}_{x} \ \{x : \ 20x - 25y \le 30, 2x + y \le 10 \\ -x + 2y \le 15, 10x + 2y \ge 15, x \text{ integer}\}.$$

The feasible set of this problem as well as its continuous relaxation are shown in Figure 8.2. The thick line is the feasible set of the continuous relaxation problem. The unique global optimal solution of the continuous relaxation problem is the point $\hat{x} = (1, 8)^\top$. This point is feasible for the discrete bilevel programming problem but it is not the global optimal solution of this problem. The unique global optimal solution of the discrete bilevel problem is found at $x^* = (2, 2)^\top$. The upper level objective function value of $\hat{x}$ is $f(\hat{x}) = -18$ and that of x^* is $f(x^*) = -22$.

□

The property that optimal solutions of the continuous relaxation of the discrete bilevel programming problem not necessarily determine an op-

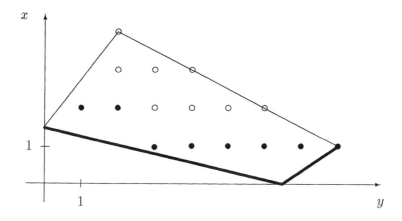

Figure 8.2.   Discrete bilevel programming problem

timal solution of the discrete problem has deep consequences for branch-and-bound algorithms [215]:

1. The optimal solution of the continuous relaxation does in general not provide a valid bound on the solution of the (mixed-)integer bilevel problem.

2. Solutions to the relaxed bilevel programming problem which are feasible for the discrete problem, can in general not be fathomed.

In [26, 215] branch-and -bound algorithms for the pure integer as well as for the mixed-integer bilevel programming problem have been given. These algorithms combine the Karush-Kuhn-Tucker conditions of the relaxed lower level problem with the enumeration scheme for forcing integrality of the variables.

8.2   CUTTING PLANE ALGORITHM

In this book we intend to give two other ideas for attacking linear bilevel programming problems with integrality constraints. First we apply a cutting plane algorithm to the discrete lower level problem with the aim to transform the discrete bilevel problem into a linear bilevel one. Second we try to illustrate how parametric discrete optimization applied to the lower level problem can be used to solve the discrete bilevel problem.

The first idea for solving discrete bilevel optimization problems using cutting planes is described for discrete lower level problems with perturbed objective functions only. It can also be applied to problems with right-hand side perturbations in the lower level problem. But then, the parametric analysis used for computing directions of descent for the linear bilevel programming problems in Step 1 of the algorithm below is more complicated. Moreover the exclusive use of a cutting plane algorithm to the lower level problem will not be successful in general, i.e. this algorithm has to be combined with branching steps of a branch-and-bound algorithm [84].

We consider the special case of the discrete bilevel programming problem where the upper level problem is a (discrete or continuous) linear problem and the lower level discrete optimization problem has parameter-independent constraints:

$$\Psi_D(y) = \operatorname*{Argmax}_x \{\langle x, y \rangle : Ax \leq a, x \geq 0, \text{ integer}\}. \tag{8.1}$$

Let the matrix A and the vector a have only integral entries and denote by

$$P = \{x \geq 0 : Ax \leq a, x \text{ integer}\}$$

the feasible set of problem (8.1). The bilevel programming problem is problem (3.2):

$$\text{``}\min_y\text{''}\{\langle d^1, x \rangle + \langle d^2, y \rangle : A^3 y = b, y \geq 0, x \in \Psi_D(y)\},$$

either in its optimistic or pessimistic variant and with or without integrality conditions. For being unique, we consider the optimistic problem without integrality constraints:

$$\min_{x,y}\{\langle d^1, x \rangle + \langle d^2, y \rangle : A^3 y = b, y \geq 0, x \in \Psi_D(y)\}. \tag{8.2}$$

If the upper level problem is also a discrete one, we have to replace the linear bilevel programming problem solved in the steps of the following algorithm by an algorithm for discrete optimization problems.

The main idea is that in the lower level problem we can replace the feasible set equivalently by its convex hull. Then, the discrete bilevel programming problem is replaced by an equivalent linear bilevel programming problem and all the results of Chapter 3 can be applied to treat this problem. The difficulty is that we do not know the convex hull of the feasible set in the lower level problem but have to compute it during the algorithm. One way to realize this idea is to use a cutting plane algorithm in the lower level. Cutting plane algorithms have proved to be a powerful tool for attacking discrete optimization problems, see e.g. [9, 14, 57, 63]. Here we want to give some ideas of how to apply the cutting plane idea to bilevel programming. The application of a cutting plane algorithm to the lower level problem has the advantage that, in each iteration of the algorithm, we have to solve only one bilevel programming problem. Moreover, the solution of this problem should give some information for the problem (8.1), (8.2) at least in each step in which an integral solution in the lower level is obtained.

First we recall the main ideas of a cutting plane algorithm for solving the lower level problem [217]. The inclusion of these ideas in an algorithm solving the discrete bilevel problem will follow later.

Consider the linear discrete programming problem (8.1) and its linear relaxation

$$\max_x \{\langle x, y \rangle : Ax \leq a, x \geq 0\} \qquad (8.3)$$

having $\Psi_D^a(y)$ as set of optimal solutions. Then, a Chvátal-Gomory cut is an inequality

$$\langle r, x \rangle \leq s$$

where $r_i = \lfloor \langle u, A^i \rangle \rfloor$ is the largest integer not exceeding the inner product of some $u \geq 0$ with the i-th column of A and $s = \lfloor \langle u, a \rangle \rfloor$ [217]. The following theorems say that, by the help of an integer point in the region of stability (cf. Definition 3.3) of an optimal solution x^* for (8.3) we can construct a Chvátal-Gomory cut cutting x^* away if and only if x^* is not integer valued. This gives one way for deriving a cutting plane algorithm for solving the problem (8.1).

DEFINITION 8.1 *Let x^* be an optimal solution of problem (8.3) for some $y = y^*$. Then, the set*

$$\mathcal{R}(x^*) = \{y : x^* \in \Psi_D^a(y)\} \qquad (8.4)$$

is the region of stability of the solution x^*.

THEOREM 8.1 *Let x^* be an optimal solution of the problem (8.3), let $y' \in \mathcal{R}(x^*)$ be integer-valued and let $u = u(y')$ be an optimal solution of the dual problem to (8.3) for y being replaced by y'. Then*

1. *$\langle r, x \rangle \leq s$ with $r_i = \lfloor \langle u, A^i \rangle \rfloor$, $i = 1, \ldots, l$ and $s = \lfloor \langle u, a \rangle \rfloor$ defines a Chvátal-Gomory cut which is satisfied by all points*

$$x \in P := \{x \geq 0 : Ax \leq a, x \text{ integer}\},$$

2. *$r \in \mathcal{R}(x^*)$,*

3. $\langle r, x^* \rangle > s$ if and only if $\langle u, a \rangle$ is not integer-valued.

THEOREM 8.2 *Let x^* be an optimal vertex for problem (8.3) which is not integer-valued. Then, there exists a Chvátal-Gomory cut $\langle r, x \rangle \leq s$ with $\langle r, x^* \rangle > s$.*

Theorems 8.1 and 8.2 suggest that we can left the procedure of computing one Chvátal-Gomory cut to the following

Oracle: Given problem (8.3) with a non-integer optimal solution x^*, compute a Chvátal-Gomory cut which is not satisfied by x^*.

Clearly, this oracle is not yet implementable in general. Moreover, by polynomial equivalence of optimization and separation [159, 217], $\mathcal{NP}$-hardness [104] of (8.1) implies $\mathcal{NP}$-completeness of the oracle itself.

The most sensitive part in the oracle is the computation of an integer vector y' in the region of stability of the optimal solution x^*. This vector has to be computed such that $\langle u, a \rangle$ is not integer-valued where u is a dual solution to x^* with y replaced by y' in (8.3). In using this approach to execute the oracle, Theorems 8.1 and 8.2 can be helpful. This method has the drawback that the computed vectors y' have often large components.

Another way to give a more clear formula for executing the oracle is to compute an approximate solution of the following mathematical programming problem in which the distance of x^* from the hyperplane $\{x : \langle r, x \rangle = s\}$ defining the cut is maximized [70]:

$$\frac{\langle u, a \rangle - \lfloor \langle u, a \rangle \rfloor}{\|r\|} \to \max$$

$$\langle r, x \rangle \leq \lfloor \langle u, a \rangle \rfloor$$
$$r_i = \lfloor \langle u, A^i \rangle \rfloor \; \forall i \qquad\qquad (8.5)$$
$$\langle u, a \rangle = \langle r, x^* \rangle$$
$$u \geq 0.$$

THEOREM 8.3 *Problem (8.5) has a finite optimal function value which is positive if and only if x^* is not integer-valued.*

The proof of this theorem shows that, in searching for an optimal solution, we can restrict us to the search on a bounded set. Since $\lfloor \langle u, A^i \rangle \rfloor$ and $\lfloor \langle u, a \rangle \rfloor$ can take only finitely many values over bounded sets of solutions u problem (8.5) decomposes into a finite number of (linear!) optimization problems. Unfortunately, a closer look on the feasible set of these linear optimization problems shows that their feasible set is in general bounded but not closed since the step function has jumps at integer values. This implies that an optimal solution of the problems (8.5) need not to exist. But remember that we introduced this problem only to give some way to improve the quality of the cutting planes and that there is no need to solve this problem up to optimality. For attacking it by means of heuristics, local search [230] in the region of stability $\mathcal{R}(x^*)$ can be used.

By use of the oracle we can define the following cutting plane algorithm for solving the discrete optimization problem (8.1):

Cutting Plane Algorithm for Discrete Linear Programs

    Input: An instance of problem (8.1).

    Output: An optimal solution.

    1: Solve the relaxed problem (8.3). Let x^* be an optimal vertex of this problem.

    2: If x^* is integer-valued, then x^* is an optimal solution of the problem (8.1).

    Else, call the oracle to compute a Chvátal-Gomory cut $\langle r, x \rangle \leq s$, add this inequality to the constraints $Ax \leq a$ (the resulting system is again denoted by $Ax \leq a$) and goto 1.

Now we come back to the bilevel programming problem. Our intention is to apply the cutting plane algorithm to this problem. Having a first look into that direction the possibility of realizing this idea seems not to be very surprising, since problem (8.1), (8.2) is equivalent to the linear bilevel programming problem

$$\langle d^1, x \rangle + \langle d^2, y \rangle \to \min_{x,y} \tag{8.6}$$

$$A^3 y = b, \ y \in \mathbb{R}^m, \ x \in \Psi_D^0(y),$$

where

$$\Psi_D^0(y) = \underset{x}{\text{Argmax}} \ \{\langle y, x \rangle : x \in \text{conv } P\}. \tag{8.7}$$

This problem has an optimistic optimal solution provided that the set

$$\{(x, y) : y \in \mathbb{R}^m, \ A^3 y = b, \ x \in \text{conv } P\}$$

is nonempty and bounded (cf. Theorem 3.3). Using the ideas forming the basis of cutting plane approaches and the theory of polyhedra, by use of sufficiently many calls of the oracle it is principally possible to compute the set conv P [217]. Then, if this set is obtained, we have to solve only one linear bilevel programming problem to obtain a (global or local) optimistic optimal solution for problem (8.1), (8.2). But this is very time-consuming. Therefore, it is suggested to use an algorithm which alternately solves a linear bilevel programming problem and calls the oracle to get a more accurate approximation of the set conv P.

Algorithm for Solving the Discrete Bilevel Optimization Problem:

      Input: Instance of problem (8.1), (8.2)

      Output: A local optimistic optimal solution.

1: Compute an optimal solution (x^*, y^*) of the linear bilevel programming problem

$$\min_{x,y}\{\langle d^1, x\rangle + \langle d^2, y\rangle : A^3 y = b, x \in \Psi_D^a(y)\},$$

where

$$\Psi_D^a(y) = \operatorname*{Argmax}_x \{\langle y, x\rangle : Ax \leq a, x \geq 0\}.$$

2: If $x^* \in \mathbf{Z}^n$, then stop.

3: Else call the oracle to get a Chvátal-Gomory cut $\langle r, x\rangle \leq s$, add this inequality to the set of inequalities $Ax \leq a$ (the resulting system of inequalities is again denoted by $Ax \leq a$), and goto 1.

For the solution of the linear bilevel programming problem in Step 1 the ideas in Chapter 3 can be used. Since changes of the feasible set of the lower level problem far away from the present iteration point can imply that global optimal solutions loose their optimality status [189], the additional effort for computing global optimal solutions of the linear bilevel problem in Step 1 makes no sense unless conv P is computed by the cutting plane algorithm.

With respect to convergence of the algorithm we have the following theorem which can be shown to be valid under the assumption that it is possible to realize all cut generation steps in the algorithm.

THEOREM 8.4 *If the cutting plane algorithm stops after a finite number of iterations at (x^*, y^*), then this point is a local optimistic optimal solution of problem (8.1), (8.2).*

REMARK 8.1 *Theoretically, cutting plane algorithms need only a finite number of cuts to construct conv P [217]. Since all vertices of conv P are integer and optimistic optimal solutions can be found at the vertices of the set $\{(x, y) : x \in \text{conv } P, A^3 y = b, y \geq 0\}$, the cutting plane algorithm will need also only a finite number of iterations to come to an end. By Theorem 8.4 this implies convergence of the algorithm.*

The above results are only valid if it is possible to compute a Chvátal-Gomory cut in each step of the algorithm cutting away a sufficiently large infeasible part N of $\{x \geq 0 : Ax \leq a\}$. By numerical reasons this need not to be true either since the volumes of the sets $N \subset \{x \geq 0 : Ax \leq a\} \setminus \text{conv } P$ tend too quickly to zero or due to the numerical impossibility to compute the cut itself. In either of these cases the algorithm stops after a finite number of iterations with an infeasible solution (x^*, y^*). In this case, the usual way out is to start a branch-and-cut procedure. We will not go into the details for such an algorithm combining branch-and-bound with the idea of a cutting plane algorithm to obtain sharper bounds. We will only give some remarks with respect to the bounding procedure and mention that the branching process should be done with respect to the upper level variable y as in algorithms for globally minimizing nonconvex functions. The reason is that branching with respect to the lower level variables can in most of the resulting subproblems not produce feasible solutions. For ideas of how to construct such a branching process, the reader is referred to [129].

8.3   INNER APPROXIMATION

If the discrete bilevel programming problem (8.1), (8.2) is solved with a branch-and-cut algorithm, then we also need an idea of how to compute valid bounds for the optimal function value of the bilevel programming problem on parts of the set $\{y \geq 0 : A^3 y = b\}$. Sometimes, solving a bilevel programming problem to global optimality is not desirable (or not possible due to NP-hardness, cf. Theorem 3.12) since this will take a large amount of time. Hence, the solution of a one-level optimization problem could be considered as being superior.

In doing that it seems to be necessary not only to approximate the feasible set of the lower level problem by means of cutting planes but also to include some kind of an approximation of the objective function. The following idea realizes this aim. This idea makes sense especially in all cases when it is easy to compute feasible solutions of the lower level problem. This is at least true if all coefficients a_{ij} are non-negative.

Let

$$V \subseteq \{x \geq 0 : Ax \leq a, \, x \in \mathbf{Z}^n\}$$

be a subset of the set of feasible solutions for the lower level problem. This subset can be generated by heuristics or approximation algorithms. Let

$$Q := \{(r, s) : \langle r, x \rangle \leq s \text{ is a Chvátal-Gomory cut generated by oracle}\}$$

be a set of cut generating vectors. Then, the optimal value of the following problem gives a bound for the optimal value of the discrete bilevel programming problem:

$$\langle d^1, x\rangle + \langle d^2, y\rangle \to \min_{x,y}$$
$$A^3 y = b, \ Ax \le a, \ x \ge 0, y \ge 0$$
$$\langle r, x\rangle \le s, \ \forall \ (r, s) \in Q,$$
$$\langle y, x\rangle \ge \langle y, x^i\rangle, \ \forall \ x^i \in V.$$

(8.8)

Clearly, the last set of inequalities in this problem implies that the x-part of an optimal solution has to be not worse than the best point in the set V with respect to the objective function of the lower level problem. It is also obvious that the optimal solution of the discrete bilevel programming problem is feasible to the problem (8.8). Hence, its optimal value is indeed a bound for the optimal value of the discrete bilevel programming problem. Moreover, if

$$V = \{x \ge 0 : Ax \le a, \ x \in \mathbf{Z}^n\}$$

(8.9)

and

$$\{x \ge 0 : Ax \le a, \langle r, x\rangle \le s, \ \forall \ (r, s) \in Q\} = \text{conv } P$$

(8.10)

then problems (8.1), (8.2) and (8.8) coincide. Hence, for increasing the accuracy of the approximation, new points in P and/or new cuts are to be computed.

With respect to the effort needed to solve problem 8.8 is is necessary to mention that this problem is a nonconvex one since the last set of constraints form nonconvex inequalities. This is not surprising since it is clearly not possible to transform a NP-hard optimization problem into a polynomially solvable one by means of a polynomial transformation.

By obvious modifications of (8.9) and (8.10) it is also possible to derive a rule for the decision if a point (x^*, y^*) generated by the algorithm is locally optimal for the discrete bilevel programming problem.

Example: Consider the problem

$$-x_1 - 2x_2 + 3y_1 + 3.2y_2 \to \min$$
$$-2 \le y_1 + y_2 \le 2, \ -2 \le y_1, y_2 \le 2, \ x \in \Psi_D(y),$$

(8.11)

where

$$\Psi_D(y) = \operatorname*{Argmax}_{x} \{\langle y, x\rangle : \ -x_1 + 3x_2 \le 3, \ x_1 - x_2 \le 1,$$
$$x_1 + x_2 \ge 2, \ x \ge 0, \ \text{integer } \}.$$

(8.12)

Let

$$V = \{(1, 1)^\top, \ (3, 2)^\top\}, \ Q = \{(-1, 1, 0)^\top\}.$$

Then, problem (8.8) reads as

$$
\begin{aligned}
&-x_1 - 2x_2 + 3y_1 + 3.2y_2 \to \min_{x,y} \\
&-2 \le y_1 + y_2 \le 2, \ -2 \le y_1, y_2 \le 2, \\
&-x_1 + 3x_2 \le 3, \ x_1 - x_2 \le 1, \ x_1 + x_2 \ge 2, \ x \ge 0, \\
&-x_1 + x_2 \ge 0, \\
&y_1 x_1 + y_2 x_2 \ge y_1 + y_2, \ y_1 x_1 + y_2 x_2 \ge 3y_1 + 2y_2.
\end{aligned}
\tag{8.13}
$$

The last two inequalities of this problem are equivalently written as

$$
y_1 x_1 + y_2 x_2 \ge \max\{y_1 + y_2, 3y_1 + 2y_2\}
$$

which can then be used to partition its feasible set into two sets:

$$
\begin{aligned}
M_i := \{(x,y): \ &-2 \le y_1 + y_2 \le 2, \ -2 \le y_1, y_2 \le 2, \ -x_1 + 3x_2 \le 3, \\
&x_1 - x_2 \le 1, \ x_1 + x_2 \ge 2, \ x \ge 0, \ -x_1 + x_2 \ge 0\} \cap L_i,
\end{aligned}
$$

where

$$
L_1 = \{(x,y): 2y_1 + y_2 \ge 0, \ y_1 x_1 + y_2 x_2 \ge 3y_1 + 2y_2\}
$$

and

$$
L_2 = \{(x,y): 2y_1 + y_2 \le 0, \ y_1 x_1 + y_2 x_2 \ge y_1 + y_2\}.
$$

The problem of minimizing the upper level objective function over M_1 has the optimal solution (x^0, y^0) with $x^0 = (3,2)^\top, y^0 = (1,-2)^\top$ and an objective function value of -10.4.

The second problem of minimizing the same function on M_2 has two isolated optimal solutions at (x^0, y^0) and (x^1, y^1) with $x^1 = (2,1)^\top, y^1 = (0,-2)^\top$ both having the same objective function value.

This shows that problem (8.8) is a nonconvex optimization problem. It can be solved by means of $|V| - 1$ optimization problems having only one nonconvex constraint. For the computation of globally optimal solutions of such problems ideas of d.c. programming can be used [274]. The application of the ideas in [274] in parallel to all the $|V| - 1$ optimization problems seems to be imaginable.

Note that the second solution is fortunately equal to the optimal solution of the discrete bilevel programming problem. The first solution is infeasible for the discrete bilevel problem.   □

8.4   KNAPSACK LOWER LEVEL PROBLEMS

In this last section we want to show how parametric discrete optimization can efficiently be applied for solving discrete bilevel optimization problems. Let $\Psi_K : \mathbb{R}_+ \to 2^{\{0,1\}^n}$ denote the solution set mapping of a right-hand side parametrized 0-1 knapsack problem

$$\Psi_K(b) = \underset{x}{\text{Argmax}} \; \{\langle c, x \rangle : \langle a, x \rangle \leq b, x \in \{0,1\}^n\}, \qquad (8.14)$$

where a, c are n-dimensional integral vectors with positive components, $b \in \mathbb{R}_+ := \{z \in \mathbb{R} : z \geq 0\}$. Then, the bilevel programming problem with knapsack constraints reads as

$$\begin{aligned} \langle d, x \rangle + fb &\to \quad \text{``}\underset{b}{\max}\text{''} \\ x \in \Psi_K(b), &\qquad b_u \leq b \leq b_o, \end{aligned} \qquad (8.15)$$

with $d \in \mathbb{R}^n$, $f \in \mathbb{R}$, $0 \leq b_u < b_o \leq \sum_{i=1}^{n} a_i$. We start our investigations with repeating from the literature some necessary details for an algorithm applied to problem (8.14) to find a description of the point-to-set mapping $\Psi_K(\cdot)$ over $[b_u, b_o]$. This can be done in pseudopolynomial time and results in a pseudopolynomial algorithm for the computation of an optimal solution of (8.14), (8.15) in three senses: Without using refined cancellation rules for unnecessary solutions we get lower optimal optimal solutions. Using different refined dropping rules, optimistic as well as pessimistic optimal solutions can be computed. An example which illustrates the difficulties if a (pessimistic or optimistic) optimal solution is searched for is also given in Subsection 8.4.2. By combination of the fully polynomial approximation scheme of [168] with the investigations of [192] on the convergence of optimal solutions of approximate problems to solutions of the original one it will be possible to construct a polynomial approximation algorithm. Other than in [192], we prove that it is not necessary to use ε-optimal solutions in the perturbed lower level problems to compute optimistic optimal solutions in our problem. Thus, the bilevel programming problem with knapsack constraints is one example for the NP-hard bilevel programming problems in which a lower optimal solution possesses an arbitrary good approximation in polynomial time.

8.4.1   SOLUTION OF KNAPSACK PROBLEMS

We start with a simple and seemingly obvious result on the computability of solutions of parametric knapsack problems.

Knapsack problems have been intensively investigated, see e.g. [202, 233] for comprehensive treatments of the problem. Parametric knapsack problems are the topic of at least the papers [50, 45]. The following result shows that the computation of one optimal solution of the knapsack problem for every right-hand side can be done in pseudopolynomial time. To present them we need the subsequent enumeration algorithm.

We organize an enumeration algorithm in the breadth: Let $M^0 = \{(0,0,\ldots,0)^\top\}$, $M^k = M^{k-1} \cup \{x + e_k : x \in M^{k-1}\}$, $k = 1,\ldots,n$, where e_k denotes the k-th unit vector. Then, by use of the bijection $i \in I$ if and only if $x_i = 1$, M^k corresponds to the power set of $\{1,\ldots,k\}$. Clearly, M^n contains all possible solutions of the 0-1 knapsack problem if no notice is taken to the constraint. Since this set is much too large, dropping rules can be used. Consider problem (8.14) for a fixed right-hand side b and the following rules for excluding elements of the sets M^k :

A: Exclude $\overline{x}$ from M^k if $\langle a, \overline{x} \rangle > b$.

B: Let $\overline{x}, \widehat{x} \in M^k$ for some k. Then, if $\langle c, \overline{x} \rangle \geq \langle c, \widehat{x} \rangle$ and $\langle a, \overline{x} \rangle \leq \langle a, \widehat{x} \rangle$, exclude $\widehat{x}$ from M^k.

C: If there exist $\overline{x}, \widehat{x} \in M^k$ for some k satisfying $\langle c, \overline{x} \rangle > \langle c, \widehat{x} \rangle$ and $\langle a, \overline{x} \rangle \leq \langle a, \widehat{x} \rangle$, then exclude $\widehat{x}$ from M^k.

Call **Algorithm 1** the algorithm implementing this enumeration idea and using Rules A and B for ruling out unnecessary points. Analogously, the procedure using Rules A and C in connection with the above enumeration idea is **Algorithm 2**. The following results are more or less obvious, we state them without proof.

THEOREM 8.5 ([168]) *Algorithm 1 computes an optimal solution for the problem (8.14) with a fixed right-hand side b with time and space complexity of $O(n \min\{b, \varphi(b)\})$, where $\varphi(b) = \max\{\langle c, x \rangle : \langle a, x \rangle \leq b, x \in \{0,1\}^n\}$.*

COROLLARY 8.1 *If the above algorithm is applied to problem (8.14) with $b_{\max} = \sum\limits_{i=1}^{n} a_i$, then for each $b \in [0, b_{\max}]$ an optimal solution is computed with a time and space complexity of $O(n \min\{b_{\max}, \sum\limits_{i=1}^{n} c_i\})$.*

COROLLARY 8.2 *Algorithm 2 computes all optimal solutions of problem (8.14) with fixed right-hand side b.*

Note that Algorithm 2 is not of pseudopolynomial complexity.

We start with considering Algorithm 1. The elements of the set M^n can be ordered with $O(\min\{b_o, \varphi(b_o)\} \log_2 \min\{b_o, \varphi(b_o)\})$ time complexity to $M^n = \{x^1, x^2, \ldots, x^p\}$ such that

$$\langle a, x^1 \rangle < \langle a, x^2 \rangle < \ldots \langle a, x^p \rangle \text{ and } \langle c, x^1 \rangle < \langle c, x^2 \rangle < \ldots \langle c, x^p \rangle,$$

where $p \leq \min\{b_o, \varphi(b_o)\}$. Then,

COROLLARY 8.3 *For $i = 1, \ldots, p - 1$, the point $x^i \in \Psi_K(b)$ if and only if $b \in [\langle a, x^i \rangle, \langle a, x^{i+1} \rangle)$. $x^p \in \Psi_K(b)$ for all $b \in [\langle a, x^p \rangle, b_o]$.*

The proof follows directly from the following property of arbitrary two different elements x^i, x^j in M^n: either

$$\langle a, x^i \rangle < \langle a, x^j \rangle \text{ and } \langle c, x^i \rangle > \langle c, x^j \rangle$$

or vice versa.

COROLLARY 8.4 *The optimal value function $\varphi : \mathbb{R}_+ \to \mathbb{R}$ defined by*

$$\varphi(b) = \max\{\langle c, x \rangle : \langle a, x \rangle \leq b, x \in \{0, 1\}^n\}$$

is piecewise constant and monotonically not decreasing. It can have jumps only at integer values of the right-hand side $b = \langle a, x^i \rangle$, $i = 1, \ldots, p$.

For a feasible solution $\overline{x}$ of (8.14) call the set

$$\mathcal{R}(\overline{x}) = \{b \in [0, \sum_{i=1}^{n} a_i] : \overline{x} \in \Psi_K(b)\}$$

again the *region of stability* of $\overline{x}$. Clearly, $\mathcal{R}(\overline{x})$ can be empty.

REMARK 8.2 *For $\overline{x} \neq (1, 1, \ldots, 1)^\top$, the set $\mathcal{R}(\overline{x})$ is a half-open interval. If $\overline{x} = (1, 1, \ldots, 1)^\top$, $\mathcal{R}(\overline{x}) = \{\langle a, \overline{x} \rangle\}$. For two points $\overline{x}, \widehat{x} \in \{0, 1\}^n$ satisfying $\mathcal{R}(\overline{x}) \cap \mathcal{R}(\widehat{x}) \neq \emptyset$ either $\mathcal{R}(\overline{x}) \subseteq \mathcal{R}(\widehat{x})$ or $\mathcal{R}(\overline{x}) \supseteq \mathcal{R}(\widehat{x})$.*

The following well-known theorem says that by using Algorithm 1 it is possible to develop a fully polynomial-time approximation scheme:

THEOREM 8.6 ([168]) *Let $\varepsilon > 0$ be arbitrary and set $c_i' = \left\lfloor \frac{n c_i}{\varepsilon c_{\max}} \right\rfloor$, $i = 1, \ldots, n$, where $c_{\max} = \max\{c_i : 1 \leq i \leq n\}$. Then, by solving the modified problem*

$$\max_x \{\langle c', x \rangle : \langle a, x \rangle \leq b, x \in \{0, 1\}^n\} \tag{8.16}$$

a solution $x^*(\varepsilon)$ is computed satisfying

$$\frac{\varphi(b) - \langle c, x^*(\varepsilon) \rangle}{\varphi(b)} \leq \varepsilon$$

with time and space complexity of $O(n^3/\varepsilon)$.

It should be mentioned that a better fully polynomial-time approximation scheme can be found in [146]. The following corollary is obtained
by combining Corollary 8.3 with Theorem 8.6 since Theorem 8.6 does
not depend on the right-hand side b. Solve problem (8.16) for $b = b_{\max}$
with $\varepsilon > 0$ and let $M^n = \{\overline{x}^1, \overline{x}^2, \ldots, \overline{x}^p\}$ with

$$\langle a, \overline{x}^1 \rangle < \langle a, \overline{x}^2 \rangle < \ldots \langle a, \overline{x}^p \rangle, \ \langle c, \overline{x}^1 \rangle < \langle c, \overline{x}^2 \rangle < \ldots \langle c, \overline{x}^p \rangle.$$

COROLLARY 8.5 *Under the above assumptions and notations, for each*
$i = 1, \ldots, p - 1$ *and* $b \in [\langle a, \overline{x}^i \rangle, \langle a, \overline{x}^{i+1} \rangle)$, *the inequality*

$$\frac{\varphi(b) - \langle c, \overline{x}^i \rangle}{\varphi(b)} \leq \varepsilon$$

holds. The time complexity for computing all solutions and describing
all regions of stability is $O\left(\frac{n^3}{\varepsilon} + \frac{n^2}{\varepsilon} \log_2 \frac{n^2}{\varepsilon}\right)$.

8.4.2   EXACT SOLUTION

We start with the following existence result for optimistic and pessimistic optimal solutions which is different from the results in the general case (cf. Theorems 5.2 and 5.3).

THEOREM 8.7 ([75]) *Let* $b_o \leq \sum\limits_{i=1}^{n} a_i$. *Then, global optimistic and pes-*
simistic optimal solutions of (8.15) exist if $f \leq 0$. *If* $f > 0$ *then either*
the (optimistic or pessimistic) global optimal solution is $(\overline{x}, b_o)$ *for some*
$\overline{x} \in \Psi(b_o)$ *or problem (8.15) has no (optimistic or pessimistic) optimal*
solution.

The proof follows from Corollary 8.4 since for fixed x its region of stabil-
ity $\mathcal{R}(x)$ is an half open interval, the function $\langle c, x(b) \rangle + fb$ is piecewise
linear with slope f and the right endpoint of each continuous part does
not belong to this part. Since this right endpoint gives the supremal
function value on this part if and only if $f > 0$ the problem can have no
optimal solution if and only if $f > 0$.

The following Theorem shows that without using an improved dropping rule in Algorithm 1 it is possible to compute a lower optimal solution.

THEOREM 8.8 ([75]) *Let* $M^n = \{x^1, \ldots, x^p\}$ *be the set of all solutions*
for problem (8.14) obtained by application of Algorithm 1. Assume that
$b_u = 0, b_o = \sum\limits_{i=1}^{n} a_i$, $f \leq 0$ *and let* $x^{i_0} \in M^n$ *satisfy*

$$\langle d, x^{i_0} \rangle + f \langle a, x^{i_0} \rangle = \max_{1 \leq i \leq p} \{ \langle d, x^i \rangle + f \langle a, x^i \rangle \}.$$

Then,

$$\langle d, x^{i_0} \rangle + f \langle a, x^{i_0} \rangle \geq \max_{b_u \leq b \leq b_o} \min_{x \in \Psi_K(b)} \{\langle d, x \rangle + fb\}.$$

REMARK 8.3 *The computation of a solution according to Theorem 8.8 has time and space complexity* $O(n \sum_{i=1}^{n} a_i)$ *in the worst case which coincides with the time and space complexity of the knapsack problem.*

Clearly, an easy modification of the last theorem can be used in the case $[b_u, b_o] \subset [0, \sum_{i=1}^{n} a_i]$. The following example shows that for computing an optimistic solution, a refined rule has to be applied in the basic enumeration process than Rule B. Similar examples can also be derived for the computation of a pessimistic solution.

Example: Let

$$c = (7, 3, 2, 2, 1)^\top, \quad a = (2, 1, 1, 2, 3)^\top.$$

Then, the enumeration algorithms results in

$$M^n = \left\{ \begin{pmatrix} 0 \\ 0 \\ 0 \\ 0 \\ 0 \end{pmatrix}, \begin{pmatrix} 0 \\ 1 \\ 0 \\ 0 \\ 0 \end{pmatrix}, \begin{pmatrix} 1 \\ 0 \\ 0 \\ 0 \\ 0 \end{pmatrix}, \begin{pmatrix} 1 \\ 1 \\ 0 \\ 0 \\ 0 \end{pmatrix}, \begin{pmatrix} 1 \\ 1 \\ 1 \\ 0 \\ 0 \end{pmatrix}, \begin{pmatrix} 1 \\ 1 \\ 1 \\ 1 \\ 0 \end{pmatrix}, \begin{pmatrix} 1 \\ 1 \\ 1 \\ 1 \\ 1 \end{pmatrix} \right\},$$

where the solutions have been ordered according to increasing objective function values. Now, take

$$d = (1, 1, 1, 2, 1)^\top, \quad f = -0.5, b_u = 0, b_0 = 5.$$

Then, $x^i = (1, 1, 1, 0, 0)^\top$ gives the best possible function value 1 for $\langle d, x \rangle + f \langle a, x \rangle$, $x \in M^n$. But $\overline{x} = (1, 1, 0, 1, 0)^\top \in \Psi_K(5)$ has $\langle d, x \rangle + f \langle a, x \rangle = 1.5 > 1$. The reason here is that for $5 \in (\sum_{i=1}^{3} a_i, \sum_{i=1}^{4} a_i)$ a new solution $\overline{x}$ becomes feasible for (8.15).   □

The refined dropping rule for Algorithm 1 is the following: The idea for this rule and the following Theorem goes back to [84].
Algorithm 3: This is Algorithm 1 with Rule B replaced by

B': Let $\widehat{x}, \overline{x} \in M^k$ for some k. Then, if $\langle c, \widehat{x} \rangle > \langle c, \overline{x} \rangle$ and $\langle a, \widehat{x} \rangle \leq \langle a, \overline{x} \rangle$, exclude $\overline{x}$ from M^k. If this is not the case, but $\langle c, \widehat{x} \rangle = \langle c, \overline{x} \rangle$, $\langle a, \widehat{x} \rangle \leq \langle a, \overline{x} \rangle$ and $\langle d, \widehat{x} \rangle \geq \langle d, \overline{x} \rangle$ exclude $\overline{x}$ from M^k.

It is obvious that Algorithm 3 again computes an element $x \in \Psi_K(b)$ for all $b \in [b_u, b_o]$ and it is also a pseudopolynomial algorithm of the time complexity $O(nb_{max})$.

THEOREM 8.9 ([75]) *Let $M^n = \{x^1, \ldots, x^p\}$ be the set of all solutions for problem (8.14) obtained by application of Algorithm 3. Assume that $b_u = 0, b_o = \sum\limits_{i=1}^{n} a_i$, $f \leq 0$ and let $x^{i_0} \in M^n$ satisfy*

$$\langle d, x^{i_0} \rangle + f\langle a, x^{i_0} \rangle = \max_{1 \leq i \leq p} \{\langle d, x^i \rangle + f\langle a, x^i \rangle\}.$$

Then, $(x^{i_0}, \langle a, x^{i_0} \rangle)$ is an optimistic optimal solution.

Using Algorithm 3 for the above Example, the solution $(1, 1, 0, 1, 0)^\top$ is no longer ruled out.

Consider now the case of a pessimistic solution and use another modification of Rule B in the enumeration algorithm:

B": Let $\widehat{x}, \overline{x} \in M^k$ for some k. Then, if $\langle c, \widehat{x} \rangle > \langle c, \overline{x} \rangle$ and $\langle a, \widehat{x} \rangle \leq \langle a, \overline{x} \rangle$, exclude $\overline{x}$ from M^k. If this is not the case, but $\langle c, \widehat{x} \rangle = \langle c, \overline{x} \rangle$, $\langle a, \widehat{x} \rangle \leq \langle a, \overline{x} \rangle$ and $\langle d, \widehat{x} \rangle \leq \langle d, \overline{x} \rangle$ exclude $\overline{x}$ from M^k.

The algorithm using Rules A and B" within the enumeration idea is called **Algorithm 4**. Rule B" again guarantees that if Algorithm 4 computes two different points $\widehat{x}, \overline{x}$ with $\langle c, \widehat{x} \rangle = \langle c, \overline{x} \rangle$ then $\langle d, \widehat{x} \rangle \neq \langle d, \widehat{x} \rangle$. Both rules B' and B" guarantee that, for each $b \in [b_u, b_o]$ there is only one point $x \in \Psi_K(b)$ in the set M^n. Since all calculations needed for evaluating the rules can be done in polynomial time $O(n)$, the overall algorithms Algorithm 3 and Algorithm 4 have the same time complexity as Algorithm 1.

THEOREM 8.10 ([75]) *For each $b \in [b_u, b_o]$ Algorithm 4 computes exactly one element $x^* \in$ Argmin $\{\langle d, x \rangle : x \in \Psi_K(b)\}$ with*

$$\langle a, x^* \rangle \leq \langle a, x \rangle \ \forall x \in \text{Argmin}_x \ \{\langle d, x \rangle : x \in \Psi_K(b)\}. \qquad (8.17)$$

COROLLARY 8.6 *Let $\{x^1, \ldots, x^p\}$ be the set of all solutions computed by Algorithm 4. Then,*

$$(x^j, \langle a, x^j \rangle) \in \text{Argmax}_i \ \{\langle d, x^i \rangle + f\langle a, x^i \rangle : 1 \leq i \leq p\}$$

is a pessimistic optimal solution for $f \leq 0$.

If $f > 0$ holds, the bilevel programming problem (8.15) has no solution unless $(x(b_o), b_o)$ is optimal but a finite supremum. Then, the only thing we can do is to compute a feasible solution which approximates the supremum as exactly as necessary. In the following theorem we will demonstrate how this can be done using as example Algorithm 3.

THEOREM 8.11 ([75]) *Let $\varepsilon > 0$ be arbitrarily chosen. Let $f > 0$ and let $M^n = \{x^1, \ldots, x^p\}$ be the points computed with Algorithm 3. Let $\langle a, x^j \rangle < \langle a, x^{j+1} \rangle$ for $j = 1, \ldots, p-1$ and $(x(b_o), b_o)$ be not an optimistic optimal solution. Then (x^{i_o}, b^*) with $x^{i_o} \in M^n$,*

$$\langle d, x^{i_o} \rangle + f\langle a, x^{i_o+1} \rangle = \max\{\langle d, x^j \rangle + f\langle a, x^{j+1} \rangle : 1 \leq j \leq p-1\}$$

and $b^* \in [\langle a, x^{i_o+1} \rangle - \varepsilon/f, \langle a, x^{i_o+1} \rangle)$ satisfies

$$\langle d, x^{i_o} \rangle + fb^* \geq \sup_{b_u \leq b \leq b_o} \max_{x \in \Psi_K(b)} \{\langle d, x \rangle + fb\} - \varepsilon.$$

The proof of this theorem follows since Algorithm 3 computes

$$x(b) \in \operatorname*{Argmax}_{x} \{\langle d, x \rangle : x \in \Psi_K(b)\}$$

for all b and since the supremal function value in (8.15) is achieved for some right endpoint of the regions of stability for one of the points computed by Algorithm 3.

8.4.3   APPROXIMATE SOLUTION

One of the main problems for knapsack problems is the question of how to construct approximation algorithms which compute feasible solutions having objective function values arbitrarily close to the optimal ones in time polynomial in the size of the problem and, if possible also, in the reciprocal of the error [146, 168]. The same seems to be difficult in bilevel programming due to the different objective functions in both levels. In the following we show that it is possible to construct a polynomial-time approximation algorithm in a somewhat weaker sense: The computed solution is in general not feasible for the bilevel programming problem but only ε-feasible for an arbitrarily small $\varepsilon > 0$. And we also show that, with decreasing infeasibility, at least a lower optimal solution can be approximated.

Let us consider the bilevel problem in a more general form than (8.14), (8.15):

$$v^k := \text{``}\max_b\text{''}\{\langle d, x \rangle + fb : x \in \Psi_K^k(b), b_u \leq b \leq b_o\}, \tag{8.18}$$

where

$$\Psi_K^k(b) = \operatorname*{Argmax}_x \left\{ \langle c^k, x \rangle : \langle a, x \rangle \leq b, x \in \{0,1\}^n \right\} \qquad (8.19)$$

with $\lim_{k \to \infty} c_i^k = c_i$ for all $i = 1, \dots, n$. It has been shown in [183] that convergence to an optimal solution of the unperturbed problem can be achieved if ε-optimal solutions in the follower's problem are used provided that the constraint set mapping of the follower's problem is a lower as well as an upper semicontinuous point-to-set mapping. But, since the constraint set mapping of discrete optimization problems is in general not lower semicontinuous (and the solution set mapping is consequently in general neither lower nor upper semicontinuous) this result cannot be applied directly to our situation. The following theorem shows that a similar result can be obtained for problem (8.15) by exploiting the finiteness of the feasible set of the follower but only with respect to the weaker optimality definition used in [131].

THEOREM 8.12 ([75]) *Consider problem (8.18) and let $f \leq 0$, $b_u = 0, b_o = \sum_{i=1}^{n} a_i$. Let $\{(x^k, b^k)\}_{k=1}^{\infty}$ be a sequence of points satisfying*

$$\langle d, x^k \rangle + f b^k \geq \max_{b_u \leq b \leq b_o} \min_{x \in \Psi_K^k(b)} \{\langle d, x \rangle + f b\}$$

and $x^k \in \Psi_K^k(b^k)$ for all k. Then, any accumulation point $(\overline{x}, \overline{b})$ of $\{(x^k, b^k)\}_{k=1}^{\infty}$ satisfies

$$\langle d, \overline{x} \rangle + f \overline{b} \geq \max_{b_u \leq b \leq b_o} \min_{x \in \Psi_K(b)} \{\langle d, x \rangle + f b\} \qquad (8.20)$$

and $\overline{x} \in \Psi_K(\overline{b})$.

REMARK 8.4 *The following remarks seem to be in order:*

1. *To get a polynomial time approximation algorithm for the bilevel programming problem we use the idea in Theorem 8.6: Let $t^k = \frac{\varepsilon^k c_{\max}}{n}$. Then, $c_i^k = t^k \left\lfloor \frac{c_i}{t^k} \right\rfloor$ converges to c_i for $\varepsilon^k \to 0$. Clearly, the common factor t^k in the objective of the lower level problem can be cancelled. The resulting algorithm is polynomial in the problem's size and in the reciprocal of ε^k.*

2. *To apply Theorem 8.12 the perturbed problems can be solved with either of the Algorithms 1, 3 or 4.*

The following example shows that the result of Theorem 8.12 is not true in general for neither the optimistic nor the pessimistic solution:

Example: Let $b_u = 1$, $b_o = 1.5$, $a_1 = a_2 = 1$, $c_1 = c_2 = 1$, $d_1 = 1$, $d_2 = 2$, $f = -1$. Then,

$$\Psi_K(b) = \{(0,1)^\top, (1,0)^\top\} \; \forall \; b \in [b_u, b_o].$$

The unique optimistic solution is $x^* = (0,1)$, $v^* = 1$ is the optimistic optimal function value. Now consider the following perturbation of the lower level objective function coefficients: $c_1^k = 1+2/k$, $c_2^k = 1+1/k$, $k = 1, 2, \ldots$. Then, $\Psi_K^k(b) = \{(1,0)^\top\}$ for all k and all $b \in [b_u, b_o]$. Hence, the sequence of optimal solutions converges to $(1,0)^\top$ having the function value $\overline{v} = 0$. And this is not optimistic optimal.

An analogous example can easily be found in the pessimistic solution's case. We should mention that this effect is not possible if ε-optimal solutions in the lower level problem are used. $\quad\Box$

THEOREM 8.13 ([75]) *Let $f > 0$ and let $(x(b_o), b_o)$ be not a pessimistic optimal solution. Let $\varepsilon > 0$ be given. Then, any accumulation point $(\overline{x}, \overline{b})$ of a sequence $\{(x^k, b^k)\}_{k=1}^{\infty}$ with*

$$\langle d, x^k\rangle + f b^k \geq \sup_{b_u \leq b \leq b_o} \min_{x \in \Psi_K^k(b)} \{\langle d, x\rangle + fb\} - \varepsilon/f \qquad (8.21)$$

and $x^k \in \Psi_K^k(b^k)$ satisfies

$$\langle d, \overline{x}\rangle + f\overline{b} \geq \sup_{b_u \leq b \leq b_o} \min_{x \in \Psi_K(b)} \{\langle d, x\rangle + fb\} - \varepsilon/f.$$

REMARK 8.5 *In distinction to Theorem 8.12 feasibility of the limit point $(\overline{x}, \overline{b})$ can in general not be shown. Under the assumptions of the last theorem we either have $\overline{x} \in \Psi_K(\overline{b})$ or $\overline{x} \in \Psi_K(\overline{b} - \delta)$ for sufficiently small $\delta > 0$.*

8.5 PROOFS

PROOF OF THEOREM 8.1:

1. By linear programming duality we have $u \geq 0$. Hence, $\langle r, x\rangle \leq s$ describes a Chvátal-Gomory cut by definition. Now, for each $x \in P$,

$$\langle r, x\rangle \leq \langle u, Az\rangle \leq \langle u, a\rangle$$

by feasibility of x and $u \geq 0$. The first part follows now by integrality of the left-hand side $\langle r, x\rangle$ of this inequality.

2. We have to show that the problem

$$\langle r, x \rangle \to \max$$
$$Ax \leq a,$$
$$x \geq 0,$$

has x^* as an optimal solution. By linear programming duality and integrality of y',

$$y_i' = \langle u, A^i \rangle = r_i$$

for each basic index i and

$$r_i = \lfloor \langle u, A^i \rangle \rfloor \leq \langle u, A^i \rangle$$

for all other indices. Hence, u defines a feasible dual variable satisfying the complementary slackness conditions. This shows the second part of the proof.

3. By $r \in \mathcal{R}(x^*)$ we have $\langle r, x^* \rangle = \langle u, Ax^* \rangle = \langle u, a \rangle$. Thus, $\langle r, x^* \rangle \leq \lfloor \langle u, a \rangle \rfloor$ if and only if $\langle u, a \rangle$ is integral.

$\square$

PROOF OF THEOREM 8.2: The region of stability $\mathcal{R}(x^*)$ is a convex cone with apex at zero and nonempty interior int $\mathcal{R}(x^*)$ [220]. Moreover, since all coefficients of the matrix A are integers, there exists an integer-valued vector $y' \in$ int $\mathcal{R}(x^*)$. Now, by linear programming duality, $\langle y', x^* \rangle = \langle u, a \rangle$ if u is an optimal solution of the dual problem to (8.3) for $y = y'$. Hence, if $\langle y', x^* \rangle$ is not an integer, then the assertion of the theorem follows. Let $\langle y', x^* \rangle$ be an integer but $x_i^* \neq \lfloor x_i^* \rfloor$ for some i. Then, by the properties of y' and $\mathcal{R}(x^*)$ there exists an integer $t > 0$ such that $t \cdot y' + e_i \in$ int $\mathcal{R}(x^*)$, where e_i denotes the i-th unit vector. But then, $\langle t \cdot y' + e_i, x^* \rangle = t \cdot \langle y', x^* \rangle + x_i^*$ is not an integer. This shows that, for $y := t \cdot y' + e_i$, the corresponding Chvátal-Gomory cut is not satisfied by x^*.

$\square$

PROOF OF THEOREM 8.3: By Theorems 8.1 and 8.2 there exist u, r, y' such that the objective function value of problem (8.5) is positive if and only if x^* is not integer-valued. Let u, r, y' be given according to Theorem 8.1 such that the objective function value of problem (8.5) is positive. Obviously, u is rational since A has integer coefficients. Hence, there exist positive integers α, β with greatest common divisor 1 such that $\langle u, a \rangle - \lfloor \langle u, a \rangle \rfloor = \frac{\alpha}{\beta}$. Now, consider the vector $u' := (\beta + 1)u$. Then,

$$\langle u', a \rangle - \lfloor \langle u', a \rangle \rfloor = (\beta + 1)\langle u, a \rangle - \lfloor (\beta + 1)\langle u, a \rangle \rfloor$$

$$= (\beta+1)\lfloor\langle u,a\rangle\rfloor + \alpha + \frac{\alpha}{\beta} - \lfloor(\beta+1)\lfloor\langle u,a\rangle\rfloor + \alpha + \frac{\alpha}{\beta}\rfloor$$

$$= \frac{\alpha}{\beta} - \lfloor\frac{\alpha}{\beta}\rfloor \leq \frac{\alpha}{\beta}.$$

Hence, by $\beta > 1$, $r_i = \lfloor\langle u,A^i\rangle\rfloor$, $i = 1,\ldots,l$, $r \neq 0$, we have

$$\frac{(\beta+1)\langle u,a\rangle - \lfloor(\beta+1)\langle u,a\rangle\rfloor}{\|\lfloor(\beta+1)u^\top A\rfloor\|} < \frac{\langle u,a\rangle - \lfloor\langle u,a\rangle\rfloor}{\|\lfloor u^\top A\rfloor\|}$$

where $\lfloor a\rfloor$ denotes the vector with the components $\lfloor a_i\rfloor$ for all i. This implies that the optimal function value of problem (8.5) is bounded. $\square$

PROOF OF THEOREM 8.4: If the algorithm stops after a finite number of iterations, the point x^* is integral.

First, if $y^* \in \text{int } \mathcal{R}(x^*)$, then by conv $P \subseteq \{x \geq 0 : Ax \leq a\}$ we have $\langle y^*, x^*\rangle > \langle y, x\rangle$ for all $x \in P \setminus \{x^*\}$ and all $\|y - y^*\| < \varepsilon$ for some $\varepsilon > 0$. This implies that

$$\{(x,y) : x \in \Psi_D(y), A^3y = b, y \geq 0\} \cap (\{(x^*,y^*)\} + \delta\mathcal{B}^n \times \varepsilon\mathcal{B}^m)$$
$$= \{(x^*,y^*) + \{0\} \times \varepsilon\mathcal{B}^m : A^3y = b, y \geq 0\}, \tag{8.22}$$

with $\delta > 0$. Hence, since each point in the right-hand side set in this inclusion is feasible for the problem in Step 1 of the algorithm and since the algorithms for solving linear bilevel programming problems compute local optimistic optimal solutions , the point (x^*,y^*) is also a local optimistic optimal solution of the discrete problem.

Second, let $y^* \notin \text{int } \mathcal{R}(x^*)$. Then, it is possible that the left-hand side in inclusion (8.22) contains points $(x^1,y^1),(x^2,y^2)$ with different but integral points x^1, y^1. But nevertheless, the point (x^*,y^*) is a local optimal optimistic solution of the auxiliary linear bilevel programming problem in Step 1 of the algorithm. This implies that $F(x^*,y^*) \leq F(x^k,y^*)$ for all integer points $x^k, k = 1,\ldots,s$ in $\Psi_D^a(y^*)$. By (discrete) parametric optimization, there is $\varepsilon > 0$ such that $\Psi_D(y) \subseteq \{x^1,\ldots,x^s\}$ for all $\|y - y^*\| < \varepsilon$. This implies that $F(x,y) \geq F(x^*,y^*)$ for all $x \in \Psi_D(y)$ and all $\|y - y^*\| < \varepsilon$. Hence, the theorem follows. $\square$

PROOF OF THEOREM 8.8: By Corollary 8.3, $x^i \in \Psi_K(b)$ for all $b \in [\langle a, x^i\rangle, \langle a, x^{i+1}\rangle)$. Hence,

$$\langle d, x^i\rangle \geq \min\{\langle d, x\rangle : x \in \Psi_K(b)\} \ \forall b \in [\langle a, x^i\rangle, \langle a, x^{i+1}\rangle).$$

This implies

$$\max\left\{\max_{1\leq i\leq p-1} \max_{b\in[\langle a,x^i\rangle,\langle a,x^{i+1}\rangle)} \langle d, x^i\rangle + fb, \langle d, x^p\rangle + f\langle a, x^p\rangle\right\}$$

$$\geq \max_{b \in [0, b_o]} \min_{x} \{\langle d, x\rangle + fb : x \in \Psi_K(b)\}.$$

Since by $f \leq 0$, the maximum on the left-hand side of this inequality is equal to $\max_{i}\{\langle d, x^i\rangle + f\langle a, x^i\rangle : 1 \leq i \leq p\}$, the theorem follows.   □

PROOF OF THEOREM 8.9: Due to Theorem 8.7 an optimal solution in the optimistic approach exists. Let this solution be $(x^*, \langle a, x^*\rangle)$. Consequently, $x^* \in \Psi_K(\langle a, x^*\rangle)$.

If $(x^{i_0}, \langle a, x^{i_0}\rangle)$ is not optimistic optimal, we have

$$\langle d, x^*\rangle + f\langle a, x^*\rangle > \langle d, x^{i_0}\rangle + f\langle a, x^{i_0}\rangle$$

and x^* has been ruled out in some iteration k. Hence, there exists a point $\widehat{x} \in M^k$ such that

$$\sum_{i=1}^{k} a_i \widehat{x}_i \leq \sum_{i=1}^{k} a_i x_i^* \text{ but } \sum_{i=1}^{k} c_i \widehat{x}_i > \sum_{i=1}^{k} c_i x_i^*$$

or

$$\sum_{i=1}^{k} a_i \widehat{x}_i \leq \sum_{i=1}^{k} a_i x_i^*, \ \sum_{i=1}^{k} c_i \widehat{x}_i = \sum_{i=1}^{k} c_i x_i^* \text{ but } \sum_{i=1}^{k} d_i \widehat{x}_i \geq \sum_{i=1}^{k} d_i x_i^*.$$

Consider the point $\overline{x}$ given by

$$\overline{x}_i = \begin{cases} \widehat{x}_i, & 1 \leq i \leq k \\ x_i^*, & k+1 \leq i \leq n \end{cases}$$

Then,

$$\langle a, \overline{x}\rangle = \sum_{i=1}^{k} a_i \widehat{x}_i + \sum_{i=k+1}^{n} a_i x_i^* \leq \langle a, x^*\rangle,$$

i.e. the point $\overline{x}$ is feasible for the lower level problem with $b = \langle a, x^*\rangle$. If the first rule is used to delete x^* we get

$$\langle c, \overline{x}\rangle = \sum_{i=1}^{k} c_i \widehat{x}_i + \sum_{i=k+1}^{n} c_i x_i^* > \langle c, x^*\rangle$$

and $x^* \notin \Psi_K(\langle a, x^*\rangle)$. Since this is not possible, the second part applies. We get $\overline{x} \in \Psi_K(\langle a, x^*\rangle)$ and

$$\langle d, \overline{x}\rangle = \sum_{i=1}^{k} d_i \widehat{x}_i + \sum_{i=k+1}^{n} d_i x_i^* \geq \langle d, x^*\rangle.$$

But $\langle d, \overline{x} \rangle > \langle d, x^* \rangle$ implies

$$\langle d, \overline{x} \rangle + f \langle a, \overline{x} \rangle \geq \langle d, \overline{x} \rangle + f \langle a, x^* \rangle > \langle d, x^* \rangle + f \langle a, x^* \rangle$$

contradicting the optimality of $(x^*, \langle a, x^* \rangle)$. Thus, $(\overline{x}, \langle a, \overline{x} \rangle)$ is also an optimistic optimal solution with $\langle d, \overline{x} \rangle + f \langle a, \overline{x} \rangle = \langle d, x^* \rangle + f \langle a, x^* \rangle$. Therefore, the same considerations of ruling out this point $\overline{x}$ in some iteration k' with $k' > k$ follow. This can only be done for $k' \leq n$. Finally, we obtain an optimistic optimal solution $\overline{x} \in M^n$ with

$$\langle d, \overline{x} \rangle + f \langle a, \overline{x} \rangle = \langle d, x^* \rangle + f \langle a, x^* \rangle > \langle d, x^{i_0} \rangle + f \langle a, x^{i_0} \rangle \geq \langle d, x^j \rangle + f \langle a, x^j \rangle$$

for all $x^j \in M^n$ and this is a contradiction. $\qquad \square$

PROOF OF THEOREM 8.10: Let $b \in [b_u, b_o]$ be fixed and assume that Algorithm 4 does not compute a solution

$$x^* \in \underset{x}{\mathrm{Argmin}} \ \{\langle d, x \rangle : x \in \Psi_K(b)\}$$

satisfying (8.17). Let x^* be the only solution with this property and assume that it has been dropped in iteration k of Algorithm 4. Then there exists a point $\widehat{x} \in \{0, 1\}^n$ such that

$$\sum_{i=1}^{k} a_i \widehat{x}_i \leq \sum_{i=1}^{k} a_i x_i^* \ \text{but} \ \sum_{i=1}^{k} c_i \widehat{x}_i > \sum_{i=1}^{k} c_i x_i^*$$

or

$$\sum_{i=1}^{k} a_i \widehat{x}_i \leq \sum_{i=1}^{k} a_i x_i^*, \sum_{i=1}^{k} c_i \widehat{x}_i = \sum_{i=1}^{k} c_i x_i^* \ \text{but} \ \sum_{i=1}^{k} d_i \widehat{x}_i \leq \sum_{i=1}^{k} d_i x_i^*.$$

Consider the point $\overline{x}$ given by

$$\overline{x}_i = \begin{cases} \widehat{x}_i, & 1 \leq i \leq k \\ x_i^*, & k+1 \leq i \leq n \end{cases}$$

Then, using the same argumentation as in the proof of Theorem 8.9 we can show that $\overline{x} \in \Psi_K(b)$. Hence, the first case is not possible, the second part applies and we get

$$\langle d, \overline{x} \rangle = \sum_{i=1}^{k} d_i \widehat{x}_i + \sum_{i=k+1}^{n} d_i x_i^* \leq \langle d, x^* \rangle.$$

By $x^* \in \underset{x}{\mathrm{Argmin}} \ \{\langle d, x \rangle : x \in \Psi_K(b)\}$ the last inequality is satisfied as an equation. But then, $\langle a, \overline{x} \rangle \leq \langle a, x^* \rangle$ which violates the uniqueness assumption for x^* at the beginning of the proof. $\qquad \square$

PROOF OF COROLLARY 8.6: If $f = 0$, the Algorithm 4 clearly computes a pessimistic optimal solution.

Let $f < 0$ and let $(x^j, \langle a, x^j \rangle)$ not be a pessimistic optimal solution. Then, by Theorem 8.7 there exists a pessimistic optimal solution (x^*, b^*) and we have

$$\langle d, x^* \rangle + f b^* > \langle d, x^j \rangle + f \langle a, x^j \rangle \tag{8.23}$$

due to

$$\langle d, x^j \rangle + f \langle a, x^j \rangle = \min\{\langle d, x \rangle + f \langle a, x^j \rangle : x \in \Psi_K(\langle a, x^j \rangle)\}$$

by Theorem 8.10. Since the existence of two points

$$\widehat{x}, \widetilde{x} \in \operatorname*{Argmin}_{x} \{\langle d, x \rangle : x \in \Psi_K(b)\}$$

with

$$\widehat{x} \in \operatorname*{Argmin}_{x} \{\langle d, x \rangle : x \in \Psi_K(b')\}, \widetilde{x} \notin \operatorname*{Argmin}_{x} \{\langle d, x \rangle : x \in \Psi_K(b')\}$$

is not possible for $b' > b$ we have $b^* = \langle a, x^* \rangle$. Assume that there is $\widehat{x} \in \operatorname*{Argmin}_{x} \{\langle d, x \rangle : x \in \Psi_K(b^*)\}$ with $\langle a, \widehat{x} \rangle < \langle a, x^* \rangle$. Then

$$\langle d, \widehat{x} \rangle + f \langle a, \widehat{x} \rangle > \langle d, \widehat{x} \rangle + f b^* = \langle d, x^* \rangle + f b^*.$$

Since this contradicts the optimality of (x^*, b^*) we have $\langle a, \widehat{x} \rangle \geq \langle a, x^* \rangle$ for all $\widehat{x}$ with $\{x^*, \widehat{x}\} \subseteq \operatorname*{Argmin}_{x} \{\langle d, x \rangle : x \in \Psi_K(b^*)\}$. Hence, by Theorem 8.10, (x^*, b^*) has computed by Algorithm 4. This implies $x^* \in \{x^1, \ldots, x^p\}$ contradicting the assumption (8.23). $\qquad\square$

PROOF OF THEOREM 8.12:

1. By $f \leq 0$ and Theorem 8.7, optimal solutions in the pessimistic sense (x^k, b^k) of problems (8.18) exist. By $a_i < \infty, i = 1, \ldots, n$, the sequence $\{(x^k, b^k)\}_{k=1}^{\infty}$ has accumulation points.

2. Take $\overline{x}^k(\overline{b}) \in \Psi_K^k(\overline{b})$ for some fixed $\overline{b}$ and let $\overline{x}(\overline{b})$ be any accumulation point of this sequence. Then,

$$\langle c^k, \overline{x}^k(\overline{b}) \rangle \geq \langle c^k, x \rangle$$

for all $x \in \{0,1\}^n, \langle a, x \rangle \leq \overline{b}$. Hence, by convergence of $\{c^k\}_{k=1}^{\infty}$, we get $\overline{x}(\overline{b}) \in \Psi_K(\overline{b})$ and, thus,

$$\langle d, \overline{x}(\overline{b}) \rangle + f \overline{b} \geq \min\{\langle d, x \rangle + f \overline{b} : x \in \Psi_K(\overline{b})\} =: v(\overline{b}). \tag{8.24}$$

3. Now, take any accumulation point $\hat{x}$ of $\{x^k\}_{k=1}^{\infty}$ and assume without loss of generality that $\{x^k\}_{k=1}^{\infty}$ itself converges to $\hat{x}$. Then, by finiteness of $\{x \in \{0,1\}^n : \langle a, x \rangle \leq b_o\}$ there exists an index k_0 such that $x^k \equiv \hat{x}$ for all $k \geq k_0$. According to Theorem 8.7 an optimistic optimal solution (x^k, b^k) of problem (8.18) satisfies $b^k = \langle a, x^k \rangle$ for all k. Hence, the accumulation point $(\hat{x}, \hat{b})$ with $\hat{b} = \langle a, \hat{x} \rangle$ of the sequence $\{x^k, b^k\}_{k=1}^{\infty}$ satisfies

$$\langle d, \hat{x} \rangle + f\hat{b} \geq \min\{\langle d, x \rangle + f\hat{b} : x \in \Psi_K(\hat{b})\} = v(\hat{b}). \qquad (8.25)$$

4. Assume that $(\hat{x}, \hat{b})$ does not satisfy the desired inequality (8.20). Then, by Theorem 8.7 there is $(\tilde{x}, \tilde{b})$ with

$$\langle d, \hat{x} \rangle + f\hat{b} < \langle d, \tilde{x} \rangle + f\tilde{b} = \max_{b_u \leq b \leq b_o} \min_{x \in \Psi_K(b)} \{\langle d, x \rangle + fb\} = v(\tilde{b}).$$

But then, for $\tilde{x}^k \in \Psi_K^k(\tilde{b})$ with

$$\tilde{x}^k \in \operatorname*{Argmin}_{x} \{\langle d, x \rangle : x \in \Psi_K^k(\tilde{b})\}$$

we derive

$$\langle d, x^k \rangle + fb^k \geq \langle d, \tilde{x}^k \rangle + f\tilde{b}$$

for all k. But since the left–hand side of this inequality is equal to $\langle d, \hat{x} \rangle + f\hat{b}$ for sufficiently large k and the right–hand side cannot get smaller than $v(\tilde{b}) = \langle d, \tilde{x} \rangle + f\tilde{b}$ in the limit by part 2 of this proof we get

$$v(\tilde{b}) \leq \langle d, \hat{x} \rangle + f\hat{b} < v(\tilde{b})$$

which gives the desired contradiction.

$$\square$$

PROOF OF THEOREM 8.13:

1. Similarly to the idea of the proof of Theorem 8.11 let (x^*, b^*) be a given point such that $\mathcal{R}(x^*) = [b_u^*, b_o^*)$ with $b^* = b_o^*$ and

$$\langle d, x^* \rangle + fb^* = \sup_{b_u \leq b \leq b_o} \min_{x \in \Psi_K(b)} \{\langle d, x \rangle + fb\}.$$

Then, there exist $\hat{b} \in [b_u^*, b_o^*)$ such that

$$\langle d, x^* \rangle + f\hat{b} \geq \langle d, x^* \rangle + fb^* - \varepsilon/f.$$

Clearly,

$$\langle d, x^* \rangle + f\widehat{b} = \min_{x \in \Psi_K(\widehat{b})} \{ \langle d, x \rangle + f\widehat{b} \}.$$

2. Using the same ideas as in the proof of Theorem 8.12 it is easy to see that any accumulation point y^* of a sequence $\{y^k\}_{k=1}^\infty$ with

$$y^k \in \operatorname*{Argmin}_{x \in \Psi_K^k(\overline{b})} \langle d, x \rangle$$

for all k satisfies $\langle d, y^* \rangle \geq \min_{x \in \Psi_K(\overline{b})} \langle d, x \rangle$.

3. Now, let $\{(x^k, b^k)\}_{k=1}^\infty$ be a sequence of points satisfying the conditions of the theorem and let $(\overline{x}, \overline{b})$ be an accumulation point of this sequence. Without loss of generality, let $(\overline{x}, \overline{b})$ be the limit point of this sequence. Then, similarly to the proof of Theorem 8.12, $x^k = \overline{x}$ for sufficiently large k and $\lim_{k \to \infty} b^k = \overline{b}$. Fix any sufficiently large $k = k^*$ such that $x^k = \overline{x}$ for all $k \geq k^*$. For each such k^* we get by (8.21)

$$\langle d, x^{k^*} \rangle + f b^{k^*} \geq \sup_{b_u \leq b \leq b_o} \min_{x \in \Psi_K^{k^*}(b)} \{ \langle d, x \rangle + fb \} - \varepsilon/f$$

$$\geq \min_{x \in \Psi_K^{k^*}(\widehat{b})} \{ \langle d, x \rangle + f\widehat{b} \} - \varepsilon/f$$

for each fixed $\widehat{b} \in [b_u^*, b_o^*)$. Passing to the limit for k^* to infinity in this inequality, the left-hand side converges and the value on the right-hand side of this inequality, for fixed $b = \widehat{b}$, cannot increase. Hence

$$\langle d, \overline{x} \rangle + f\overline{b} \geq \min_{x \in \Psi_K(\widehat{b})} \{ \langle d, x \rangle + f\widehat{b} \} - \varepsilon/f = \langle d, x^* \rangle + f\widehat{b} - \varepsilon/f$$

by part 1. Since this is true for all $\widehat{b} \in [b_u^*, b_o^*)$ sufficiently close to b_o^* it is also satisfied for $b = b_o^*$ and we get the assertion of the theorem.

$\square$

References

[1] S. Addoune. *Optimisation à deux niveaux : Conditions d'optimalité, approximation et stabilité.* PhD thesis, Université de Bourgogne, Département de Mathématique, 1994.

[2] E. Aiyoshi and K. Shimizu. Hierarchical decentralized systems and its new solution by a barrier method. *IEEE Transactions on Systems, Man, and Cybernetics,* 11:444–449, 1981.

[3] E. Aiyoshi and K. Shimizu. A solution method for the static constrained Stackelberg problem via penalty method. *IEEE Transactions on Automatic Control,* 29:1111–1114, 1984.

[4] G. Anandalingam. A mathematical programming model of decentralized multi-level systems. *Journal of the Operational Research Society,* 39:1021–1033, 1988.

[5] G. Anandalingam and V. Apprey. Multi-level programming and conflict resolution. *European Journal of Operational Research,* 51:233–247, 1991.

[6] G. Anandalingam and T.L. Friesz (eds.). Hierarchical optimization. *Annals of Operations Research,* 34, 1992.

[7] G. Anandalingam and D. White. A solution method for the linear static Stackelberg problem using penalty functions. *IEEE Transactions on Automatic Control,* 35:1170–1173, 1990.

[8] K.J. Arrow, L. Hurwicz, and H. Uzawa. Constraint qualifications in maximization problems. *Naval Research Logistics Quarterly,* 8:175–191, 1961.

[9] N. Ascheuer, L.F. Escudero, M. Grötschel, and M. Stoer. A cutting plane approach to the sequential ordering problem (with applications to job scheduling in manufacturing). *SIAM Journal on Optimization,* 3:25–42, 1993.

[10] J.-P. Aubin and I. Ekeland. *Applied Nonlinear Analysis.* Wiley – Interscience, J. Wiley & Sons, New York, 1984.

[11] C. Audet, P. Hansen, B. Jaumard, and G. Savard. Links between linear bilevel and mixed 0-1 programming problems. *Journal of Optimization Theory and Applications,* 93:273–300, 1997.

283

[12] A. Auslender. Regularity theorems in sensitivity theory with nonsmooth data. In J. Guddat, H.Th. Jongen, F. Nožička, and B. Kummer, editors, *Parametric Optimization and Related Topics*, pages 9–15. Akademie Verlag, Berlin, 1987.

[13] A. Auslender and R. Cominetti. First and second order sensitivity analysis of nonlinear programs under directional constraint qualification condition. *Optimization*, 21:351–363, 1990.

[14] E. Balas, S. Ceria, and G. Cornuejols. A lift-and-project cutting plane algorithm for mixed 0–1 programs. *Mathematical Programming*, 58:295–324, 1993.

[15] B. Bank. Qualitative Stabilitätsuntersuchungen rein- und gemischt-ganzzahliger linearer parametrischer Optimierungsproblems. *Seminarberichte Humboldt-Universität zu Berlin, Sektion Mathematik*, 6, 1978.

[16] B. Bank. *Stability analysis in pure and mixed-integer linear programming*. Number 23 in Lecture Notes in Control and Information Sciences. Springer-Verlag, Berlin, 1980.

[17] B. Bank, J. Guddat, D. Klatte, B. Kummer, and K. Tammer. *Non-Linear Parametric Optimization*. Akademie-Verlag, Berlin, 1982.

[18] B. Bank and R. Mandel. *Parametric Integer Optimization*. Number 39 in Mathematical Research. Akademie Verlag, Berlin, 1988.

[19] J.F. Bard. A grid search algorithm for the linear bilevel programming problem. In *Proceedings of the 14th Annual Meeting of the American Institute for Decision Science*, pages 256–258, 1982.

[20] J.F. Bard. An algorithm for the general bilevel programming problem. *Mathematics of Operations Research*, 8:260–272, 1983.

[21] J.F. Bard. Coordination of a multidivisional organization through two levels of management. *OMEGA*, 11:457–468, 1983.

[22] J.F. Bard. An investigation of the linear three level programming problem. *IEEE Transactions on Systems, Man, and Cybernetics*, 14:711–717, 1984.

[23] J.F. Bard. Optimality conditions for the bilevel programming problem. *Naval Research Logistics Quarterly*, 31:13–26, 1984.

[24] J.F. Bard. Convex two-level optimization. *Mathematical Programming*, 40:15–27, 1988.

[25] J.F. Bard. Some properties of the bilevel programming problem. *Journal of Optimization Theory and Applications*, 68:371–378, 1991.

[26] J.F. Bard. *Practical Bilevel Optimization: Algorithms and Applications*. Kluwer Academic Publishers, Dordrecht, 1998.

[27] J.F. Bard and J. Falk. An explicit solution to the multi-level programming problem. *Computers and Operations Research*, 9:77–100, 1982.

[28] J.F. Bard and J. Moore. An algorithm for the discrete bilevel programming problem. *Naval Research Logistics*, 39:419–435, 1992.

[29] J.F. Bard, J. Plummer, and J.C. Sourie. Determining tax credits for converting nonfood crops to biofuels: an application of bilevel programming. In A. Migdalas, P.M. Pardalos, and P. Värbrand, editors, *Multilevel Optimization: Algorithms and Applications*, pages 23–50. Kluwer Academic Publishers, Dordrecht, 1998.

[30] R. Baumgart and K. Beer. A practicable way for computing the directional derivative of the optimal value function in convex programming. *Optimization*, 25:379–389, 1992.

[31] M.S. Bazaraa, H.D. Sherali, and C.M. Shetty. *Nonlinear Programming: Theory and Algorithms*. J. Wiley & Sons, New York, 1992.

[32] E.M. Bednarczuk and W. Song. Contingent epiderivative and its application to set-valued optimization. *Control and Cybernetics*, 27:375–386, 1998.

[33] K. Beer. *Lösung großer linearer Optimierungsaufgaben*. Deutscher Verlag der Wissenschaften, Berlin, 1977.

[34] O. Ben-Ayed. *Bilevel linear programming: analysis and application to the network design problem*. PhD thesis, University of Illinois at Urbana-Champaign, 1988.

[35] O. Ben-Ayed. A bilevel linear programming model applied to Tunisian interegional High way network design problem. *Revue Tunesienne d'Economie et de Gestion*, V:234–277, 1990.

[36] O. Ben-Ayed and C. Blair. Computational difficulties of bilevel linear programming. *Operations Research*, 38:556–560, 1990.

[37] O. Ben-Ayed, C. Blair, D. Boyce, and L. LeBlanc. Construction of a real-world bilevel linear programming model of the highway design problem. *Annals of Operations Research*, 34:219–254, 1992.

[38] O. Ben-Ayed, D. Boyce, and C. Blair. A general bilevel linear programming formulation of the network design problem. *Transportation Research*, 22 B:311–318, 1988.

[39] H. Benson. On the structure and properties of a linear multilevel programming problem. *Journal of Optimization Theory and Applications*, 60:353–373, 1989.

[40] D.P. Bertsekas. *Constrained Optimization and Lagrange Multipliers*. Academic Press, New York, 1982.

[41] W. Bialas and M. Karwan. Multilevel linear programming. Technical Report 78-1, Operations Research Program, State University of New York at Buffalo, 1978.

[42] W. Bialas and M. Karwan. On two-level optimization. *IEEE Transactions on Automatic Control*, 27:211–214, 1982.

[43] W. Bialas and M. Karwan. Two-level linear programming. *Management Science*, 30:1004–1020, 1984.

[44] J. Bisschop, W. Candler, J. Duloy, and G. O'Mara. The indus basin model: a special application of two-level linear programming. *Mathematical Programming Study*, 20:30–38, 1982.

[45] C.E. Blair. Sensitivity analysis for knapsack problems: A negative result. *Discrete Applied Mathematics*, 81:133–139, 1998.

[46] J.F. Bonnans and A. Shapiro. *Perturbation Analysis of Optimization Problems.* Springer-Verlag, New York et al., 2000.

[47] J. Bracken and J. McGill. Mathematical programs with optimization problems in the constraints. *Operations Research*, 21:37–44, 1973.

[48] J. Bracken and J. McGill. Defense applications of mathematical programs with optimization problems in the constraints. *Operations Research*, 22:1086–1096, 1974.

[49] J. Bracken and J.T. McGill. Production and marketing decisions with multiple objectives in a competitive environment. *Journal of Optimization Theory and Applications*, 24:449–458, 1978.

[50] R.E. Burkard and U. Pferschy. The inverse-parametric knapsack problem. *European Journal of Operational Research*, 83(2):376–393, 1995.

[51] J.V. Burke. An exact penalization viewpoint of constrained optimization. *SIAM Journal on Control and Optimization*, 29:968–998, 1991.

[52] H.I. Calvete and C. Gale. On the quasiconcave bilevel programming problem. *Journal of Optimization Theory and Applications*, 98:613–622, 1998.

[53] H.I. Calvete and C. Gale. The bilevel linear/linear fractional programming problem. *European Journal of Operational Research*, 114:188–197, 1999.

[54] W. Candler and R. Norton. Multilevel programming and development policy. Technical Report 258, World Bank Staff, Washington D.C., 1977.

[55] W. Candler and R. Townsley. A linear two-level programming problem. *Computers and Operations Research*, 9:59–76, 1982.

[56] R. Cassidy, M. Kirby, and W. Raike. Efficient distribution of resources through three levels of government. *Management Science*, 17:462–473, 1971.

[57] S. Ceria, C. Cordier, H. Marchand, and L.A. Wolsey. Cutting planes for integer programs with general integer variables. *Mathematical Programming*, 81:201–214, 1998.

[58] R.W. Chaney. Piecewise C^k functions in nonsmooth analysis. *Nonlinear Analysis, Methods & Applications*, 15:649–660, 1990.

[59] Y. Chen and M. Florian. The nonlinear bilevel programming problem: a general formulation and optimality conditions. Technical Report CRT-794, Centre de Recherche sur les Transports, 1991.

[60] F.H. Clarke. A new approach to Lagrange multipliers. *Mathematics of Operations Research*, 1:165–174, 1976.

[61] F.H. Clarke. *Optimization and Nonsmooth Analysis*. John Wiley & Sons, New York, 1983.

[62] M.M. Čoban, P.S. Kenderov, and J.P. Revalski. Generic well-posedness of optimization problems in topological spaces. *Mathematika*, 36:301–324, 1989.

[63] W. Cook, R. Kannan, and A. Schrijver. Chvátal closures for mixed integer programming problems. *Mathematical Programming*, 47:155–174, 1990.

[64] B. Cornet. Sensitivity analysis in optimization. Technical Report CORE Discussion Paper No. 8322, Université Catholique de Louvain, Louvain-de-Neuve, Belgium, 1983.

[65] J.W. Danskin. The theory of max–min with applications. *SIAM Journal of Applied Mathematics*, 14:641–664, 1966.

[66] S. Dempe. A simple algorithm for the linear bilevel programming problem. *Optimization*, 18:373–385, 1987.

[67] S. Dempe. A necessary and a sufficient optimality condition for bilevel programming problems. *Optimization*, 25:341–354, 1992.

[68] S. Dempe. Directional differentiability of optimal solutions under Slater's condition. *Mathematical Programming*, 59:49–69, 1993.

[69] S. Dempe. On the leader's dilemma and a new idea for attacking bilevel programming problems. Technical report, Technische Universität Chemnitz, Fachbereich Mathematik, 1993. http://www.mathe.tu-freiberg.de/~dempe.

[70] S. Dempe. Discrete bilevel optimization problems. Technical report, TU Chemnitz, 1995. http://www.mathe.tu-freiberg.de/~dempe.

[71] S. Dempe. On generalized differentiability of optimal solutions and its application to an algorithm for solving bilevel optimization problems. In D.-Z. Du, L. Qi, and R.S. Womersley, editor, *Recent advances in nonsmooth optimization*, pages 36–56. World Scientific Publishers, Singapore, 1995.

[72] S. Dempe. Applicability of two-level optimization to issues of environmental policy. Technical report, Institut für Wirtschftsinformatik, Universität Leipzig, 1996. http://www.mathe.tu-freiberg.de/~dempe.

[73] S. Dempe. An implicit function approach to bilevel programming problems. In A. Migdalas, P.M. Pardalos, and P. Värbrand, editors, *Multilevel Optimization: Algorithms and Applications*, pages 273–294. Kluwer Academic Publishers, Dordrecht, 1998.

[74] S. Dempe. A bundle algorithm applied to bilevel programming problems with non-unique lower level solutions. *Computational Optimization and Applications*, 15:145–166, 2000.

[75] S. Dempe and K. Richter. Bilevel programming with knapsack constraints. *Central European Journal of Operations Research*, 8:93–107, 2000.

[76] S. Dempe and H. Schmidt. On an algorithm solving two-level programming problems with nonunique lower level solutions. *Computational Optimization and Applications*, 6:227–249, 1996.

[77] S. Dempe and T. Unger. Generalized PC^1 functions and their application to bilevel programming. *Optimization*, 46:311–326, 1999.

[78] S. Dempe and S. Vogel. The subdifferential of the optimal solution in parametric optimization. *Optimization*. to appear.

[79] V.F. Dem'yanov, C. Lemarechal, and J. Zowe. Approximation to a set–valued mapping i. a proposal. *Applied Mathematics and Optimization*, 14:203–214, 1986.

[80] V.F. Demyanov and A.M. Rubinov. On quasidifferentiable mappings. *Mathematische Operationsforschung und Statistik, series optimization*, 14:3–21, 1983.

[81] X. Deng. Complexity issues in bilevel linear programming. In A. Migdalas, P.M. Pardalos, and P. Värbrand, editors, *Multilevel Optimization: Algorithms and Applications*, pages 149–164. Kluwer Academic Publishers, Dordrecht, 1998.

[82] L. DeSilets, B. Golden, Q. Wang, and R. Kumar. Predicting salinity in the Chesapeake Bay using backpropagation. *Computers and Operations Research*, 19:277–285, 1992.

[83] M.E. Dyer and L.G. Proll. An algorithm for determining all extreme points of a convex polytope. *Mathematical Programming*, 12:81–96, 1985.

[84] J. Eckardt. Zwei-Ebenen-Optimierung mit diskreten Aufgaben in der unteren Ebene. Master's thesis, TU Bergakademie Freiberg, Fakultät für Mathematik und Informatik, 1998.

[85] T. Edmunds and J.F. Bard. Algorithms for nonlinear bilevel mathematical programming. *IEEE Transactions on Systems, Man, and Cybernetics*, 21:83–89, 1991.

[86] T. Edmunds and J.F. Bard. An algorithm for the mixed-integer nonlinear bilevel programming problem. *Annals of Operations Research*, 34:149–162, 1992.

[87] T.A. Edmunds and J.F. Bard. A decomposition technique for discrete time optimal control problems with an application to water resources management. *Math. Comput. Modelling*, 13:61–78, 1990.

[88] K.-H. Elster, editor. *Modern Mathematical Methods of Optimization*. Akademie-Verlag, Berlin, 1993.

[89] I.I. Eremin. The penalty method in convex programming. *Soviet Mathematical Doklady*, 8:459–462, 1966.

[90] I.I. Eremin, V.D. Masurov, and N.N. Astav'ev. *Improper Problems of Linear and Convex Programming*. Nauka, Moscow, 1983. (in Russian).

[91] F.I. Ereshko and A.S. Zlobin. Algorithm for centralized resource allocation between active subsystems. *Economika i matematičeskie metody*, (4):703–713, 1977. (in Russian).

[92] R. Ewert. *Wirtschaftsprüfung und asymmetrische Information*. Springer Verlag, Berlin, 1990.

[93] R. Ewert and A. Wagenhofer. *Interne Unternehmensrechnung*. Springer Verlag, Berlin, 1993.

[94] J.F. Falk and J. Liu. On bilevel programming, Part I: General nonlinear cases. *Mathematical Programming*, 70:47–72, 1995.

[95] P. Ferrari. A model of urban transport management. *Transportation Research, Part B*, 33:43–61, 1999.

[96] A.V. Fiacco. *Introduction to Sensitivity and Stability Analysis in Nonlinear Programming*. Academic Press, New York, 1983.

[97] A.V. Fiacco and G.P. McCormick. *Nonlinear Programming: Sequential Unconstrained Minimization Techniques*. John Wiley & Sons, New York, 1968.

[98] A. Fischer. A special Newton-type optimization method. *Optimization*, 24:269–284, 1992.

[99] J. Fortuny-Amat and B. McCarl. A representation and economic interpretation of a two-level programming problem. *Journal of the Operational Research Society*, 32:783–792, 1981.

[100] T. Friesz, R. Tobin, H. Cho, and N. Mehta. Sensivity analysis based heuristic algorithms for mathematical programs with variational inequality constraints. *Mathematical Programming*, 48:265–284, 1990.

[101] M. Fukushima and J.-S. Pang. Convergence of a smoothing continuation method for mathematical programs with complementarity constraints. In *Ill-posed Variational Problems and Regularization Techniques*, number 477 in Lecture Notes in Economics and Mathematical Systems. Springer, Berlin et al., 1999.

[102] J. Fülöp. On the equivalence between a linear bilevel programming problem and linear optimization over the efficient set. Technical Report WP 93-1, Laboratory of Operations Research and Decision Systems, Computer and Automation Institute, Hungarian Academy of Sciences, 1993.

[103] M.R. Garey and D.S. Johnson. "Strong" NP-completeness results: Motivation, examples and implications. *Journal of the ACM*, 25:499–508, 1978.

[104] M.R. Garey and D.S Johnson. *Computers and Intractability: A Guide to the Theory of NP-Completeness*. W.H. Freeman and Co., San Francisco, 1979.

[105] J. Gauvin. A necessary and sufficient regularity condition to have bounded multipliers in nonconvex programming. *Mathematical Programming*, 12:136–139, 1977.

[106] J. Gauvin and F. Dubeau. Differential properties of the marginal function in mathematical programming. *Mathematical Programming Study*, 19:101–119, 1982.

[107] J. Gauvin and R. Janin. Directional behaviour of optimal solutions in nonlinear mathematical programming. *Mathematics of Operations Research*, 14:629–649, 1988.

[108] J. Gauvin and R. Janin. Directional Lipschitzian optimal solutions and directional derivative of the optimal value function in nonlinear mathematical programming. *Analyse non linéaire. Ann. Inst. H. Poincaré Anal. Non Linéaire*, 6:305–324, 1989.

[109] H. Gfrerer. Hölder continuity of solutions of perturbed optimization problems under mangasarian-fromowitz constraint qualification. In J. Guddat et al., editor, *Parametric Optimization and Related Topics*, pages 113–127. Akademie-Verlag, Berlin, 1987.

[110] R. Gibbons. *A Primer in Game Theory*. Harvester Wheatsheaf, New York et al., 1992.

[111] B. Gollan. On the marginal function in nonlinear programming. *Mathematics of Operations Research*, 9:208–221, 1984.

[112] S.J. Grossman and O.D. Hart. An analysis of the principal-agent problem. *Econometrica*, 51:7–45, 1983.

[113] J. Guddat, F. Guerra Vasquez, and H.Th. Jongen. *Parametric Optimization: Singularities, Pathfollowing and Jumps*. John Wiley & Sons, Chichester and B.G. Teubner, Stuttgart, 1990.

[114] W.W. Hager. Lipschitz continuity for constrained processes. *SIAM Journal on Control and Optimization*, 17:321–328, 1979.

[115] S.-P. Han and O.L. Mangasarian. Exact penalty functions in nonlinear programming. *Mathematical Programming*, 17:251–269, 1979.

[116] P. Hansen, B. Jaumard, and G. Savard. New branch-and-bound rules for linear bilevel programming. *SIAM Journal on Scientific and Statistical Computing*, 13:1194–1217, 1992.

[117] P.T. Harker and J.-S. Pang. Existence of optimal solutions to mathematical programs with equilibrium constraints. *Operations Research Letters*, 7:61–64, 1988.

[118] J. Hertz, A. Krogh, and R.G. Palmer. *Introduction to the theory of neural computation*. Addison-Wesley, Redwood City, California, 1991.

[119] G. Hibino, M. Kainuma, and Y. Matsuoka. Two-level mathematical programming for analyzing subsidy options to reduce greenhouse-gas emissions. Technical Report WP-96-129, IIASA, Laxenburg, Austria, 1996.

[120] J.-B. Hiriart-Urruty. Approximate first-order and second-order derivatives of a marginal function in convex optimization. *Journal of Optimization Theory and Applications*, 48:127–140, 1986.

[121] J.-B. Hiriart-Urruty. From convex to nonconvex optimization. Necessary and sufficient conditions for global optimality. In F.H. Clarke et al., editor, *Nonsmooth Optimization and Related Topics*, pages 219–240. Plenum Press, New York, 1989.

[122] J.-P. Hiriart-Urruty. Refinement of necessary optimality conditions in nondifferentiable programming. *Applied Mathematics and Optimization*, 5:63–82, 1979.

[123] J.-P. Hiriart-Urruty and C. Lemarechal. *Convex Analysis and Minimization Algorithms, Vol. 1*. Springer-Verlag, Berlin et. al., 1993.

[124] J.-P. Hiriart-Urruty and C. Lemarechal. *Convex Analysis and Minimization Algorithms, Vol. 2*. Springer-Verlag, Berlin et. al., 1993.

[125] M.W. Hirsch. *Differential Topology*. Springer, Berlin et al., 1994.

[126] A.J. Hoffman. On approximate solutions of systems of linear inequalities. *J. Res. Nat. Bur. Standards*, 49:263–265, 1952.

[127] W.W. Hogan. Directional derivatives of extremal-value functions with applications to the completely convex case. *Operations Research*, 21:188–209, 1973.

[128] L. Hola. Most of the optimization problems have unique solution. *C.R. Acad. Bulgare Sci.*, 42:5–8, 1989.

[129] R. Horst and H. Tuy. *Global Optimization: Deterministic Approaches*. Springer Verlag, Berlin, 1990.

[130] J.P. Ignizio. *Linear Programming in Single – & Multiple – Objective Systems*. Prentice – Hall Internat. Ser. Indust. Syst. Engin., Englewood Cliffs, 1982.

[131] Y. Ishizuka and E. Aiyoshi. Double penalty method for bilevel optimization problems. *Annals of Operations Research*, 34:73–88, 1992.

[132] J. Jahn and R. Rank. Contingent epiderivatives and set-valued optimization. *ZOR – Mathematical Methods of Operations Research*, 46(2), 1997.

[133] R.G. Jeroslow. The polynomial hierarchy and a simple model for competitive analysis. *Mathematical Programming*, 32:146–164, 1985.

[134] F. Jia, F. Yang, and S.-Y. Wang. Sensitivity analysis in bilevel linear programming. *Systems Science and Mathematical Sciences*, 11:359–366, 1998.

[135] H. Jiang and D. Ralph. QPECgen, a MATLAB generator for mathematical programs with quadratic objectives and affine variational inequality constraints. *Computational Optimization and Applications*, 13:25–59, 1999.

[136] H. Jiang and D. Ralph. Smooth SQP methods for mathematical programs with nonlinear complementarity constraints. *SIAM Journal on Optimization*, 10:779–808, 2000.

[137] K. Jittorntrum. Solution point differentiability without strict complementarity in nonlinear programming. *Mathematical Programming Study*, 21:127–138, 1984.

[138] H.Th. Jongen, P. Jonker, and F. Twilt. *Nonlinear Optimization in R^n. I. Morse Theory, Chebyshev Approximation.* P. Lang, Frankfurt/M. et al., 1983.

[139] H.Th. Jongen, P. Jonker, and F. Twilt. *Nonlinear Optimization in R^n. II. Transversality, Flows, Parametric Aspects.* P. Lang, Frankfurt/M. et al., 1986.

[140] H.Th. Jongen, T. Möbert, and K. Tammer. On iterated minimization in nonconvex optimization. *Mathematics of Operations Research*, 11:679–691, 1986.

[141] H.Th. Jongen and G.-W. Weber. Nonlinear optimization: Characterization of structural optimization. *Journal of Global Optimization*, 1:47–64, 1991.

[142] H.Th. Jongen and G.-W. Weber. Nonconvex optimization and its structural fronties. In *Modern Methods of Optimization, Bayreuth, 1990*, number 378 in Lecture Notes in Economics and Mathematical Systems, pages 151–203, 1992.

[143] C. Kanzow. Some noninterior continuation methods for linear complementarity problems. *SIAM Journal on Matrix Analysis and Applications*, 17:851–868, 1996.

[144] N. Karmarkar. A new polynomial-time algorithm for linear programming. *Combinatorica*, 4:373–395, 1984.

[145] K. Kassicieh. Evaluation of incentive structure strategies in resource allocation problems. *Large Scale Systems*, 10:193–202, 1986.

[146] H. Kellerer and U. Pferschy. A new fully polynomial approximation scheme for the knapsack problem. *Lecture Notes in Computer Science*, 1444:123–134, 1998.

[147] P.S. Kenderov and R.E. Lucchetti. Generic well-posedness of supinf problems. *Bulletin of the Australean Mathematical Society*, 54:5–25, 1996.

[148] L.G. Khachiyan. A polynomial algorithm in linear programming. *Dokl. Akad. Nauk SSSR*, 244:1093–1096, 1975. (in Russian).

[149] K.-P. Kistner and M. Switalski. Hierarchical production planning: necessity, problems, and methods. *Zeitschrift für Operations Research*, 33:199–212, 1989.

[150] K.C. Kiwiel. *Methods of Descent for Nondifferentiable Optimization.* Springer - Verlag, Berlin, 1985.

[151] K.C. Kiwiel. Restricted step and Levenberg-Marquardt techniques in proximal bundle methods for nonconvex nondifferentiable optimization. *SIAM Journal on Optimization*, 6:227–249, 1996.

[152] D. Klatte. Upper Lipschitz behavior of solutions to perturbed $C^{1,1}$ programs. *Mathematical Programming.* to appear.

[153] D. Klatte and B. Kummer. Stability properties of infima and optimal solutions of parametric optimization problems. In V.F. Demyanov, editor, *Nondifferentiable Optimization: Motivations and Applications, Proceedings of the IIASA Workshop, Sopron, 1984.* IIASA, Laxenburg, 1984.

[154] D. Klatte and B. Kummer. Strong stability in nonlinear programming revisited. *Journal of the Australian Mathematical Society, Ser. B*, 40:336–352, 1999.

[155] D. Klatte and K. Tammer. On second-order sufficient optimality conditions for $C^{1,1}$-optimization problems. *Optimization*, 19:169–179, 1088.

[156] V. Klee. Separation and support properties of convex sets – a survey. In A.V. Balakrishnan, editor, *Control Theory and the Calculus of Variations*, pages 235–303. Academic Press, New York, London, 1969.

[157] M. Kojima. Strongly stable stationary solutions in nonlinear programs. In S.M. Robinson, editor, *Analysis and Computation of Fixed Points*, pages 93–138. Academic Press, New York, 1980.

[158] A.A. Korbut and Yu.Yu. Finkel'shteĭn. *Diskrete Optimierung*. Akademie-Verlag, Berlin, 1971.

[159] B. Korte and J. Vygen. *Combinatorial Optimization: Theory and Algorithms*. Springer Verlag, Berlin et al., 2000.

[160] M. Kočvara and J.V. Outrata. A nondifferentiable approach to the solution of optimum design problems with variational inequalities. In P. Kall, editor, *System Modelling and Optimization (Proc. 15. IFIP Conference on System Modelling and Optimization, Zürich, 1991)*, pages 364–373, 1992.

[161] M. Kočvara and J.V. Outrata. A numerical solution of two selected shape optimization problems. In *System Modelling and Optimization (Proc.16. IFIP Conference on System Modelling and Optimization, Compiegne, 1993*.

[162] H.W. Kuhn and A.W. Tucker. Nonlinear programming. In *Proceedings of the 2nd Berkelay Symposium on Mathematical Statistics and Probability*, pages 481–492. University of California Press, Berkelay, 1951.

[163] B. Kummer. Newton's method for non-differentiable functions. In *Advances in Mathematical Optimization*, volume 45 of *Mathematical Research*. Akademie-Verlag, Berlin, 1988.

[164] H.P. Künzi and W. Krelle. *Nichtlineare Programmierung*. Springer-Verlag, Berlin, 1962.

[165] J. Kyparisis. On uniqueness of Kuhn-Tucker multipliers in nonlinear programming. *Mathematical Programming*, 32:242–246, 1985.

[166] H. Laux. *Risiko, Anreiz und Kontrolle*. Springer Verlag, Berlin, 1990.

[167] H. Laux and H.Y. Schenk-Mathes. *Lineare und Nichtlineare Anreizsysteme*. Physica-Verlag, 1992.

[168] E.L. Lawler. Fast approximation algorithms for knapsack problems. *Mathematical Operations Research*, 4:339–356, 1979.

[169] C. Lemarechal. A view of line searches. In *Optimization and Optimal Control*, Lecture Notes in Control and Information Sciences. Springer Verlag, Berlin et al., 1981.

[170] C. Lemaréchal. Nondifferentiable optimization. In *Handbooks in Operations Research and Management Science*, volume 1. North Holland, Amsterdam, 1989.

[171] E.S. Levitin. *Perturbation Theory in Mathematical Programming and its Applications*. J. Wiley & Sons, Chichester, 1994.

[172] A.B. Levy. Lipschitzian multifunctions and a Lipschitzian inverse mapping theorem. *Mathematics of Operations Research*, 26:105–118, 2001.

[173] M.B. Lignola and J. Morgan. Existence and approximation results for minsup problems. In *Proceedings of the Operations Research 90*, pages 157–164. Springer-Verlag, Berlin, 1990.

[174] M.B. Lignola and J. Morgan. Topological existence and stability for Stackelberg problems. Technical Report 47-1991, Dipartimento di Matematica ed Applicazioni "R. Caccioppoli", Universita di Napoli, 1991.

[175] M.B. Lignola and J. Morgan. Convergences of marginal functions with dependent constraints. *Optimization*, 23:189–213, 1992.

[176] J. Liu. Sensitivity analysis in nonlinear programs and variational inequalities via continuous selections. *SIAM Journal on Control and Optimization*, 33(4), 1995.

[177] Y.-H. Liu and T.H. Spencer. Solving a bilevel linear program when the inner decision maker controls few variables. *European Journal of Operational Research*, 81:644–651, 1995.

[178] K. Lommatsch. Begriffe und Ergebnisse der linearen parametrischen Optimierung. In K. Lommatsch, editor, *Anwendungen der linearen parametrischen Optimierung*, pages 5–22. Akademie-Verlag, Berlin, 1979.

[179] P. Loridan and J. Morgan. Approximate solutions for two-level optimization problems. In K. Hoffman, J. Hiriart-Urruty, C. Lemarechal and J. Zowe, editor, *Trends in Mathematical Optimization*, volume 84 of *International Series of Numerical Mathematics*, pages 181–196. Birkhäuser Verlag, Basel, 1988.

[180] P. Loridan and J. Morgan. New results on approximate solutions in two-level optimization. *Optimization*, 20:819–836, 1989.

[181] P. Loridan and J. Morgan. ε-regularized two-level optimzation problems: approximation and existence results. In *Optimization – Fifth French-German Conference (Varez)*, pages 99–113. Lecture Notes in Mathematics, Springer Verlag, Berlin, No. 1405, 1989.

[182] P. Loridan and J. Morgan. Regularization for two-level optimization problems. In *Advances in Optimization. Proceedings of the 6th French-German Conference on Optimization, Lambrecht*, pages 239–255. Springer Verlag, Berlin, 1991.

[183] P. Loridan and J. Morgan. Least-norm regularization for weak two-level optimization problems. In *Optimization, Optimal Control and Partial Differential Equations*, volume 107 of *International Series of Numerical Mathematics*, pages 307–318. Birkhäuser Verlag, Basel, 1992.

[184] P. Loridan and J. Morgan. On strict ε-solutions for a two-level optimization problem. In *Proceedings of the International Conference on Operations Research 90*, pages 165–172. Springer Verlag, Berlin, 1992.

[185] P. Loridan and J. Morgan. Weak via strong Stackelberg problem: New results. *Journal of Global Optimization*, 8:263–287, 1996.

[186] R. Lucchetti, F. Mignanego, and G. Pieri. Existence theorem of equilibrium points in Stackelberg games with constraints. *Optimization*, 18:857–866, 1987.

[187] B. Luderer. Über der Äquivalenz nichtlinearer Optimierungsaufgaben. *Wissenschaftliche Zeitschrift der TH Karl-MarxStadt*, 26:257–258, 1983.

[188] Z.-Q. Luo, J.-S. Pang, and D. Ralph. *Mathematical Programs with Equilibrium Constraints*. Cambridge University Press, Cambridge, 1996.

[189] C.M. Macal and A.P. Hurter. Dependence of bilevel mathematical programs on irrelevant consrtaints. *Computers and Operations Research*, 24:1129–1140, 1997.

[190] K. Malanowski. Differentiability with respect to parameters of solutions to convex programming problems. *Mathematical Programming*, 33:352–361, 1985.

[191] L. Mallozzi and J. Morgan. ε-mixed strategies for static continuous-kernel Stackelberg problems. *Journal of Optimization Theory and Applications*, 78:303–316, 1993.

[192] L. Mallozzi and J. Morgan. Weak Stackelberg problem and mixed solutions under data perturbations. *Optimization*, 32:269–290, 1995.

[193] O.L. Mangasarian. Mathematical programming in neural networks. Technical report, Computer Science Department, University of Wisconsin, 1992.

[194] O.L. Mangasarian. Misclassification minimization. *Journal of Global Optimization*, 5:309–323, 1994.

[195] O.L. Mangasarian and S. Fromowitz. The Fritz John necessary optimality condition in the presence of equality and inequality constraints. *Journal of Mathematical Analysis and Applications*, 17:37–47, 1967.

[196] O.L. Mangasarian and J.-S. Pang. Exact penalty functions for mathematical programs with linear complementarity constraints. *Optimization*, 42:1–8, 1997.

[197] P. Marcotte. Network optimization with continuous control parameters. *Transportation Science*, 17:181–197, 1983.

[198] P. Marcotte. Network design problem with congestion effects: a case of bilevel programming. *Mathematical Programming*, 34:142–162, 1986.

[199] P. Marcotte. A note on a bilevel programming algorithm by LeBlanc and Boyce. *Transportation Research*, 22 B:233–237, 1988.

[200] P. Marcotte and G. Marquis. Efficient implementation of heuristics for the continuous network design problem. *Annals of Operations Research*, 34:163–176, 1992.

[201] P. Marcotte and G. Savard. A note on the Pareto optimality of solutions to the linear bilevel programming problem. *Computers and Operations Research*, 18:355–359, 1991.

[202] S. Martello and P. Toth. *Knapsack Problems*. J. Wiley & Sons, Chichester, 1990.

[203] R. Mifflin. Semismooth and semiconvex functions in constrained optimization. *SIAM Journal on Control and Optimization*, 15:959–972, 1977.

[204] A. Migdalas. Bilevel programming in traffic planning: Models, methods and Challenge. *Journal of Global Optimization*, 7:381–405, 1995.

[205] A. Migdalas, P.M. Pardalos, and P. Värbrand (eds.). *Multilevel Optimization: Algorithms and Applications*. Kluwer Academic Publishers, Dordrecht, 1998.

[206] V.S. Mikhalevich, A.M. Gupal, and V.I. Norkin. *Methods of Nonconvex Optimization*. Nauka, Moscow, 1987. (in Russian).

[207] T. Miller, T. Friesz, and R. Tobin. Heuristic algorithms for delivered price spatially competitive network facility location problems. *Annals of Operations Research*, 34:177–202, 1992.

[208] T.C. Miller, T.L. Friesz, and R.L. Tobin. *Equilibrium Facility Location on Networks*. Springer, Berlin et al., 1996.

[209] L.I. Minchenko and P.P. Sakolchik. Hölder behavior of optimal solutions and directional differentiability of marginal functions in nonlinear programming. *Journal of Optimization Theory and Applications*, 90:555–580, 1996.

[210] L.I. Minchenko and T.V. Satsura. Calculation of directional derivatives in max-min-problems. *Computational Mathematics and Mathematical Physics*, 37:16–20, 1997.

[211] L.I. Minchenko and A. Volosevich. Strongly differentiable multifunctions and directional differentiability of marginal functions. Technical report, Belarussian State University of Informatics and Radioelectronics, Minsk, 1999.

[212] K. Mizukami and H. Xu. Closed-loop Stackelberg strategies for linear-quadratic descriptor systems. *Journal of Optimization Theory and Applications*, 74:151–170, 1992.

[213] D.A. Molodtsov. The solution of a certain class of non-antagonistic games. *Žurnal Vyčislitel'noĭ Matematiki i Matematičeskoĭ Fiziki*, 16:1451–1456, 1976.

[214] J. Moore. *Extensions to the multilevel linear programming problem*. PhD thesis, Department of Mechanical Engineering, University of Texas, Austin, 1988.

[215] J. Moore and J.F. Bard. The mixed integer linear bilevel programming problem. *Operations Research*, 38:911–921, 1990.

[216] S. Narula and A. Nwosu. An algorithm to solve a two-level resource control preemptive hierarchical programming problem. In P. Serafini, editor, *Mathematics of multiple-objective programming*. Springer-Verlag, Berlin, 1985.

[217] G.L. Nemhauser and L.A. Wolsey. *Integer and Combinatorial Optimization*. John Wiley & Sons, New York, 1988.

[218] Y. Nesterov and A. Nemirovskii. *Interior-Point Polynomial Algorithms in Convex Programming*. SIAM, Philadelphia, 1994.

[219] M.G. Nicholls. Aluminium production modelling – a non-linear bi-level programming approach. *Operations Research*, 43:208–218, 1995.

[220] F. Nožička, J. Guddat, H. Hollatz, and B. Bank. *Theorie der linearen parametrischen Optimierung*. Akademie-Verlag, Berlin, 1974.

[221] W. Oeder. *Ein Verfahren zur Lösung von Zwei-Ebenen-Optimierungsaufgaben in Verbindung mit der Untersuchung von chemischen Gleichgewichten*. PhD thesis, Technische Universität Karl-Marx-Stadt, 1988.

[222] J.M. Ortega and W. Rheinboldt. *Iterative Solutions of Nonlinear Equations in Several Variables*. Academic Press, New York, 1970.

[223] J. Outrata. On the numerical solution of a class of Stackelberg problems. *ZOR - Mathematical Methods of Operations Research*, 34:255–277, 1990.

[224] J. Outrata, M. Kočvara, and J. Zowe. *Nonsmooth Approach to Optimization Problems with Equilibrium Constraints*. Kluwer Academic Publishers, Dordrecht, 1998.

[225] J. Outrata and J. Zowe. A numerical approach to optimization problems with variational inequality constraints. *Mathematical Programming*, 68:105–130, 1995.

[226] J.V. Outrata. On the numerical solution of a class of Stackelberg problems. *Zeitschrift für Operations Research*, 34:255 – 277, 1990.

[227] J.V. Outrata. Optimality conditions for a class of mathematical programs with equilibrium constraints. Technical report, Academy of Sciences of the Czech Republic, 1997.

[228] G. Owen. *Game Theory*. Saunders Company, Philadelphia, 1968.

[229] J.-S. Pang and M. Fukushima. Complementarity constraint qualifications and simplified B-stationarity conditions for mathematical programs with equilibrium constraints. *Computational Optimization and Applications*, 13:111–136, 1999.

[230] C.H. Papadimitriou and K. Steiglitz. *Combinatorial Optimization: Algorithms and Complexity*. Prentice-Hall, Englewood Cliffs, 1982.

[231] T. Petersen. *Optimale Anreizsysteme*. Gabler Verlag, Wiesbaden, 1989.

[232] A.C. Pigou. *The economics of welfare*. London, 1923.

[233] D. Pisinger and P. Toth. Knapsack problems. In D.Z. Du and P. Pardalos, editors, *Handbook of Combinatorial Optimization*, pages 1–89. Kluwer, Dordrecht, 1998.

[234] J.C. Pomerol. The Lagrange multiplier set and the generalized gradient set of the marginal function in a differentiable program in a Banach space. *Journal of Optimization Theory and Applications*, 38:307–317, 1982.

[235] D. Ralph and S. Dempe. Directional derivatives of the solution of a parametric nonlinear program. *Mathematical Programming*, 70:159–172, 1995.

[236] P. Recht. *Generalized Derivatives: An Approach to a New Gradient in Nonsmooth Optimization*, volume 136 of *Mathematical Systems in Economics*. Anton Hain, Meisenheim, 1993.

[237] R. Rees. The theory of principal and agent. part 1. *Bulletin of Economic Research*, 37:3–26, 1985.

[238] R. Rees. The theory of principal and agent. part 2. *Bulletin of Economic Research*, 37:75–95, 1985.

[239] K. Richter. An EOQ repair and waste disposal model. Technical report, Europa-Universität Viadrina Frankfurt(Oder), 1994.

[240] S.M. Robinson. Generalized equations and their solutions, part II: Applications to nonlinear programming. *Mathematical Programming Study*, 19:200–221, 1982.

[241] R.T. Rockafellar. *Convex analysis*. Princeton University Press, Princeton, 1970.

[242] R.T. Rockafellar. Lagrange multipliers and subderivatives of optimal value functions in nonlinear programming. *Mathematical Programming Study*, 17:28–66, 1982.

[243] R.T. Rockafellar. Directional differentiability of the optimal value function in a nonlinear programming problem. *Mathematical Programming Study*, 21:213–226, 1984.

[244] R.T. Rockafellar. Lipschitzian properties of multifunctions. *Nonlinear Analysis, Theory, Methods and Applications*, 9:867–885, 1985.

[245] R.T. Rockafellar. Maximal monotone relations and the second derivatives of nonsmooth functions. *Ann. Inst. H. Poincare Anal. Non Lineaire*, 2:167–184, 1985.

[246] J. Rückmann. *Einparametrische nichtkonvexe Optimierung: Strukturuntersuchungen und eine Verallgemeinerung des Einbettungsprinzips*. PhD thesis, Technische Hochschule Leipzig, 1988.

[247] K.S. Sagyngaliev. Coordinated resource allocation in three-level active system. *Izvestija Akademii Nauk SSSR, Avtomatika i Telemechanika*, (10):81–88, 1986. (in Russian).

[248] H. Scheel. Ein Straffunktionsansatz für Optimierungsprobleme mit Gleichgewichtsrestriktionen. Master's thesis, Institut für Statistik und Mathematische Wirtschaftstheorie der Universität Karlsruhe, Germany, 1994.

[249] H. Scheel and S. Scholtes. Mathematical programs with equilibrium constraints: stationarity, optimality, and sensitivity. Technical report, Research Papers in Management Studies, The Judge Institute of Management Studies, University of Cambridge, 1998. to appear in Mathematics of Operations Research.

[250] H. Schmidt. *Zwei-Ebenen-Optimierungsaufgaben mit mehrelementiger Lösung der unteren Ebene.* PhD thesis, Fakultät für Mathematik, Technische Universität Chemnitz-Zwickau, 1995.

[251] S. Scholtes. Introduction to piecewise differentiable equations. Technical report, Universität Karlsruhe, Institut für Statistik und Mathematische Wirtschaftstheorie, 1994. No. 53/1994.

[252] S. Scholtes and M. Stöhr. Exact penalization of mathematical programs with equilibrium constraints. *SIAM Journal on Control and Optimization*, 37:617–652, 1999.

[253] S. Scholtes and M. Stöhr. How stringent is the linear independence assumption for mathematical programs with stationarity constraints? Technical report, University of Cambridge, Department of Engineering and Judge Institute of Management Studies, Cambridge, 1999.

[254] H. Schramm. *Eine Kombination von bundle- und trust-region-Verfahren zur Lösung nichtdifferenzierbarer Optimierungsprobleme.* Bayreuther Mathematische Schriften, Bayreuth, No. 30, 1989.

[255] H. Schramm and J. Zowe. A version of the bundle idea for minimizing a nonsmooth function: conceptual idea, convergence analysis, numerical results. *SIAM Journal on Optimization*, 2:121–152, 1992.

[256] A. Shapiro. Perturbation theory of nonlinear programs when the set of optimal solutions is not a singleton. *Applied Mathematics and Optimization*, 18:215–229, 1988.

[257] A. Shapiro. Sensitivity analysis of nonlinear programs and differentiability properties of metric projections. *SIAM Journal Control Optimization*, 26:628–645, 1988.

[258] J.W. Shavlik, R.J. Mooney, and G.G. Towell. Symbolic and neural network learning algorithms: An experimental comparison. *Machine Learning*, 6:111–143, 1991.

[259] H. Sherali. A multiple leader Stackelberg model and analysis. *Operations Research*, 32:390–404, 1984.

[260] H.D. Sherali, A.L. Soyster, and F.H. Murphy. Stackelberg-Nash-Cournot equilibria: Characterizations and Computations. *Operations Research*, 31:253–276, 1983.

[261] K. Shimizu, Y. Ishizuka, and J.F. Bard. *Nondifferentiable and Two–Level Mathematical Programming.* Kluwer Academic Publishers, Dordrecht, 1997.

[262] K. Shimizu and M. Lu. A global optimization method for the Stackelberg problem with convex functions via problem transformations and concave programming. *IEEE Transactions on Systems, Man, and Cybernetics*, 25:1635–1640, 1995.

[263] M. Simaan. Stackelberg optimization of two-level systems. *IEEE Transactions on Systems, Man, and Cybernetics*, 7:554–557, 1977.

[264] P.K. Simpson. *Artifical neural systems*. Pergamon Press, New York, 1990.

[265] W.R. Smith and R.W. Missen. *Chemical Reaction Equilibrium Analysis: Theory and Algorithms*. John Wiley & Sons, New York, 1982.

[266] V.F. Solohovic. Unstable extremal problem and geometric properties of Banach spaces. *Soviet Mathematical Doklady*, 11:1470–1472, 1970.

[267] H.v. Stackelberg. *Marktform und Gleichgewicht*. Springer-Verlag, Berlin, 1934. engl. transl.: The Theory of the Market Economy, Oxford University Press, 1952.

[268] S. Sternberg. *Lectures on Differential Geometry*. Prentice Hall, Englewood Cliffs, 1964.

[269] M. Stöhr. *Nonsmooth trust region methods and their applications to mathematical programs with equilibrium constraints*. Shaker Verlag, Aachen, 2000.

[270] M. Studniarski. Necessary and sufficient conditions for isolated local minimum of nonsmooth functions. *SIAM Journal Control Optimization*, 24:1044 – 1049, 1986.

[271] S. Suh and T. Kim. Solving nonlinear bilevel programming models of the equilibrium network design problem: a comparative review. *Annals of Operations Research*, 34:203–218, 1992.

[272] T. Tanino and T. Ogawa. An algorithm for solving two-level convex optimization problems. *International Journal of Systems Science*, 15:163–174, 1984.

[273] M.C. Thierry, M. Salomon, J.A.E.E. Van Nunen, and L.N. Van Wassenhove. Strategic production and operations management issues in product recovery management. Technical report, INSEAD, Fontainebleau, France, 1994.

[274] H. Tuy. D.c. optimization: theory, methods and algorithms. In R. Horst and P.M. Pardalos, editors, *Handbook of Global Optimization*, pages 149–216. Kluwer Academic Publishers, Dordrecht, 1995.

[275] H. Tuy. Bilevel linear programming, multiobjective programming, and monotonic reverse convex programming. In A. Migdalas, P.M. Pardalos, and P. Värbrand, editors, *Multilevel Optimization: Algorithms and Applications*, pages 295–314. Kluwer Academic Publishers, Dordrecht, 1998.

[276] H. Tuy, A. Migdalas, and P. Värbrand. A quasiconcave minimization method for solving linear two-level programs. *Journal of Global Optimization*, 4:243–263, 1994.

[277] A.N. Tykhonov. Methods for the regularization of optimal control problems. *Soviet Mathematical Doklady*, 6:761–763, 1965.

[278] L.N. Vicente and P.H. Calamai. Bilevel and multilevel programming: A bibliography review. *Journal of Global Optimization*, 5(3):291–306, 1994.

[279] L.N. Vicente, G. Savard, and J.J. Judice. The discrete linear bilevel programming problem. *Journal of Optimization Theory and Applications*, 89:597–614, 1996.

[280] S. Vogel. Untersuchungen zur Berechnung des verallgemeinerten Jacobians für die optimale Lösung parametrischer nichtlinearer Optimierungsaufgaben. Master's thesis, Technische Universität Chemnitz-Zwickau, 1996.

[281] S. Wang and F.A. Lootsma. A hierarchical optimization model of resource allocation. *Optimization*, 28:351–365, 1994.

[282] D.E. Ward. Exact penalties and sufficient conditions for optimality in nonsmooth optimization. *Journal of Optimization Theory and Applications*, 57:485–499, 1988.

[283] J. Weimann. *Umweltökonomik*. Springer-Verlag, Berlin, 1991.

[284] U. Wen. *Mathematical methods for multilevel linear programming*. PhD thesis, Department of Industrial Engineering, State University of New York at Buffalo, 1981.

[285] U. Wen. The "Kth-Best" algorithm for multilevel programming. Technical report, Department of Operations Research, State University of New York at Buffalo, 1981.

[286] U. Wen. A solution procedure for the resource control problem in two-level hierarchical decision processes. *Journal of Chinese Institute of Engineers*, 6:91–97, 1983.

[287] U. Wen and W. Bialas. The hybrid algorithm for solving the three-level linear programming problem. *Computers and Operations Research*, 13:367–377, 1986.

[288] U.-P. Wen and S.-T. Hsu. Efficient solutions for the linear bilevel programming problem. *European Journal of Operational Research*, 62:354–362, 1991.

[289] R.E. Wendell and A.P. Hurter Jr. Minimization of a non-separable objective function subject to disjoint constraints. *Operations Research*, 24:643–657, 1976.

[290] R.J.-B. Wets. On inf-compact mathematical programs. In *Fifth conference on optimization techniques, part 1, Lect. Notes Comput. Sci.*, volume 3, pages 426–436. Springer-Verlag, Berlin, 1973.

[291] D.J. White. Multilevel programming, rational reaction sets, and efficient solutions. *Journal of Optimization Theory and Applications*, 87:727–746, 1995.

[292] D.J. White and G. Anandalingam. A penalty function approach for solving bi-level linear programs. *Journal of Global Optimization*, 3:397–419, 1993.

[293] P. Wolfe. A method of conjugate subgradients for minimizing nondifferentiable convex functions. *Mathematical Programming Study*, 3:145–173, 1975.

[294] C. Xu and T. Chen. Incentive strategies with many followers. *Acta Automatica Sinica*, 17:577–581, 1991. in chinese.

[295] J.J. Ye. Constraint qualifications and necessary optimality conditions for optimization problems with variational inequalities. *SIAM Journal on Optimization*, 10:943–962, 2000.

[296] J.J. Ye and X.Y. Ye. Necessary optimality conditions for optimization problems with variational inequality constraints. *Mathematics of Operations Research*, 22:977–997, 1997.

[297] J.J. Ye and D.L. Zhu. Optimality conditions for bilevel programming problems. *Optimization*, 33:9–27, 1995. with correction in Optimization 39(1997), pp. 361-366.

[298] Y. Yuan. On the convergence of a new trust region algorithm. *Numerische Mathematik*, 70:515–539, 1995.

[299] W.I. Zangwill. Non-linear programming via penalty functions. *Management Science*, 13:344–358, 1967.

[300] W.I. Zangwill. *Nonlinear Programming; A Unified Approach*. Prentice Hall, Englewood Cliffs, 1969.

[301] R. Zhang. Problems of hierarchical optimization in finite dimensions. *SIAM Journal on Optimization*, 4:521–536, 1994.

[302] J. Zowe. The BT-algorithm for minimizing a nonsmooth functional subject to linear constraints. In F.H. Clarke et al., editor, *Nonsmooth Optimization and Related Topics*, pages 459–480. Plenum Press, New York, 1989.

Notations

2^A	power set of a set A, i.e. family of all subsets of a set A
$A + B$	Minkowski sum $A + B = \{z = x + y$ for some $x \in A,\ y \in B\}$
$A - B$	Minkowski sum of A and $(-1) \cdot B$
C^*	dual cone to cone C (see Definition 5.6)
E	Unit matrix of appropriate dimension
$E\Lambda(x, y)$	vertex set of $\Lambda(x, y)$
e^p	p-dimensional vector of 1's: $e^p = (1, \ldots, 1)^\top \in \mathbb{R}^p$
e_i	ith unit vector
$F(X)$	image of a set X via a function F: $F(X) = \bigcup_{x \in X} F(x)$.
$I(x, y)$	set of active constraints at $x \in M(y)$ (cf. pages 62, 63)
$I_z(x^0)$	index set of active selection functions of a PC^1–function (cf. 71)
$I_z^0(x^0)$	index set of strongly active selection functions of a PC^1–function (cf. 73)
$J(\lambda)$	index set of positive components of λ (cf. page 66)
$K_{\Psi(y^0)}^0$	a tangent cone to the optimal solution set
$K_C(x^0)$	contingent cone at $x^0 \in \operatorname{cl} C$ to set C (see Definition 4.6)
$K_{\Psi(y)}^0(x, \lambda, \mu)$	tangent cone to $\Psi(y^0)$ (cf. page 224)
$QP(\lambda, \mu, r)$	quadratic problem for computing the directional derivative of the optimal solution function (cf. (4.20))
$S(r)$	linear problem for computing the directional derivative of the optimal solution function (cf. (4.21))
$U_\delta(y)$	δ neighborhood of a point y
$V_\varepsilon(x)$	ε neighborhood of a point x (cf. page 63)
$x^I(\cdot)$	selection function of the PC^1-function $x(\cdot)$
$z = (x, y)$	abbreviation used at many places
$\alpha_r(x^0)$	radial limit of a generalized PC^1-function (cf. Definition 5.11)
conv A	convex hull of a set A
int A	set of inner points of a set A
$\Lambda(x, y)$	set of Lagrange multipliers of an optimization problem (cf. page 62)
λ, μ	lower level multiplier vectors
rg A	Rank of the matrix A
$\nabla f(x)$	gradient of a function $f : \mathbb{R}^n \to \mathbb{R}$, the gradient is a row vector
$\nabla^2 f(x)$	Hessian matrix of $f : \mathbb{R}^n \to \mathbb{R}$
$\nabla_{xy}^2 f(x, y)$	matrix of second order mixed derivatives of $f : \mathbb{R}^n \times \mathbb{R}^m \to \mathbb{R}$
$\nabla_x f(x, y)$	gradient of $f : \mathbb{R}^n \times \mathbb{R}^m \to \mathbb{R}$ with respect to the x variables
$\nu = (\lambda, \mu)$	abbreviation used at many places
$\Psi(\cdot)$	solution set mapping of the lower level problem (cf. pages 1, 62)
$\Psi_D(\cdot)$	solution set mapping of a discrete lower level problem (cf. (8.1))
$\Psi_L(\cdot)$	solution set mapping of the linear lower level problem (cf. pages 21, 27, 40)
$\Psi_\varepsilon(y)$	ε-optimal solution set mapping of problem 4.1 (cf. page 88)
$\mathbb{R}^n$	n-dimensional Euclidean vector space
$\mathbb{R}_{++}$	set of positive numbers $\mathbb{R}_{++} = \{z : z > 0\}$
$\mathbb{R}_+$	set of nonnegative numbers $\mathbb{R}_+ = \{z : z \geq 0\}$

$\mathrm{Supp}(z, z^i)$	support set of an active selection function of a PC^1–function (cf. 71)
$\mathbf{Z}^n$	space of integral n-dimensional vectors
$\varrho(x, A)$	distance of a point x from a set A (cf. (4.9))
$\mathcal{B}^l$	unit sphere in $\mathbb{R}^l$
$\mathcal{I}(\lambda)$	family of all index sets satisfying conditions (C1) and (C2) (cf. page 78)
$\mathcal{RP}(y^0)$	reachable part of the set of optimal solutions (c.f. (7.3))
$\mathcal{R}(x)$	region of stability for a solution of a linear program (cf. page 260)
$\mathcal{R}(\mathbf{B})$	region of stability for a solution of a linear program (cf. page 40)
$d\alpha(x^0, \cdot)$	radial-directionally derivative of a generalized PC^1-function (cf. Definition 5.11)
grph Γ	graph of a point-to-set mapping (cf. page 24)
$C^l(\mathbb{R}^r, \mathbb{R}^t)$	space of l times continuously differential functions mapping $\mathbb{R}^r$ to $\mathbb{R}^t$
$\|\cdot\|_\infty$	L_∞ norm
$\|\cdot\|_1$	L_1 norm
$\|\cdot\|_2$	Euclidean norm

Index

305

Nonconvex Optimization and Its Applications

Nonconvex Optimization and Its Applications

Nonconvex Optimization and Its Applications

44. A. Rubinov: *Abstract Convexity and Global Optimization.* 2000
    ISBN 0-7923-6323-X
45. R.G. Strongin and Y.D. Sergeyev: *Global Optimization with Non-Convex Constraints.* 2000
    ISBN 0-7923-6490-2
46. X.-S. Zhang: *Neural Networks in Optimization.* 2000   ISBN 0-7923-6515-1
47. H. Jongen, P. Jonker and F. Twilt: *Nonlinear Optimization in Finite Dimensions.* Morse Theory, Chebyshev Approximation, Transversability, Flows, Parametric Aspects. 2000   ISBN 0-7923-6561-5
48. R. Horst, P.M. Pardalos and N.V. Thoai: *Introduction to Global Optimization.* 2nd Edition. 2000   ISBN 0-7923-6574-7
49. S.P. Uryasev (ed.): *Probabilistic Constrained Optimization.* Methodology and Applications. 2000   ISBN 0-7923-6644-1
50. D.Y. Gao, R.W. Ogden and G.E. Stavroulakis (eds.): *Nonsmooth/Nonconvex Mechanics.* Modeling, Analysis and Numerical Methods. 2001   ISBN 0-7923-6786-3
51. A. Atkinson, B. Bogacka and A. Zhigljavsky (eds.): *Optimum Design 2000.* 2001
    ISBN 0-7923-6798-7
52. M. do Rosário Grossinho and S.A. Tersian: *An Introduction to Minimax Theorems and Their Applications to Differential Equations.* 2001   ISBN 0-7923-6832-0
53. A. Migdalas, P.M. Pardalos and P. Värbrand (eds.): *From Local to Global Optimization.* 2001   ISBN 0-7923-6883-5
54. N. Hadjisavvas and P.M. Pardalos (eds.): *Advances in Convex Analysis and Global Optimization.* Honoring the Memory of C. Caratheodory (1873-1950). 2001
    ISBN 0-7923-6942-4
55. R.P. Gilbert, P.D. Panagiotopoulos[†] and P.M. Pardalos (eds.): *From Convexity to Nonconvexity.* 2001   ISBN 0-7923-7144-5
56. D.-Z. Du, P.M. Pardalos and W. Wu: *Mathematical Theory of Optimization.* 2001
    ISBN 1-4020-0015-4
57. M.A. Goberna and M.A. López (eds.): *Semi-Infinite Programming.* Recent Advances. 2001   ISBN 1-4020-0032-4
58. F. Giannessi, A. Maugeri and P.M. Pardalos (eds.): *Equilibrium Problems: Nonsmooth Optimization and Variational Inequality Models.* 2001   ISBN 1-4020-0161-4
59. G. Dzemyda, V. Šaltenis and A. Žilinskas (eds.): *Stochastic and Global Optimization.* 2002   ISBN 1-4020-0484-2
60. D. Klatte and B. Kummer: *Nonsmooth Equations in Optimization.* Regularity, Calculus, Methods and Applications. 2002   ISBN 1-4020-0550-4
61. S. Dempe: *Foundations of Bilevel Programming.* 2002   ISBN 1-4020-0631-4
62. P.M. Pardalos and H.E. Romeijn (eds.): *Handbook of Global Optimization, Volume 2.* 2002   ISBN 1-4020-0632-2